Patrick Doherty, Witold Łukaszewicz, Andrzej Skowron, Andrzej Szałas

Knowledge Representation Techniques

Studies in Fuzziness and Soft Computing, Volume 202

Editor-in-chief
Prof. Janusz Kacprzyk
Systems Research Institute
Polish Academy of Sciences
ul. Newelska 6
01-447 Warsaw
Poland
E-mail: kacprzyk@ibspan.waw.pl

Further volumes of this series can be found on our homepage: springer.com

Patrick Doherty
Witold Łukaszewicz
Andrzej Skowron
Andrzej Szałas

Knowledge Representation Techniques

A Rough Set Approach

Professor Patrick Doherty
Department of Computer
and Information Science
Linköping University, Sweden
E-mail: patdo@ida.liu.se

Professor Andrzej Skowron
Institute of Mathematics
Warsaw University
Warsaw, Poland
E-mail: skowron@mimuw.edu.pl

Professor Witold Łukaszewicz
The University of Economics
and Computer Science
Olsztyn, Poland
E-mail: witlu@ida.liu.se

Professor Andrzej Szałas
The University of Economics
and Computer Science
Olsztyn, Poland
and
Department of Computer
and Information Science
Linköping University, Sweden
E-mail: andsz@ida.liu.se

ISSN print edition: 1434-9922
ISSN electronic edition: 1860-0808

ISBN 978-3-642-07012-9 e-ISBN 978-3-540-33519-1

Springer is a part of Springer Science+Business Media
springer.com

Softcover reprint of the hardcover 1st edition 2006

Cover design: Erich Kirchner, Heidelberg

Contents

Part III From Sensors to Relations

Part I

Introduction and Preliminaries

1

Introduction

1.1 Background

The basis for the material in this book centers around research done in an ongoing long-term project which focuses on the development of highly autonomous unmanned aerial vehicle systems.[1] The actual platform which serves as a case study for the research in this book will be described in detail later in this chapter. Before doing that, a brief background of the motivations behind this research will be provided. One of the main research topics in the project is knowledge representation and reasoning and its use in UAV platforms. A very strong constraint has been placed on the nature of research done in the project where theoretical results, to the greatest extent possible, should serve as a basis for tractable reasoning mechanisms for use in a fully deployed autonomous UAV operating under soft real-time constraints associated with the types of mission scenarios envisioned. Considering that much of the work with knowledge representation in this context focuses on application domains where one can only hope for an incomplete characterization of such domains, this methodological constraint has proven to be quite challenging since, in essence, the focus is on tractable approximate and nonmonotonic reasoning systems. As is well known, until recently, nonmonotonic formalisms have had a notorious reputation for lack of tractable and scalable reasoning systems. At an early stage, a decision was made to investigate a number of standard nonmonotonic reasoning approaches and their combination with approximate reasoning techniques based on the use of rough set theory, or at the very least, guided by intuitions from rough set theory. In addition, a decision was also made to deal seriously with the sense/reasoning gap associated with most state-of-the-art robotic systems where it is often the case that high-level reasoning systems are not strongly grounded in the sensory data continually generated by sensor platforms. Pragmatically, an effort has been made to

[1] UAV is an acronym for Unmanned Aerial Vehicles.

P. Doherty et al.: *Knowledge Representation Techniques*, Studfuzz **202**, 3–16 (2006)
www.springerlink.com

instantiate any new results in the context of traditional relational database or deductive database technology, with the intention of integrating such systems on our experimental UAV platforms as stand-alone services which can be used by other autonomous functionality integrated in the UAV platforms. An attempt has also been made to show how one could ground such reasoning mechanisms in such systems through the use of machine learning techniques for generating rough relations or classifiers which can then be embedded as part of the knowledge representations stored in extended database format.

This book contains a cohesive, self-contained collection of many of the theoretical and applied research results that have been achieved so far, and which for the most part pertain to nonmonotonic and approximate reasoning systems developed for the experimental UAV platforms. The work is far from complete and the longer term goals have not yet been reached. That being said, this book provides a foundation for continuing research along the lines sketched above. Throughout the book, it is more or less assumed that the background paradigm or framework in which approximate and nonmonotonic reasoning systems can be embedded is that of an agent or multi-agent framework. Although agent architectures are not developed in detail, it is assumed that each agent will have one or more knowledge bases containing both static and dynamic information about itself and its surrounding environment. From this perspective, an agent can be either a software agent (or softbot), or an actual physical artifact. In both cases, one would have to deal with the sense/reasoning gap in order for knowledge representation and inference mechanisms to be of practical use in such systems. In the following sections, intelligent artifacts and the sense/reasoning gap are discussed in further detail and an actual physical artifact used in our experimentation is described.

1.2 Intelligent Artifacts and Agents

The use of intelligent artifacts, both at the workplace and in the home, is becoming increasingly more pervasive due to a number of factors which include the accessibility of the Internet/World-Wide-Web to the broad masses, the drop in price and increase in capacity of computer processors and memory, and the integration of computer technology with telecommunications. Intelligent artifacts are man-made physical systems containing computational equipment and software that provide them with capabilities for receiving and comprehending sensory data, for reasoning, and for performing rational action in their environment. The spectrum of capabilities and the sophistication of an artifact's ability to interface to its environment and reason about it varies with the type of artifact, its intended tasks, the complexity of the environment in which it is embedded, and its ability to adapt its models of the environment at different levels of knowledge abstraction. Representative examples of intelligent artifacts ranging from less to more complex would be mobile telephones, personal digital assistants (PDAs), softbots on the World-Wide-Web, collections of distributed communicating artifacts which serve as components of smart homes, mobile robots, unmanned aerial vehicles, and many more.

One unifying conceptual framework that can be used to view these increasingly more complex integrated computer systems is as societies of agents (virtually and/or physically embedded in their respective environments) with the capacity to acquire information about their environments, structure the information and interpret it as knowledge, and use this knowledge in a rational manner to enhance goal-directed behavior which is used to achieve tasks and to function robustly in their dynamic and complex environments.

1.3 Knowledge Representation

An essential component in agent architectures is the agent's knowledge representation component which includes a variety of knowledge and data repositories with associated inference mechanisms. The knowledge representation component is used by the agent to provide it with models of its embedding environment and of its own and other agent capabilities, in addition, to reasoning efficiently about them. It is becoming increasingly important to move away from the notion of a single knowledge representation mechanism with one knowledge source and inference method to multiple forms of knowledge representation with several inference methods. This viewpoint introduces an interesting set of complex issues related to the merging of knowledge from disparate sources and the use of adjudication or conflict resolution policies to provide coherence of knowledge sources (see also Chapter 9).

Due to the embedded nature of these agent societies in complex dynamic environments, it is also becoming increasingly important to take seriously the gap between access to low-level sensory data and its fusion and integration with more qualitative knowledge structures. These signal-to-symbol transformations should be viewed as an on-going process with a great deal of feedback between the levels of processing. In addition, because the embedding environments are often as complex and dynamic as those faced by humans, the knowledge representations which are used as models of the environment must necessarily be partial, elaboration tolerant and approximate in nature.

Figure 1.1 provides a schematic of the sense/reasoning gap. In any robotic system or complex agent system, there is a continual flow of data throw the system and much processing along the way. As sensor data is input into an agent system, more traditional sensor fusion techniques are used to construct quantitative models used for navigation or to control manipulators. For sophisticated autonomous behavior, these models are not adequate due to the limited temporal horizons assumed and the limited predictive or anticipatory capability which can be derived from such models. Intelligent behavior

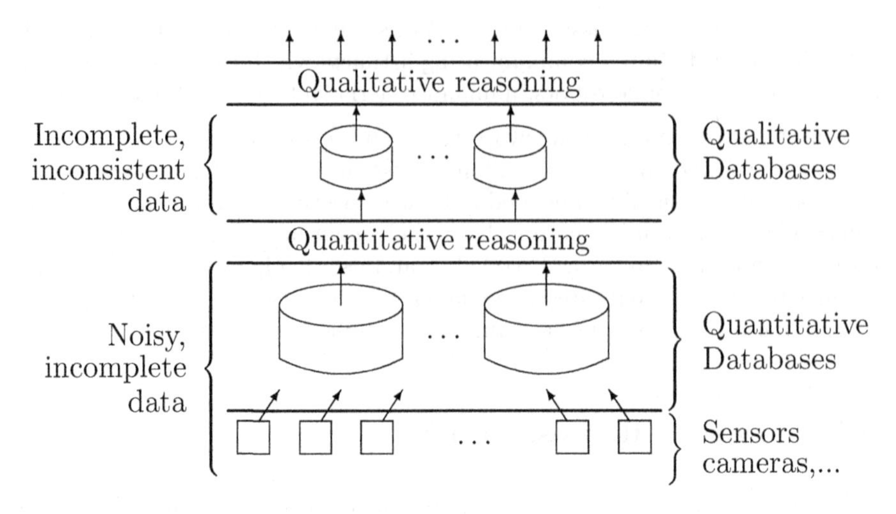

Fig. 1.1. Sense/reasoning gap.

requires more complex qualitative models which are constructed from properties of entities and relations between entities. Such entities can be physical or epistemic. Generally, neither complete information about entities, nor sufficient and necessary conditions for relations are present in the models that can be constructed from noisy and incomplete sensor data. So there is a flow of data from and into quantitative to combined quantitative/qualitative to qualitative structures which is a necessary part of any robotic architecture.

Research in this book is structured along the lines depicted in the figure, one part which takes us from sensor data to relations of a quantitative/qualitative nature and another part which takes us from these relations to more complex qualitative knowledge representations useful for robotic agents. In related work, more pragmatic issues which deal with the practical distributed data flow management through complex robotic architectures has been investigated, but will not be covered in this book.

Concept Acquisition and Fluid Knowledge Structures

Just as an agent's surrounding environment is in flux, so are many of its knowledge structures. For example approximate relations may initially be defined with very weak sufficient and/or necessary conditions, but as the agent learns more about its environment these conditions may be tightened and even sometimes loosened. The grounding of approximate concepts and relations is contextual and based on the quality and quantity of sensory data at hand. Part of the approach pursued in this book, is to develop a framework for the specification, implementation and management of fluid knowledge structures containing both quantitative and qualitative components, where the knowledge structures are grounded in the embedding environments in which they are used. We assume a fine granularity as a basis for concept acquisition, grounding and knowledge structure design which is the result of using intuitions from rough set theory.

Approximate reasoning can be made (self-)adaptive by applying machine learning techniques and tuning various parameters with the goal to minimize the sizes of boundary regions of relations. To do this, assume that certain concepts which we call primitive concepts have been acquired through a machine learning process where learning samples are provided from sensor data and approximations of concepts are induced from the data. One particularly interesting approach to this is the use of rough set based supervised machine learning techniques. It is important to emphasize that the induced concepts are approximate in nature and fluid in the sense that additional machine learning may modify the concept. In other words, concepts are inherently contextual and subject to elaboration and change in a number of ways. Primitive concepts may change as new sensor data is acquired and fused with existing data through diverse processes associated with particular sensory platforms. At some point, constraints associated with other more abstract concepts having dependencies with primitive concepts may influence the definition of the primitive concept.

As an example of these ideas, take a situation involving an UAV operating over a road and traffic environment. In this case, the meaning of concepts such as *fast* or *slow*, *small* or *large vehicle*, *near*, *far*, or *between*, will have a meaning different from that in another application with other temporal and spatial constraints. Assuming these primitive concepts as given and that they are continually re-grounded in changes in the operational environment via additional machine learning or sensor fusion, we would then like to use these primitive concepts as the *ur*-elements in our knowledge representation structures. Since these *ur*-elements are inherently approximate, contextual and elaboration tolerant in nature, any knowledge structure containing these concepts should also inherit or be influenced by these characteristics. In fact, there are even more primitive *ur*-elements in the system we envision which can be used to define the primitive concepts themselves if a specific concept learning policy based on rough sets is used. These are the elementary sets used in rough set theory to define contextual approximations to sets. This book investigates a number of techniques for grounding concepts and relations through the use of machine learning techniques and using primitive concepts and relations to construct more abstract approximate knowledge structures.

In the following sections, we will describe an experimental UAV platform in detail which has been used in our experimentation and which has been the driving force in developing the techniques in this book.

1.4 The WITAS UAV Experimental Platform

One of the long-term goals for the use of the research results described in this book is to use them as a basis for specifying, constructing and managing a particular class of approximate knowledge structures in intelligent artifacts. In our current research, the particular artifact we use as an experimental platform is an unmanned aerial vehicle flying over operational environments populated by traffic. In such scenarios, knowledge about both the environment below and the unmanned aerial vehicle agent's own epistemic state must be acquired in a timely manner in order for the knowledge to be of use to the agent while achieving its goals. Consequently, the results must provide for an efficient implementation of both the knowledge structures themselves and the inference mechanisms used to query these structures for information.

WITAS (pronounced vee-tas) is an acronym for the Wallenberg Information Technology and Autonomous Systems Laboratory at Linköping University, Sweden. The WITAS UAV Project was a long term project (1997-2005) with the goal of designing, specifying and implementing the IT subsystem for an intelligent autonomous aircraft and embedding it in an actual platform. Although the WITAS UAV project is finished, related work in the area will continue to be pursued. We have been using a Yamaha RMAX VTOL (vertical take-off and landing system) developed by Yamaha Motor Company Ltd., as a platform of choice.

An important part of the project involved identifying core functionalities required for the successful development of such systems and doing basic and applied research in the areas identified. The topics associated with this book fall under the umbrella of one such core functionality: approximate knowledge structures and their associated inference mechanisms.

The project encompassed the design of a command and control system for a UAV and its integration in a suitable deliberative/reactive architecture; the design of high-level cognitive tasks, intermediate reactive behaviors, low-level control-based behaviors and their integration with each other; the integration of sensory capabilities with the command and control architecture, in particular the use of an active vision system; the development of hybrid, mode-based low-level control systems to supervise and schedule control behaviors; the signal-to-symbol conversions from sensory data to qualitative structures used in mediating choice of actions and synthesizing plans to attain operational mission goals; and the development of the systems architecture for the physical UAV platform.

In addition the project also encompassed the design and development of the necessary tools and research infrastructure required to achieve the goals of the project. This included the development of model-based distributed simulation tools and languages used in the concurrent engineering required to move incrementally from software emulation and simulation to the actual hardware components used in the final product.

The operational environment used is over widely varying geographical terrain with traffic networks and vehicle interaction of varying degrees of density. Possible applications are emergency services assistance, monitoring and surveillance, use of a UAV as a mobile sensory platform in an integrated real-time traffic control system and photogrammetry applications. Figure 1.2 shows a bird's eye view of one of the operational environments and test flight areas used in the project.

Fig. 1.2. Revinge Emergency Services Training Area in southern Sweden.

Much effort has gone into the development of useful ground control station interfaces which encourage the idea of *push-button missions*, letting the system itself plan and execute complex missions with as little effort as possible required from the ground operator other than stating mission goals at a high-level of abstraction and monitoring the execution of the ensuing mission.

An example of such a push-button mission that has been used as an application scenario in our research is a combined monitoring/surveillance and photogrammetry mission out in the field in an urban area with the goal of investigating facades of building structures and gathering both video sequences and photographs of building facades. For this experiment, we have used the Yamaha RMAX helicopter system as a platform. Let's assume the operational environment is in an urban area with a complex configuration of building and road structures. A number of these physical structures are of interest since one has previously observed suspicious behavior and suspects the possibility of terrorist activity. The goal of the mission is to investigate a number of these buildings and acquire video and photos from each of the building's facades. It is assumed the UAV has a 3D model of the area and a GIS with building and road structure information on-line.

The ground operator would simply mark building structures of interest on a map display and press a button to generate a complete multi-segment mission that flies to each building, moves to waypoints to view each facade, positions the camera accordingly and begins to relay video and/or photographs. The motion plans generated are also guaranteed to be collision-free from static obstacles. If the ground operator is satisfied with the generated mission, he or she simply clicks a confirm button and the mission begins. During the mission, the ground operator has the possibility of suspending the mission to take a closer look at interesting facades of buildings, perhaps taking a closer look into windows or openings and then continuing the mission. This mission has been successfully executed robustly and repeatedly from take-off to landing using the RMAX.

The UAV experimental platform offers an ideal environment for experimentation with the knowledge representation framework we propose, because the system architecture is rich with different types of knowledge representation structures, the operational environment is quite complex and dynamic, and signal-to-symbol transformations of data are an integral part of the architecture. In addition, much of the knowledge acquired by the UAV will be necessarily approximate in nature. In several parts of the book, we will use examples from this application domain to describe and motivate some of our techniques. In the following sections, we provide a more detailed description of the hardware and software components which make up the UAV platform and which make such complex missions possible to execute autonomously.

1.4.1 The Hardware Platform

The WITAS UAV platform [48] is a slightly modified Yamaha RMAX helicopter (Figure 1.3). It has a total length of 3.6 m (including main rotor) and is powered by a 21 hp two-stroke engine with a maximum takeoff weight of 95 kg. The helicopter has a built-in attitude sensor (YAS) and an attitude control system (YACS). The hardware platform developed during the WITAS UAV project is integrated with the Yamaha platform as shown in Figure 1.4. It contains three PC104 embedded computers. The primary flight control (PFC)

Fig. 1.3. The WITAS RMAX helicopter in an urban environment.

system runs on a PIII (700Mhz), and includes a wireless Ethernet bridge, a GPS receiver, and several additional sensors including a barometric altitude sensor. The PFC is connected to the YAS and YACS, an image processing computer and a computer for deliberative capabilities. The image processing (IPC) system runs on the second PC104 embedded computer (PIII 700MHz), and includes a color CCD camera mounted on a pan/tilt unit, a video transmitter and a recorder (miniDV). The deliberative/reactive (DRC) system runs on the third PC104 embedded computer (Pentium-M 1.4GHz) and executes all high-end autonomous functionality. Network communication between computers is physically realized with serial line RS232C and Ethernet. Ethernet is mainly used for CORBA applications (see below), remote login and file transfer, while serial lines are used for hard real-time networking.

1.4.2 The Software Architecture

A hybrid deliberative/reactive software architecture has been developed for the UAV . Conceptually, it is a layered, hierarchical system with deliberative, reactive and control components, although the system can easily support both vertical and horizontal data and control flow. Figure 1.5 presents the functional layer structure of the architecture and emphasizes its reactive-concentric nature. Reactive task procedures (TPs) can call both deliberative and flight control services concurrently.

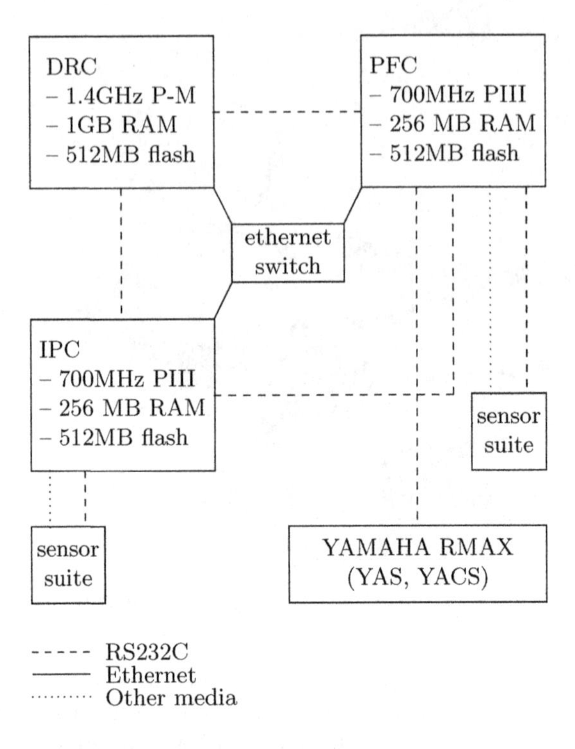

Fig. 1.4. On-board hardware schematic.

The software implementation is based on CORBA (Common Object Request Broker Architecture), which is often used as middleware for object-based distributed systems. It enables different objects or components to communicate with each other regardless of the programming languages in which they are written, their location on different processors or the operating systems they running on. A component can act as a client, a server or as both.

The functional interfaces to components are specified via the use of IDL (Interface Definition Language). The majority of the functionalities which are part of the architecture can be viewed as CORBA objects or collections of objects, where the communication infrastructure is provided by CORBA facilities and other services such as real-time and standard event channels. This architectural choice provides us with an ideal development environment and versatile run-time system with built-in scalability, modularity, software relocatability on various hardware configurations, performance (real-time event channels and schedulers), and support for plug-and-play software modules.

With respect to timing characteristics, the architecture can be divided into two layers:

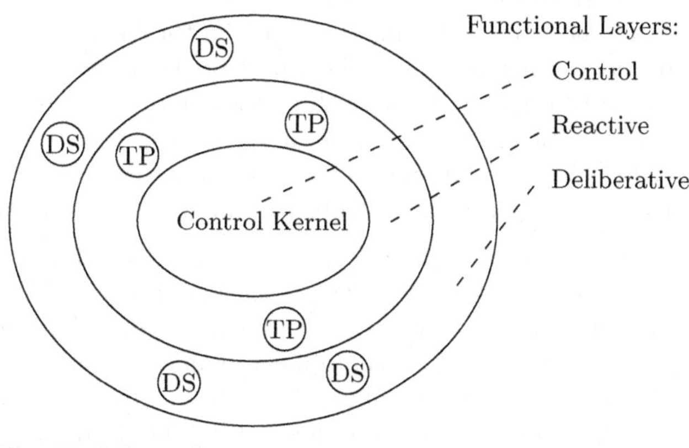

Fig. 1.5. Functional structure of the architecture.

1. the hard real-time part, which mostly deals with hardware and control laws (also referred to as the Control Kernel)
2. the non real-time part, which includes deliberative services of the system (also referred to as the High-level system) [2].

All three computers in our UAV platform (i.e., PFC, IPC and DRC) have both hard and soft real-time components but the processor time is assigned to them in different proportions.

On one extreme, the PFC runs mostly hard real-time tasks with only minimum user space applications (e.g., SSH daemon for remote login). On the other extreme, the DRC uses the real-time part only for device drivers and real-time communication. The majority of processor time is spent on running the deliberative services. The deliberative services include, among others, a Path Planner, a Task Procedure Execution Module, a Helicopter Server which encapsulates the Control Kernel (CK) of the UAV system, a Task Planner, a Chronicle Recognition System, DyKnow (a data stream manager), A Geographic Information System (GIS), and an Approximate Deductive Database System. The latter is presented in this book.

The CK is a distributed real-time runtime environment and is used for accessing the hardware, implementing continuous control laws, and control mode switching. Moreover, the CK coordinates the real-time communication between all three on-board computers as well as between CKs of other robotic

[2] Note that distinction between the Control Kernel and the High-level system is done based mainly on the timing characteristics and it does not exclude, for example, placing some deliberative services (e.g., prediction) in the Control Kernel.

systems. In our case, we perform multi-platform missions with two identical RMAX helicopter platforms developed in the WITAS UAV project. The CK is implemented using C code. This part of the system uses the Real-Time Application Interface (RTAI) which provides industrial-grade real time operating system functionality. RTAI is a hard real-time extension to a standard Linux kernel (Debian in our case) and has been developed at the Department of the Aerospace Engineering of Politecnico di Milano (DIAPM).

The high-level part of the system has reduced timing requirements and is responsible for coordinating the execution of reactive Task Procedures (TPs). This part of the system uses CORBA as its distribution backbone. A TP is a high-level procedural execution component which provides a computational mechanism for achieving different robotic behaviors by using both deliberative and control components in a highly distributed and concurrent manner.

The control and sensing components of the system are accessible for TPs through the Helicopter Server which in turn uses an interface provided by the Control Kernel. A TP can initiate one of the autonomous control flight modes available in the UAV (i.e., take off, vision-based landing, hovering, dynamic path following or reactive flight modes for interception and tracking). The high-level deliberative services are accessible to TPs in a client-server relationship, where the deliberative services such as the approximate database service are wrapped as CORBA servers and have specific IDL interfaces for access and use. Additionally, TPs can control the payload of the UAV platform which currently consists of the video camera mounted on a pan-tilt unit. TPs receive data delivered by the PFC and IPC computers, i.e., helicopter state and camera system state (including image processing results), respectively. The Helicopter Server on one side uses CORBA to be accessible by TPs or other components of the system, on the other side it communicates through shared memory with the HCSM based interface running in the real-time part of the DRC software.

One of the challenges when working with such complex artifacts is to take seriously all the hard and soft realtime constraints present in such systems in addition to the noisy and incomplete sensor data which has to be integrated with other qualitative data present in the system. Integrating knowledge representation and reasoning components with such systems is a real challenge and often influences the way one thinks of such functionalities from the theory stage to application. The results in this book have been very much influenced by this perspective.

1.5 Book Structure

The book consists of three parts.

Part I (Chapters 2-5) collects background material needed in the rest of the book. In Chapter 2 the reader will find a summary of set-theoretic notions, a presentation of classical (propositional, first- and second-order) logic, together with fixpoint calculus and three-valued propositional logic, and a brief exposition of basic notions of computational complexity. Chapter 3 provides a short introduction to rough sets theory. This theory can be viewed as a foundation of many solutions developed in this book. Chapter 4 is devoted to relational and deductive databases. In particular, we discuss here various queries languages, including Datalog and Datalog with negation. Finally, Chapter 5 is a brief presentation of non-monotonic reasoning, with emphasis put on two of the most prominent non-monotonic formalisms, namely default logic and circumscription.

Part II (Chapters 6-11) provides a number of techniques that can be used to represent knowledge on the basis of incomplete, imprecise and (sometimes) incorrect information. It is assumed that this information is given in the form of particular relations, extracted from sensor observations and various high-level rules provided by an expert or by machine learning methods. In Chapter 6, we study *rough knowledge databases*. These can be regarded as generalizations of classical databases, where classical relations have been replaced by the rough ones. In Chapter 7, we show how crisp and rough knowledge can be combined. We introduce here a notion of *approximation transducer* which provides a means of generating an approximate relation (the output) in terms of other approximate relations (the input) using a logical theory specifying relationships between the input and the output. In Chapter 8, we discuss important logical concepts of *weakest sufficient and strongest necessary conditions.* These notions can be used in many practical applications including building communication interfaces between agents using different languages, information hiding, knowledge compilation and abduction. Chapter 9 presents the CAKE methodology which provides a means for constructing and visualizing complex inference patterns associated with rough relations and default reasoning.[3]. In Chapter 10, we illustrate how CAKE can be used to formalize a subset of default logic. Finally, in Chapter 11, we provide a small case study, based on the WITAS UAV application domain, to illustrate various knowledge representation and reasoning techniques presented in the earlier chapters of Part II.

Part III of the book (Chapters 12-15) is concerned with low-level knowledge representation techniques and concepts, i.e., techniques and concepts which are useful while extracting relations from sensor data. Chapter 12 introduces the idea of an *information granule.* This concept forms a basis for *granular*

[3] CAKE is an acronym for *Computer Aided Knowledge Engineering.*

computing which, in turn, provides a bridge between data analysis tools and logic-based approaches to knowledge representation. In Chapter 13, we discuss *tolerance spaces* which allow us to transform quantitative representations of concepts based on a notion of similarity into quantitative representations of concepts. Chapter 14 reviews a rough set approach to machine learning. Finally, Chapter 15 is devoted to a case study showing a UAV learning process.

1.6 Related Work

Information about the WITAS UAV project and its continuation can be found at the following websites [216, 215]. [45] provides an overview on the project as pertains to knowledge representation and UAVs and [47] provides an earlier general overview of the project. [48, 122] describe the software architecture used on the RMAX and [39, 123] describe the path following and vision-based landing modes used by the RMAX, respectively. [156, 155, 235] describe work with sample-based motion planning for UAVs.

Acknowledgments

The research and writing of this book has been supported in part by the Wallenberg Foundation under the WITAS UAV Project (1997 - 2205). Witold Łukaszewicz, Andrzej Skowron and Andrzej Szałas have also been supported by KBN grants 8 T11C 009 19 and 8T11C 025 19 in addition to MNiI grants 3T11C 002 26 and 3T11C 023 29.

We would also like to acknowledge the Department of Computer and Information Science of Linköping University, Sweden and the University of Economics and Computer Science in Olsztyn, Poland for providing additional financial support and a stimulating research environments.

2

Basic Notions

2.1 BNF Notation

We define the syntax of various logical languages using BNF notation with some commonly used additions. Elements (words) of the defined language are called *terminal symbols. Syntactic categories*, i.e., sets of well-formed expressions are represented by *non-terminal symbols* and denoted by $\langle Name \rangle$, where $Name$ is the name of a category. Syntactic categories are defined over non-terminal and terminal symbols using rules of the form:

$$\langle S \rangle ::= E_1 \mid\mid E_2 \mid\mid \ldots \mid\mid E_k$$

meaning that $\langle S \rangle$ is to be the least set of words containing only terminal symbols and formed according to the expression E_1 or E_2 or ... or E_k. Notation $\{E\}$ is used to indicate that expression E can be repeated 0 or more times and $[E]$ denotes that expression E is optional, i.e., might or might not occur.

Note also that we use notation $\langle X \rangle$, where X is a set of elements, to denote the syntactic category consisting of all elements of X.

Example 2.1.1. Assume we want to define arithmetic expressions containing the variable x, the addition symbol $+$ and parentheses "(" and ")." Terminal symbols are then $x, +, (,)$. We use one non-terminal symbol, $\langle Expr \rangle$, representing well-formed expressions. The following rule defines the syntactic category $\langle Expr \rangle$, i.e., the set of all well-formed expressions:

$$\langle Expr \rangle ::= x \mid\mid \langle Expr \rangle + \langle Expr \rangle \{+\langle Expr \rangle\} \mid\mid (\langle Expr \rangle).$$

Now, for instance, $(x+x+x)+x$ is a well-formed expression, but $x+\langle Expr \rangle$ is not, since $\langle Expr \rangle$ is not a terminal symbol. □

P. Doherty et al.: *Knowledge Representation Techniques*, Studfuzz **202**, 17–38 (2006)
www.springerlink.com

2.2 Sets, Relations, Functions

We assume that the reader is familiar with the algebra of sets. As usual, we write:

- $\emptyset$ to denote the *empty set*
- ω to denote the set of natural numbers
- $\mathcal{R}$ to denote the set of real numbers
- $e \in A$ to mean that an individual e *belongs* to (or is a *member*, or is an *element* of) the set A
- $A \subseteq B$ to mean that A is a *subset* of B (A is *included in* B).

The set of all subsets of a set A is called the *powerset* of A and is denoted by $\text{Pow}(A)$. By Dom we denote a given *universe*. The *cardinality* of a set A is denoted by $|A|$.[1] *Sequences* (*tuples*) of elements $a_1, \ldots, a_k$ are denoted by $\langle a_1, \ldots, a_k \rangle$.

We also use the standard notation:

- $-A \stackrel{\text{def}}{=} \{a : a \notin A\}$ to denote the *complement* of A
- $A - B \stackrel{\text{def}}{=} \{a : a \in A \text{ and } a \notin B\}$ to denote the *difference* of A and B,
- $A \cup B \stackrel{\text{def}}{=} \{a : a \in A \text{ or } a \in B\}$ to denote the *union* of A and B,
- $A \cap B \stackrel{\text{def}}{=} \{a : a \in A \text{ and } a \in B\}$ to denote the *intersection* of A and B,
- $A_1 \times \ldots \times A_k \stackrel{\text{def}}{=} \{\langle a_1, \ldots, a_k \rangle : a_1 \in A_1, \ldots, a_k \in A_k\}$ to denote the *Cartesian product* of sets $A_1, \ldots, A_k$, where k is a natural number. By A^k we denote the Cartesian product $\underbrace{A \times \ldots \times A}_{k \text{ times}}$.

By a *multiset* we understand a set that can contain many copies of the same elements.

By a *covering of a set* A we understand a family $\{A_i\}_{i \in I}$ of non-empty subsets of A, such that $A = \bigcup_{i \in I} A_i$. A covering $\{A_i\}_{i \in I}$ is called a *partition of the set* A if and only if for all $i, j \in I$, if $i \neq j$ then $A_i \cap A_j = \emptyset$.

For any natural number k, a k-argument *relation* over the sets $A_1, \ldots, A_k$ is any subset of the Cartesian product $A_1 \times \ldots \times A_k$. A *total relation* over the sets $A_1, \ldots, A_k$ is the whole Cartesian product $A_1 \times \ldots \times A_k$. By a k-argument *function* f from $A_1 \times \ldots \times A_k$ to a set A, denoted by $f : A_1 \times \ldots \times A_k \longrightarrow A$,

[1] Through this book we mainly deal with finite sets, where cardinality of a set is simply the number of elements of the set.

we mean a relation over $A_1 \times \ldots \times A_k \times A$ such that for any $\langle a_1, \ldots, a_k \rangle \in A_1 \times \ldots \times A_k$ there is exactly one $a \in A$ such that $\langle a_1, \ldots, a_k, a \rangle \in f$. We call such an a a *value* of f and denote it by $f(a_1, \ldots, a_k)$.

By a *composition of binary relations* R_1 and R_2 we mean the relation

$$R_1; R_2 \stackrel{\text{def}}{=} \{\langle a, b \rangle \mid \text{ there is } c \text{ such that } \langle a, c \rangle \in R_1 \text{ and } \langle c, b \rangle \in R_2\}.$$

Many relations, defined on a set, say A, may satisfy some important properties. Among those properties, perhaps the most frequently occurring are specified below, where the symbol $\leq$ stands for the considered relation,

- *reflexivity*: for all $x \in A$, $x \leq x$
- *symmetry*: for all $x, y \in A$, if $x \leq y$ then $y \leq x$
- *anti-symmetry*: for all $x, y \in A$, if $x \leq y$ and $y \leq x$ then $x = y$
- *transitivity*: for all $x, y, z \in A$, if $x \leq y$ and $y \leq z$ then $x \leq z$
- *linearity*: for all $x, y \in A$, $x \leq y$ or $y \leq x$.

By an *equivalence relation* on a set A we understand any binary relation R which is reflexive, symmetric and transitive. Equivalence relations provide us with a partition of a given set into subsets, called *equivalence classes* in such a way that each equivalence class contains elements equivalent w.r.t. a given relation. Let $\cong$ be an equivalence relation on A. Then for $a \in A$, by $[a]_{\cong}$ we denote the equivalence class of all elements equivalent to a, i.e., the set $\{e \in A : e \cong a\}$. By $A/\cong$ we denote the set of all equivalence classes of elements of A, i.e., $A/\cong \stackrel{\text{def}}{=} \{[a]_{\cong} \mid a \in A\}$.

Example 2.2.1. The most common equivalence relation is the equality relation $=$. Its equivalence classes contain singletons.

Equivalence relations are often used in order to abstract from irrelevant features of objects. For instance, in order to classify persons by age, one can define the following equivalence relation $\sim$ on the set Per of persons (assuming that a function $age : Per \longrightarrow \omega$ is given), where $p, p' \in Per$:

$$p \sim p' \text{ if and only if } age(p) = age(p').$$

Equivalence classes of relation $\sim$ consist of persons of the same age, belonging to Per. □

Among relations, orderings play an important rôle. By a *partial order* on a set A we shall understand any binary relation $\leq$ which is reflexive, anti-symmetric and transitive. By $a < b$ we shall mean that $a \leq b$ and $a \neq b$. The order $<$ is often referred to as *strict partial order*. If a partial order is linear, then we say that it is a *linear order* on A. A strict partial order satisfying the linearity condition is called a *strict linear order.*

We often use the notion of the *transitive closure* of a binary relation R, denoted by $\text{TC}(R)$, and defined as the least transitive relation containing R, i.e., the least relation containing R and closed under the following condition:

$$\left.\begin{array}{c}\text{TC}(R)(x,y)\\ \text{and}\\ \text{TC}(R)(y,z)\end{array}\right\}\text{ imply } \text{TC}(R)(x,z).$$

It is easily observed that the transitive closure over finite domains can alternatively be defined by means of compositions of relation R with itself as follows:

$$\text{TC}(R) = \bigcup_{k>1} \underbrace{R;\ldots;R}_{k \text{ times}}.$$

The following example illustrates notions of orderings and transitive closure.

Example 2.2.2.

1. The inclusion relation $\subseteq$ is a partial order on sets.
2. The standard relation $\leq$ defined on real numbers is a linear order.
3. Consider a set *Per* containing persons. One can define a binary relation *Parent* on *Per*, such that $Parent(x, y)$ means that x is a parent of y. The transitive closure, $\text{TC}(Parent)$ defines the ancestor relation on set *Per*. □

By the *reflexive closure* of a binary relation R we understand relation

$$R \cup \{\langle x, x\rangle \mid x \in \text{DOM}\},$$

and by the *symmetric closure* of R, the relation

$$R \cup \{\langle y, x\rangle \mid \langle x, y\rangle \in R\}.$$

2.3 Metric Spaces

One important concept used in this book is that of a distance. The concept of distance is formally introduced by means of metric spaces.

A *metric space* is a pair $\langle A, \delta\rangle$, where A is a set and δ is a function

$$\delta : A \times A \longrightarrow \mathcal{R}$$

which, for all $x, y, z \in A$, satisfies:

$\delta(x, y) \geq 0$
$\delta(x, y) = 0$ if and only if $x = y$ (*reflexivity*)
$\delta(x, y) = \delta(y, x)$ (*symmetry*)
$\delta(x, z) \leq \delta(x, y) + \delta(y, z)$ (triangle law).

Any function δ satisfying the above properties is called a *metric* for A and $\delta(x, y)$ is called the *distance* between x and y.

If δ is not required to satisfy the triangle law and satisfies the first three laws, then $\langle A, \delta \rangle$ is called a *semi-metric space.* In this case δ is called a *semi-metric* for A and $\delta(x, y)$ is called the *semi-distance* between x and y.

Example 2.3.1.

1. An example of a metric space is $\langle \mathcal{R}, \delta \rangle$, where $\mathcal{R}$ is the set of reals, and $\delta(x, y) \stackrel{\text{def}}{=} |x - y|$, where $|u|$ stands for the absolute value of u.
2. Another example is $\langle \mathcal{R} \times \mathcal{R}, \delta_E \rangle$, where

$$\delta_E(\langle x, y \rangle, \langle x', y' \rangle) \stackrel{\text{def}}{=} \sqrt{(x - x')^2 + (y - y')^2}.$$

The metric δ_E is called the *Euclidean metric.*

3. Consider $M = \langle \{\mathsf{red}, \mathsf{green}, \mathsf{blue}\}, \delta \rangle$, where δ is given by

$$\delta(x, y) \stackrel{\text{def}}{=} \begin{cases} 0.0 \text{ for } x = y \\ 1.0 \text{ for } x, y \in \{\mathsf{blue}, \mathsf{green}\} \text{ and } x \neq y \\ 2.0 \text{ for } x, y \in \{\mathsf{red}, \mathsf{blue}\} \text{ and } x \neq y \\ 4.0 \text{ for } x, y \in \{\mathsf{red}, \mathsf{green}\} \text{ and } x \neq y. \end{cases}$$

Then M is a semi-metric space and is not a metric space, since in this case $\delta(\mathsf{red}, \mathsf{blue}) + \delta(\mathsf{blue}, \mathsf{green}) \not\geq \delta(\mathsf{red}, \mathsf{green})$, which violates the triangle law. □

2.4 Computational Complexity

We consider *time complexity* and *space complexity* with the intuition that the time complexity refers to the time spent during the computing process and space complexity refers to the amount of memory used during the computing process. We assume the sequential machine model.[2]

[2] Observe that all reasonable models considered in the literature have the same computational power with regards to the complexity classes we are interested in.

In order to measure complexity, we need a parameter describing the input size. We shall always assume that the input size is a natural number. Let $n \in \omega$ be an input size and let $f : \omega \longrightarrow \omega$ be a function. We shall deal with the following complexity classes:

- $\text{DTIME}(f(n))$, $\text{NTIME}(f(n))$ - problems solvable by deterministic (respectively nondeterministic) algorithms in time $\leq f(n)$
- $\text{DSPACE}(f(n))$, $\text{NSPACE}(f(n))$ - problems solvable by deterministic (respectively nondeterministic) algorithms in space $\leq f(n)$.

The most important complexity classes considered in this book are:

$$\begin{aligned}
\text{PTIME} &\stackrel{\text{def}}{=} \bigcup_{k \in \omega} \text{DTIME}(n^k) \\
\text{NPTIME} &\stackrel{\text{def}}{=} \bigcup_{k \in \omega} \text{NTIME}(n^k) \\
\text{PSPACE} &\stackrel{\text{def}}{=} \bigcup_{k \in \omega} \text{DSPACE}(n^k) \\
\text{LOGSPACE} &\stackrel{\text{def}}{=} \text{DSPACE}(\log(n)) \\
\text{NLOGSPACE} &\stackrel{\text{def}}{=} \text{NSPACE}(\log(n)).
\end{aligned}$$

Observe that

$$\text{LOGSPACE} \subseteq \text{NLOGSPACE} \subseteq \text{PTIME} \subseteq \text{NPTIME} \subseteq \text{PSPACE}.$$

It is, however, unknown whether the inclusions are proper. Classes containing *complements* of the problems of a given class are denoted using the prefix CO- preceding the class name. For example, CO-NPTIME denotes the class of complements of problems from NPTIME.

Classes PTIME and LOGSPACE are considered *tractable*. However, the classes NPTIME and PSPACE are hypothesized to be *intractable* since no deterministic polynomial time algorithms for the latter classes are known.

By an *oracle for a problem* P we mean a querying mechanism giving answers to instances of P. Each call to the oracle is regarded as a single step. If C, C' are complexity classes, then by $C[C']$ we shall denote the class of problems with complexity in class C, provided that an oracle for C' is given. The complexity classes forming the *polynomial hierarchy*, $\Delta_k^P, \Sigma_k^P, \Pi_k^P$, for $k \in \omega$, are defined as follows:

$$\begin{aligned}
\Delta_0^P = \Sigma_0^P = \Pi_0^P &\stackrel{\text{def}}{=} \text{PTIME} \\
\Delta_{k+1}^P &\stackrel{\text{def}}{=} \text{PTIME}[\Sigma_k^P] \\
\Sigma_{k+1}^P &\stackrel{\text{def}}{=} \text{NPTIME}[\Sigma_k^P]
\end{aligned}$$

$$\Pi^P_{k+1} \stackrel{\text{def}}{=} \text{co-}\Sigma^P_{k+1}.$$

It can be proved that, for all $k \in \omega$, $\Sigma^P_k, \Pi^P_k, \Delta^P_k \subseteq$ PSPACE. It is also known that NPTIME, co-NPTIME $\subseteq \Sigma^P_2 \cap \Pi^P_2$.

We say that a problem Q is *polynomially reducible* to a problem P if and only if there is a deterministic polynomial time algorithm, Alg, which translates data d for Q into data $Alg(d)$ for P in such a way that d satisfies problem Q if and only if $Alg(d)$ satisfies problem P. Given a complexity class C we say that a problem P is *C-hard* if all other problems in C are polynomially reducible to P. P is called *C-complete* if it is C-hard and $P \in C$.

The complexity classes considered so far are examples of classes of so-called computable problems. Namely, we define *computable problems* as problems for which algorithms exist that compute answers to the problems and always terminate. However, in many areas of knowledge representation one deals with even more complex problems. A particularly important class is that of *partially computable problems*, i.e., problems for which there are algorithms answering TRUE if the given data satisfy the problem and, in the opposite case providing the answer FALSE or no answer at all (e.g., looping forever).[3] We further say that a problem is *uncomputable* if it is not computable.[4]

In this book we shall always provide tractable machinery for solving the considered knowledge engineering problems. In fact, one of our primary goals is to keep the complexity as low as possible, but still provide an expressive and powerful formalism for knowledge representation and applications.

2.5 Propositional Calculus

2.5.1 Introduction and Definitions

Let V_0 be a set of *propositional variables* (or *Boolean variables*), i.e., variables representing truth values TRUE, FALSE, standing for true and false, respectively. The set {TRUE, FALSE} is denoted by BOOL. We further assume that truth values are ordered, FALSE $\leq$ TRUE, and use min(...) and max(...) to denote the minimum and maximum of a given set of truth values.

We build *propositional formulas* (*sentences*) from truth values and propositional variables by applying propositional connectives $\neg, \wedge, \vee, \rightarrow, \equiv$, standing for *negation*, *conjunction*, *disjunction*, *implication* and *equivalence*, respectively. The set of propositional formulas is denoted by F_0. More formally, the syntax of propositional formulas is defined by the following rules:

[3] Of course, any computable problem is also partially computable.

[4] In particular, partially computable problems might be uncomputable.

$$\langle F_0 \rangle ::= \langle \textsc{Bool} \rangle \mid\mid \langle V_0 \rangle \mid\mid \neg \langle F_0 \rangle \mid\mid \langle F_0 \rangle \wedge \langle F_0 \rangle \mid\mid \langle F_0 \rangle \vee \langle F_0 \rangle \mid\mid \\ \langle F_0 \rangle \rightarrow \langle F_0 \rangle \mid\mid \langle F_0 \rangle \equiv \langle F_0 \rangle \mid\mid (\langle F_0 \rangle)$$

To make formulas easier to comprehend, the brackets ( ) will be sometimes replaced by [ ] or { }.

The semantics of propositional formulas is given by assigning truth values to propositional variables and then calculating values of formulas. Let

$$v : V_0 \longrightarrow \textsc{Bool}$$

be such an assignment (called a *valuation of propositional variables*). Then v is extended to define the *truth value* of propositional formulas as follows:

$$\begin{aligned} v(\neg A) &= \begin{cases} \textsc{True} \text{ if } v(A) = \textsc{False} \\ \textsc{False} \text{ otherwise} \end{cases} \\ v(A \wedge B) &= \min(v(A), v(B)) \\ v(A \vee B) &= \max(v(A), v(B)) \\ v(A \rightarrow B) &= \textsc{True} \text{ if and only if } v(A) \leq v(B) \\ v(A \equiv B) &= \textsc{True} \text{ if and only if } v(A) = v(B). \end{aligned}$$

A propositional formula A is *satisfiable* if there is a valuation v such that $v(A) = \textsc{True}$. It is a *tautology* if for all valuations v we have $v(A) = \textsc{True}$.

By a *propositional literal*, we understand either a propositional variable, or its negation, or a truth value from $\{\textsc{True}, \textsc{False}\}$. A literal is positive if it is TRUE or a variable, and is negative if it is FALSE or the negation of a variable. A *propositional term* is a conjunction of literals in which no propositional variable appears more than once. A *propositional clause* is any disjunction of propositional literals. A *propositional Horn clause* is a clause with at most one positive literal. We say that a formula is in *conjunctive normal form*, CNF, if it is a conjunction of clauses. It is in *disjunctive normal form*, DNF, if it is a disjunction of terms.[5] a formula is in *negation normal form*, NNF, if any negation occurs only in literals.

Any formula can be equivalently transformed into CNF, DNF and NNF. The transformation, into CNF or DNF may exponentially increase the size of the formula, while the transformation into NNF may increase or decrease the size of the formula by a constant factor.

[5] Note that we can always restrict ourselves to terms in which no variable appears more than once: each repeated occurrence of a variable p can be removed from a term, whereas any term including p and $\neg p$ can be replaced by FALSE.

2.5.2 Complexity of Reasoning

Theorem 2.5.1. *The problem of checking satisfiability of propositional formulas is* NPTIME-*complete. Checking whether a formula is a propositional tautology is a* CO-NPTIME-*complete problem.* □

Theorem 2.5.2. *The problem of checking satisfiability of propositional Horn clauses is in* PTIME. □

2.5.3 Prime Implicants

In this section, we introduce the idea of a *prime implicant* which plays a crucial role in Chapter 14.

Definition 2.5.3. *Let t be a term different from* FALSE *and suppose that A is a formula. We say that t is an* implicant *of A if and only if the formula $t \to A$ is a tautology. An implicant t of A is said to be* prime *if and only if no proper subterm of t is an implicant of A.* □

To provide a method of computing prime implicants, we need some terminology. We say that a term t_1 *absorbs* a term t_2 if and only if either t_1 is TRUE, or t_2 is FALSE, or t_1 is a subterm of t_2. For instance, the term a absorbs the term $a \wedge l$.

Let A be a formula in DNF. We write ABS(A) to denote the formula obtained from A by deleting all absorbed terms. A and ABS(A) are equivalent.

Two terms are said to have an *opposition* if and only if one of them contains a literal p and the other the literal $\neg p$. For instance, the terms $\neg p \wedge q$ and $p \wedge r$ have a single opposition, in the variable p.

Suppose that two terms, t_1 and t_2, have exactly one opposition. Then the *consensus* of t_1 and t_2, written $cons(t_1, t_2)$, is the term obtained from the conjunction $t_1 \wedge t_2$ by deleting the opposed variables, as well as any repeated variables. For example, $cons(\neg p \wedge q, p \wedge r)$ is $q \wedge r$.

Let A be a formula. The *Blake canonical form* of A, written BCF(A), is the formula obtained from A by the following construction:

1. replace A by its disjunctive normal form; denote the resulting formula by B
2. repeat as long as possible:

   if B contains a pair t_1 and t_2 of terms whose consensus exists and no term of B is a subformula of $cons(t_1, t_2)$, then
   $B := B \vee cons(t_1, t_2)$

3. take $\textsc{Abs}(B)$. This is $\textsc{Bcf}(A)$.

Proposition 2.5.4.

1. *Formulas A and* $\textsc{Bcf}(A)$ *are equivalent.*
2. $\textsc{Bcf}(A)$ *is the disjunction of all prime implicants of A.* □

The problem of transforming a formula into $\textsc{Bcf}$ is, in general, of a high complexity. In particular, the problem of finding a prime implicant is NPTIME-hard. Moreover, there can be exponentially many prime implicants for a given formula.

Example 2.5.5. Let A be

$$[(\neg p \lor \neg q \lor r) \land (\neg p \lor r \lor s)] \to (\neg p \land s).$$

The $\textsc{Dnf}$ form of A, denoted by B, is

$$(p \land q \land \neg r) \lor (p \land \neg r \land \neg s) \lor (\neg p \land s).$$

After performing Step (2) from the above construction, we get

$$(\neg p \land s) \lor (p \land q \land \neg r) \lor (p \land \neg r \land \neg s) \lor (q \land s \land \neg r). \tag{2.1}$$

Since $\textsc{Abs}(2.1) \equiv (2.1)$, the formula (2.1) is the Blake canonical form of A. Accordingly, prime implicants for formula A are:

$$\neg p \land s,\ p \land q \land \neg r,\ p \land \neg r \land \neg s,\ q \land s \land \neg r.$$

□

2.6 Predicate Calculus

2.6.1 Introduction and Definitions

Let V_{I} be a set of *individual variables* representing values of some domain. In order to define the language and semantics of *predicate calculus* (or, in other words, *first-order logic*) we assume that we are given a set of function symbols $\textsc{Fun} = \{f_i : i \in I\}$ and a set of relation symbols $\textsc{Rel} = \{R_j : j \in J\}$, where I, J are some finite sets. Functions and relations may have arguments. The number of arguments is called the *arity* of the function or relation, respectively. Functions and relations of arity 0 are called *individual constants* (or *constants*) and *Boolean constants*, respectively. The set of individual constants is denoted by CONST. Symbols of arity 1 are usually called *unary* and of arity 2 are called *binary*.[6] The set of function symbols and relation symbols together with their arities is called the *signature* or *vocabulary.*

Functional expressions in predicate calculus are represented by terms. We define the set of *terms*, denoted by TERMS, by the following rule:

$$\langle \text{TERMS} \rangle ::= \langle \text{CONST} \rangle \mid\mid \langle V_\text{I} \rangle \mid\mid \langle \text{FUN} \rangle([\langle \text{TERMS} \rangle \{, \langle \text{TERMS} \rangle\}])$$

where the number of terms which are arguments of the function symbol $\langle \text{FUN} \rangle$ above is equal to the arity of the function symbol.

Terms without variables are called *ground terms.*

Atomic formulas are defined by means of the the following syntax rule:

$$\langle \text{ATOMIC FORMULA} \rangle ::= \langle \text{REL} \rangle([\langle \text{TERMS} \rangle \{, \langle \text{TERMS} \rangle\}])$$

where the number of terms which are arguments of the relation symbol of $\langle \text{REL} \rangle$ above is equal to the arity of the relation symbol.

Atomic formulas without variables are called *ground formulas.*

Formulas of predicate calculus, denoted by F_I, are now defined by means of the following rule:

$$\begin{aligned} \langle F_\text{I} \rangle ::= \ & \langle \text{BOOL} \rangle \mid\mid \langle \text{ATOMIC FORMULA} \rangle \mid\mid \\ & \neg \langle F_\text{I} \rangle \mid\mid \langle F_\text{I} \rangle \wedge \langle F_\text{I} \rangle \mid\mid \langle F_\text{I} \rangle \vee \langle F_\text{I} \rangle \mid\mid \langle F_\text{I} \rangle \rightarrow \langle F_\text{I} \rangle \mid\mid \langle F_\text{I} \rangle \equiv \langle F_\text{I} \rangle \mid\mid \\ & \forall \langle V_\text{I} \rangle . \langle F_\text{I} \rangle \mid\mid \exists \langle V_\text{I} \rangle . \langle F_\text{I} \rangle \mid\mid (\langle F_\text{I} \rangle) \end{aligned}$$

The semantics of first-order formulas is given by a valuation of individual variables together with an interpretation of function symbols and relation symbols as functions and relations, respectively. The interpretation of function symbols and relation symbols is defined by *relational structures* of the form

$$\langle \text{DOM}, \{f_i^{\text{DOM}} : i \in I\}, \{R_j^{\text{DOM}} : j \in J\} \rangle,$$

where:

- DOM is a non-empty set, called the *domain* or *universe* of the relational structure
- f_i^{DOM} denotes a function corresponding to the function symbol f_i, for $i \in I$
- R_j^{DOM} denotes a relation corresponding to the relation symbol R_j, for $j \in J$.

[6] Observe that in the case of binary relations or functions we often use traditional infix notation. For instance we write $x \leq y$ rather than $\leq (x, y)$.

For the sake of simplicity, in the rest of the book we often abbreviate f_i^{DOM} and R_j^{DOM} by f_i and R_j, respectively.

For a given signature Sig, by $\text{STRUC}[Sig]$ we denote the class of all relational structures built over the signature Sig.

Let $v : V_\mathsf{I} \longrightarrow \text{DOM}$ be a valuation of individual variables. By v_a^x we denote the valuation obtained from v by assigning value a to variable x and leaving all other values of variables unchanged, i.e.,

$$v_a^x(z) = \begin{cases} a & \text{if } z = x \\ v(z) & \text{otherwise} \end{cases}$$

The valuation v is extended to provide values of terms as follows:

$$v(f(t_1, \ldots, t_k)) = f^{\text{DOM}}(v(t_1), \ldots, v(t_k)),$$

where $f \in \text{FUN}$ is a k-argument function symbol and $t_1, \ldots, t_k \in \text{TERMS}$. Then v is extended to define the *truth value* of first-order formulas as follows:

$$\begin{aligned}
v(R(t_1, \ldots, t_k)) &= R^{\text{DOM}}(v(t_1), \ldots, v(t_k)) \\
v(\neg A) &= \begin{cases} \text{TRUE} & \text{if } v(A) = \text{FALSE} \\ \text{FALSE} & \text{otherwise} \end{cases} \\
v(A \land B) &= \min(v(A), v(B)) \\
v(A \lor B) &= \max(v(A), v(B)) \\
v(A \to B) &= \text{TRUE if and only if } v(A) \leq v(B) \\
v(A \equiv B) &= \text{TRUE if and only if } v(A) = v(B) \\
v(\forall x.A(x)) &= \min(\{v_a^x(A(x)) : a \in \text{DOM}\}) \\
v(\exists x.A(x)) &= \max(\{v_a^x(A(x)) : a \in \text{DOM}\}),
\end{aligned}$$

where $R \in \text{REL}$ is a k-argument relation and A, B are formulas.

A first-order formula A is *satisfiable* if there is a relational structure

$$M = \langle \text{DOM}, \{f_i^{\text{DOM}} : i \in I\}, \{R_j^{\text{DOM}} : j \in J\} \rangle$$

and a valuation $v : V_\mathsf{I} \longrightarrow \text{DOM}$ such that its extension to F_I satisfies $v(A) = \text{TRUE}$. Formula A is *valid* in a relational structure M if for all valuations $v : V_\mathsf{I} \longrightarrow \text{DOM}$, $v(A) = \text{TRUE}$. In such a case we also say that M is a *model* for A. Formula A is a *tautology* if for all relational structures of suitable signature and all valuations v we have $v(A) = \text{TRUE}$. For a set of formulas $F \subseteq F_\mathsf{I}$ and a formula $A \in F_\mathsf{I}$, by *entailment* (or a *semantic consequence relation*),[7] denoted by $F \models A$, we mean that A is satisfied in any relational

[7] We also use the term "entailment" in more general context, but always assume that $G \models B$ means that formula B is true in all models of the set of formulas G, where formulas, validity and the notion of models depend on a given logic.

structure which is a model of all formulas of F. By $Cn(F)$ we mean the set of all semantic consequences of F, i.e.,

$$Cn(F) = \{A \in F_{\mathrm{I}} : F \models A\}.$$

A first-order formula is called *open* if it does not contain quantifiers. A variable occurrence is *free* in a formula if it is not bound by a quantifier, otherwise it is called *bound*. A formula is called *closed* (or a *sentence*) if it does not contain free occurrences of variables. Any set of sentences is called a *first-order theory*, or *theory*, for short.

In knowledge engineering applications it is usually assumed that the considered sets of sentences are finite. A finite theory is identified with the conjunction of its sentences.

A formula is in the *prenex normal form*, PNF, if all its quantifiers are in its prefix, i.e., if it is of the form $Q_1 x_1 \ldots . Q_k x_k . A$, where $Q_1, \ldots, Q_k \in \{\forall, \exists\}$ and A is an open formula. Any formula can be equivalently transformed into PNF. The transformation into PNF may increase or decrease the size of the formula by a constant factor.

By a *universal formula* we mean a formula in the prenex normal form, without existential quantifiers. A set of universal formulas is called a *universal theory*. By a *first-order literal* (or *literal*, for short) we understand an atomic formula or its negation. A *ground literal* is a literal without variables. A *first-order clause* is any, possibly empty, disjunction of first-order literals, preceded by a possibly empty prefix of universal quantifiers. A literal is *positive* if it is an atomic formula and is *negative* if it is the negation of an atomic formula. A relation symbol R *occurs positively* (respectively *negatively*) in a formula A if it appears under an even (respectively odd) number of negations.[8] A formula A is *positive w.r.t. relation symbol* R iff all occurrences of R in A are positive. A formula A is *negative w.r.t. relation symbol* R iff all occurrences of R in A are negative.

A relation symbol R is *similar to a formula* A if and only if the arity of R is equal to the number of free variables of A.

A *first-order Horn clause*, or *Horn clause*, for short, is a clause with at most one positive literal.

Semi-Horn formulas are defined by the following syntax rule:

$$\langle \text{SEMI-HORN FORMULA} \rangle ::= \langle \text{ATOMIC FORMULA} \rangle \rightarrow \langle F_{\mathrm{I}} \rangle \;||\; \tag{2.2}$$

$$\langle F_{\mathrm{I}} \rangle \rightarrow \langle \text{ATOMIC FORMULA} \rangle \tag{2.3}$$

[8] It is assumed here that all implications of the form $p \rightarrow q$ are substituted by $\neg p \vee q$ and all equivalences of the form $p \equiv q$ are substituted by $(\neg p \vee q) \wedge (\neg q \vee p)$.

where the formula $\langle F_{\text{I}} \rangle$ is an arbitrary classical first-order formula positive w.r.t. relation symbol represented by $\langle$ATOMIC FORMULA$\rangle$, and the only terms allowed in the atomic formula are variables. The atomic formula is called the *head* of the formula and the first-order formula is called the *body* of the formula. Semi-Horn formulas are assumed to be implicitly universally quantified, i.e., any free variable is bounded by an implicit universal quantifier. *Semi-Horn rules* (or *rules*, for short), are semi-Horn formulas in which the only terms are constant and variable symbols.

If the head of a rule contains a relation symbol R, we call the rule *semi-Horn w.r.t. R*. If the body of a rule does not contain the relation symbol appearing in its head, the rule is called *nonrecursive*. A conjunction of (nonrecursive) rules w.r.t. a relation symbol R is called a (*nonrecursive*) *semi-Horn theory w.r.t. R*.

We often consider slight variations of the syntax of semi-Horn formulas and rules. Namely, we also allow formulas of the form:

$$\langle \text{ATOMIC FORMULA} \rangle \{, \neg \langle \text{ATOMIC FORMULA} \rangle\} \rightarrow \langle F_{\text{I}} \rangle \tag{2.4}$$

$$\langle F_{\text{I}} \rangle \rightarrow \langle \text{ATOMIC FORMULA} \rangle \{, \neg \langle \text{ATOMIC FORMULA} \rangle\} \tag{2.5}$$

where the comma on the lefthand side of the formula (2.4) denotes the conjunction $\wedge$ and on the righthand side of the formula (2.5) denotes the disjunction $\vee$. This convention comes from sequent notation. Formulas of the forms (2.4) and (2.5) can equivalently be presented in the form of semi-Horn formulas respectively as follows:

$$\langle \text{ATOMIC FORMULA} \rangle \rightarrow \langle F_{\text{I}} \rangle \{ \vee \langle \text{ATOMIC FORMULA} \rangle\}$$

$$\langle F_{\text{I}} \rangle \{ \wedge \langle \text{ATOMIC FORMULA} \rangle\} \rightarrow \langle \text{ATOMIC FORMULA} \rangle.$$

2.6.2 Complexity of Reasoning

Using predicate calculus as a practical reasoning tool is somewhat questionable, because of the complexity of the logic. Existing first-order theorem provers solve the reasoning problem partially and exploit the fact that checking whether a first-order formula is a tautology is only partially computable.

The following theorem quotes the most important facts on the complexity of general first-order reasoning.

Theorem 2.6.1.

1. *The problem of checking whether a given first-order formula is a tautology is uncomputable but is partially computable.*
2. *The problem of checking whether a given first-order formula is satisfiable, is not partially computable.* □

Fortunately, when fixing a finite domain relational structure, one ends up in a tractable situation, as stated in the following theorem.

Theorem 2.6.2. *Assume we are given any fixed first-order formula. Then checking whether it is valid in a given finite domain relational structure is in* PTIME *and* LOGSPACE *w.r.t. the size of the domain.* □

If one would like to investigate properties of first-order formulas valid in all finite domain structures, one would end up in quite a complex situation, as we have the following theorem.

Theorem 2.6.3.

1. *The problem of checking whether a first-order formula is valid in all finite domain structures is not partially computable.*
2. *The problem of checking whether there is a finite domain structure satisfying a given first-order formula is partially computable but is not computable.* □

2.7 Second-order Logic

2.7.1 Introduction and Definitions

In knowledge representation it is often necessary to formulate properties using phrases of the form "there is a relation," "for any relation," i.e., to use quantifiers over relations. Such quantifiers are called *second-order quantifiers* and are allowed in *second-order logic.*

Second-order logic is an extension of the predicate calculus obtained by admitting second-order quantifiers. In order to define this logic we have to add variables representing relations. The set of *relational variables* is denoted by V_{II}.[9]

Formulas of second-order logic, denoted by F_{II}, are defined by means of the following rules.

$$\begin{aligned}\langle F_{\mathrm{II}}\rangle ::= \; & \langle \mathrm{BOOL}\rangle \;||\; \langle V_{\mathrm{II}}\rangle \;||\; \langle F_{\mathrm{I}}\rangle \;||\; \neg\langle F_{\mathrm{II}}\rangle \;||\; \langle F_{\mathrm{II}}\rangle \wedge \langle F_{\mathrm{II}}\rangle \;||\; \langle F_{\mathrm{II}}\rangle \vee \langle F_{\mathrm{II}}\rangle \;|| \\ & \langle F_{\mathrm{II}}\rangle \to \langle F_{\mathrm{II}}\rangle \;||\; \langle F_{\mathrm{II}}\rangle \equiv \langle F_{\mathrm{II}}\rangle \;||\; \forall\langle V_{\mathrm{I}}\rangle.\langle F_{\mathrm{II}}\rangle \;||\; \exists\langle V_{\mathrm{I}}\rangle.\langle F_{\mathrm{II}}\rangle \;|| \\ & \forall\langle V_{\mathrm{II}}\rangle.\langle F_{\mathrm{II}}\rangle \;||\; \exists\langle V_{\mathrm{II}}\rangle.\langle F_{\mathrm{II}}\rangle \;||\; (\langle F_{\mathrm{II}}\rangle)\end{aligned}$$

[9] Relational variables are sometimes called *second-order variables.* In second-order logic one can also consider function variables, but we shall not use this possibility here.

By an *existential fragment of second-order logic* we shall mean the set of second-order formulas, whose second-order quantifiers can only be existential and appear in front of the formula.

The semantics of second-order logic is an extension of the semantics of the predicate calculus. We then only have to provide the semantics for relational variables and second-order quantifiers.

Let $R \in \text{REL}$ be a k-argument relation symbol. Assume we are given a relational structure $M = \langle \text{DOM}, \{f_i^{\text{DOM}} : i \in I\}, \{R_j^{\text{DOM}} : j \in J\} \rangle$,. Denote by $\text{REL}(M)$ the set of all relations over DOM. Let $v' : V_{\text{I}} \longrightarrow \text{DOM}$ be a valuation of individual variables and $v'' : V_{\text{II}} \longrightarrow \text{REL}(M)$ be a valuation assigning relations to relational variables. Valuations v', v'' can be extended to the valuation assigning truth values to second-order formulas, $v : F_{\text{II}} \longrightarrow \text{BOOL}$, as follows, assuming that first-order connectives and quantifiers are defined as in Section 2.6:

$$v(\forall X.A(X)) = \min(\{v_S^X(A(X))) : S \subseteq \text{DOM}^k\})$$
$$v(\exists X.A(X)) = \max(\{v_S^X(A(X)) : S \subseteq \text{DOM}^k\}),$$

where X is a k-argument relational variable and by v_S^X we denote the valuation obtained from v by assigning value S to variable X and leaving all other values of variables unchanged.

2.7.2 Complexity of Reasoning

Theorem 2.7.1. *Both checking whether a second-order formula is satisfiable or whether it is a tautology are not partially computable problems.*[10] □

Theorem 2.7.2. *Given a second-order formula, checking its satisfiability and validity over a given finite domain relational structure is* PSPACE*-complete. If the formula belongs to the existential fragment of second-order logic, then checking its satisfiability over a given finite domain relational structure is* NPTIME*-complete.* □

2.8 Fixpoint Calculus

2.8.1 Introduction and Definitions

In many contexts, in particular in the theory of relational and deductive databases and knowledge representation, fixpoint calculus is considered to

[10] In fact, the problem is even much more complex than partially computable problems or their complements.

be one of the most important tools which maintain a good balance between expressiveness and complexity. Fixpoint formulas allow one to define many notions reflecting computer science phenomena in a concise way.

Formulas of the fixpoint calculus, denoted by F_{X}, are defined by means of the following rules.

$$\begin{aligned}\langle F_{\mathsf{X}}\rangle ::= &\ \langle F_{\mathsf{I}}\rangle \ || \\ &\ \text{LFP}\ \langle V_{\mathsf{II}}\rangle[(\langle V_{\mathsf{I}}\rangle\{,\langle V_{\mathsf{I}}\rangle\})].\langle F_{\mathsf{X}}\rangle \text{ where } \langle F_{\mathsf{X}}\rangle \text{ is positive w.r.t. } \langle V_{\mathsf{II}}\rangle \ || \\ &\ \text{GFP}\ \langle V_{\mathsf{II}}\rangle[(\langle V_{\mathsf{I}}\rangle\{,\langle V_{\mathsf{I}}\rangle\})].\langle F_{\mathsf{X}}\rangle \text{ where } \langle F_{\mathsf{X}}\rangle \text{ is positive w.r.t. } \langle V_{\mathsf{II}}\rangle \ || \\ &\ \neg\langle F_{\mathsf{X}}\rangle \ || \ \langle F_{\mathsf{X}}\rangle \wedge \langle F_{\mathsf{X}}\rangle \ || \ \langle F_{\mathsf{X}}\rangle \vee \langle F_{\mathsf{X}}\rangle \ || \ \langle F_{\mathsf{X}}\rangle \to \langle F_{\mathsf{X}}\rangle \ || \\ &\ \langle F_{\mathsf{X}}\rangle \equiv \langle F_{\mathsf{X}}\rangle \ || \ \forall\langle V_{\mathsf{I}}\rangle.\langle F_{\mathsf{X}}\rangle \ || \ \exists\langle V_{\mathsf{I}}\rangle.\langle F_{\mathsf{X}}\rangle \ || \ (\langle F_{\mathsf{X}}\rangle)\end{aligned}$$

For the sake of simplicity we often write LFP R. (GFP R.), where R is a relation symbol rather than a second-order relational variable.

Let $T(X)$ be a fixpoint formula with second-order variable X. The semantics of LFP $X(\bar{x}).T(X)$ and GFP $X(\bar{x}).T(X)$ is the least and the greatest fixpoint of $T(X)$, i.e., the least and the greatest relation $X(\bar{x})$ such that $X(\bar{x}) \equiv T(X)$. Since T is assumed positive w.r.t. $X(\bar{x})$, such fixpoints exist. More precisely, given a relational structure

$$\langle \text{DOM}, \{f_i^{\text{DOM}} : i \in I\}, \{R_j^{\text{DOM}} : j \in J\}\rangle,$$

any valuation $v : V_{\mathsf{I}} \longrightarrow \text{DOM}$ can be extended to a valuation

$$v' : F_{\mathsf{X}} \longrightarrow \text{POW}(\text{DOM})$$

as follows, assuming that the cases of first-order connectives and quantifiers are defined as in Section 2.6:

$$\begin{aligned} v'(\text{LFP}\,X(\bar{x}).A(X)) &= \text{ the least (w.r.t. } \subseteq)) \text{ relation } S \text{ such that} \\ &\quad\ S(\bar{x}) \equiv v_S^X(A(X)) \\ v'(\text{GFP}\,X(\bar{x}).A(X)) &= \text{ the greatest (w.r.t. } \subseteq)) \text{ relation } S \text{ such that} \\ &\quad\ S(\bar{x}) \equiv v_S^X(A(X)). \end{aligned}$$

Example 2.8.1. The transitive closure of a binary relation R can be defined by the following fixpoint formula:

$$\text{TC}(R)(x,y) \equiv \text{LFP}\,X(x,y).[R(x,y) \vee \exists z.(R(x,z) \wedge X(z,y))].$$

Consider now the following example, where we are given a unary relation $Wise$ and a binary relation *Colleague* defined on a set *Per* of persons and suppose we want to calculate the relation $Wisest$ as the greatest relation satisfying the following constraint, meaning that $Wisest$ are those who are wise and have only wisest colleagues:

$\forall x.[Wisest(x) \rightarrow (Wise(x) \wedge \forall y.(Colleague(x,y) \rightarrow Wisest(y)))]$.

The $Wisest$ relation is defined by the following fixpoint formula:

$\text{GFP}\, X(x).[Wise(x) \wedge \forall y.(Colleague(x,y) \rightarrow X(y))]$. □

It is sometimes convenient to define more than one relation by means of fixpoint equations. This gives rise to *simultaneous fixpoints* defined by allowing many relations as arguments of fixpoint operators. The syntax is then modified by assuming new syntax rules for fixpoint operators and leaving the other rules unchanged. The new rules for fixpoints are the following:

$\text{LFP}\, \langle V_{\text{II}} \rangle [(\langle V_{\text{I}} \rangle \{, \langle V_{\text{I}} \rangle \})] \{, \langle V_{\text{II}} \rangle [(\langle V_{\text{I}} \rangle \{, \langle V_{\text{I}} \rangle \})] \}.\langle F_{\text{X}} \rangle \;||$
$\text{GFP}\, \langle V_{\text{II}} \rangle [(\langle V_{\text{I}} \rangle \{, \langle V_{\text{I}} \rangle \})] \{, \langle V_{\text{II}} \rangle [(\langle V_{\text{I}} \rangle \{, \langle V_{\text{I}} \rangle \})] \}.\langle F_{\text{X}} \rangle$

where $\langle F_{\text{X}} \rangle$ is positive w.r.t. all relational variables in $\langle V_{\text{II}} \rangle \{, \langle V_{\text{II}} \rangle \}$.

Let $T(X_1, \ldots, X_n) : \underbrace{F_{\text{X}} \times \ldots F_{\text{X}}}_{n\ times} \longrightarrow \underbrace{F_{\text{X}} \times \ldots F_{\text{X}}}_{n\ times}$. Given a relational structure $\langle \text{DOM}, \{f_i^{\text{DOM}} : i \in I\}, \{R_j^{\text{DOM}} : j \in J\} \rangle$, any valuation $v : V_{\text{I}} \longrightarrow \text{DOM}$ can be extended as follows:

$v(\text{LFP}\, X_1(\bar{x}_1), \ldots, X_n(\bar{x}_n).T(X_1, \ldots, X_n)) = \langle S_1, \ldots, S_n \rangle$
where $S_1, \ldots, S_n$ are the least (w.r.t. $\subseteq$)) relations such that
$\langle S_1(\bar{x}), \ldots, S_n(\bar{x}) \rangle \equiv v_{S_1,\ldots,S_n}^{X_1,\ldots,X_n}(T(X_1, \ldots, X_n))$
$v(\text{GFP}\, X_1(\bar{x}_1), \ldots, X_n(\bar{x}_n).T(X_1, \ldots, X_n)) = \langle S_1, \ldots, S_n \rangle$
where $S_1, \ldots, S_n$ are the greatest (w.r.t. $\subseteq$)) relations such that
$\langle S_1(\bar{x}), \ldots, S_n(\bar{x}) \rangle \equiv v_{S_1,\ldots,S_n}^{X_1,\ldots,X_n}(T(X_1, \ldots, X_n))$,

where $v_{S_1,\ldots,S_n}^{X_1,\ldots,X_n}(T(X_1, \ldots, X_n))$ denotes the tuple of values $v(T(X_1, \ldots, X_n))$, in which, for $1 \leq i \leq n$, X_i is interpreted as S_i.

Since T is assumed positive w.r.t. $X_1, \ldots, X_n$, such fixpoints exist.

2.8.2 Complexity of Reasoning

Theorem 2.8.2. *Both checking whether a fixpoint formula is satisfiable or whether it is a tautology are not partially computable problems.*[11] □

Theorem 2.8.3. *Given a fixpoint formula without function symbols, checking its satisfiability or validity over a given finite domain relational structure is in* PTIME. □

[11] In fact, as in the case of the second-order logic, the problem is much more complex than partially computable problems or their complements.

2.9 Second-Order Quantifier Elimination

Theorem 2.9.2, formulated below, allows us to eliminate second-order quantifiers from formulas which are in the form appearing on the left-hand side of the equivalences (2.6), (2.7). Observe, that in the context of databases one remains in a tractable framework, since fixpoint formulas over finite domains are computable in polynomial time (and space).

Let $B(X)$ be a second-order formula, where X is a k-argument relational variable and let $C(\bar{x})$ be a first-order formula with free variables $\bar{x} = \langle x_1, \ldots, x_k \rangle$. Then by $B[X(\bar{t}) := C(\bar{x})]$ we mean the formula obtained from $B(X)$ by substituting each occurrence of X of the form $X(\bar{t})$ in $B(X)$ by $C(\bar{t})$, renaming the bound variables in $C(\bar{x})$ with fresh variables.

Example 2.9.1. Let $B(X) \equiv \forall z.[X(y,z) \vee X(f(y), g(x,z))]$, where X is a relational variable and let $C(x,y) \equiv \exists z.R(x,y,z)$. Then

$$B[X(t_1,t_2) := C(x,y)]$$

is defined by

$$\forall z.[\underbrace{\exists z'.R(y,z,z')}_{C'(y,z)} \vee \underbrace{\exists z'.R(f(y),g(x,z),z')}_{C'(f(y),g(x,z))}],$$

where $C'(x,y)$ is obtained from $C(x,y)$ by renaming the bound variable z with z'. □

Let $A(\bar{x})$ be a formula with free variables $\bar{x}$ and $\bar{t}$ be a tuple of terms. Then by $A(\bar{x})[\bar{t}]$ we mean the application of $A(\bar{x})$ to terms $\bar{t}$.

Theorem 2.9.2. Assume that formula A is a first-order formula positive w.r.t. X.

- if B is a first-order formula negative w.r.t. X then

$$\exists X.\forall \bar{y}.[A(X) \to X(\bar{y})] \wedge [B(X)] \equiv B[X(\bar{t}) := \text{LFP}\, X(\bar{y}).A(X)[\bar{t}]] \tag{2.6}$$

- if B is a first-order formula positive w.r.t. X then

$$\exists X.\forall \bar{y}.[X(\bar{y}) \to A(X)] \wedge [B(X)] \equiv B[X(\bar{t}) := \text{GFP}\, X(\bar{y}).A(X)[\bar{t}]]. \tag{2.7}$$

□

Observe that, whenever formula A in Theorem 2.9.2 does not contain X, the resulting formula is easily reducible to a first-order formula, as in this case both $\text{LFP}\, X(\bar{y}).A$ and $\text{GFP}\, X(\bar{y}).A$ are equivalent to A.

Example 2.9.3. Consider the following second-order formula:

$$\exists X.\forall x.\forall y.[(S(x,y) \vee X(y,x)) \rightarrow X(x,y)] \wedge [\neg X(a,b) \vee \forall z.(\neg X(a,z))] \tag{2.8}$$

According to Theorem 2.9.2, equivalence (2.6), formula (2.8) is equivalent to:

$$\begin{aligned}&\neg \text{LFP}\, X(x,y).(S(x,y) \vee X(y,x))[a,b] \vee \\ &\qquad \forall z.(\neg \text{LFP}\, X(x,y).(S(x,y) \vee X(y,x))[a,z]).\end{aligned} \tag{2.9}$$

Observe that the definition of the least fixpoint appearing in (2.9) is obtained on the basis of the first conjunct of (2.8). The successive lines of (2.9) represent substitutions of $\neg X(a,b)$ and $\forall z.(\neg X(a,z))$ of (2.8) by the obtained definition of the fixpoint. □

2.10 Three-Valued Logic

In this book we often deal with unknown values due to the incompleteness and uncertainty of information. Thus, in many contexts we allow the use of three truth values TRUE, FALSE and UNKNOWN. The set of the three values is denoted by 3-BOOL. We assume that the values are ordered by the *truth ordering* FALSE $\leq$ UNKNOWN $\leq$ TRUE.[12].

The syntax of 3-valued formulas is similar to that of the propositional and predicate calculus. The only exception is that the syntactic category ⟨BOOL⟩ is now to be replaced by ⟨3-BOOL⟩.

The semantics of three-valued propositional formulas is given by assigning truth values to propositional variables and then calculating values of formulas. Let V_3 denote the set of *three-valued propositional variables* and let

$$v : V_3 \longrightarrow \text{3-BOOL}$$

be such an assignment of three-valued truth values to variables. Then v is extended to define the *truth value of* 3*-valued propositional formulas* as follows:[13]

[12] In many contexts the so-called *information ordering* is more suitable. In this ordering UNKNOWN $\leq$ FALSE, UNKNOWN $\leq$ TRUE and FALSE and TRUE are not comparable.

[13] There are many known three-valued logics, where the definitions of truth values of propositional connectives are different (for references see Section 2.11). Here we present a well-known version of the three-valued logic, based on Kleene logic, which is used in the book.

$$v(\neg A) = \begin{cases} \text{TRUE} & \text{if } v(A) = \text{FALSE} \\ \text{UNKNOWN} & \text{if } v(A) = \text{UNKNOWN} \\ \text{FALSE} & \text{if } v(A) = \text{TRUE} \end{cases} \tag{2.10}$$
$$v(A \wedge B) = \min(v(A), v(B))$$
$$v(A \vee B) = \max(v(A), v(B))$$
$$v(A \rightarrow B) = \text{TRUE if and only if } v(A) \leq v(B)$$
$$v(A \equiv B) = \text{TRUE if and only if } v(A) = v(B),$$

where min, max are defined according to the truth ordering.

The syntax of three-valued predicate calculus is that of two-valued predicate calculus.

Let us now define the semantics of the three-valued predicate calculus. By a *three-valued interpretation* we shall understand a *three-valued relational structure* $\langle \text{DOM}, \{f_i^{\text{DOM}} : i \in I\}, \{R_j^{\text{DOM}} : j \in J\}\rangle$, where:

- DOM is a non-empty set, called the *domain* or *universe* of the relational structure
- for $i \in I$, f_i^{DOM} denotes a function corresponding to the function symbol f_i
- for $j \in J$, R_j^{DOM} denotes a *three-valued relation* corresponding to the relation symbol R_j, where by a k-argument three-valued relation we mean a function from $\text{DOM}^k \longrightarrow$ 3-BOOL.

The semantics can now be defined as in the case of predicate calculus, where min and max are understood w.r.t. to the extended truth ordering.

The complexity of reasoning in three-valued logic is the same as the reasoning in the corresponding two-valued logics.

2.11 Bibliographic Notes

In this chapter basic concepts and results that are used throughout the book are presented. For a more comprehensive discussion of the topics concerning logics the reader is referred, e.g., to [68, 69]. A good exposition of Boolean reasoning can be found in [30].

Computational complexity is covered in many books, including, e.g., [78, 89, 92, 147].

We present the syntax of various logical languages in the widely accepted Backus-Naur BNF form. For details see, e.g., [117].

Theorem 2.5.1 on NPTIME-completeness of satisfiability of propositional formulas is due to Cook. The complexity of tractable fragments of propositional calculus is studied, e.g., in [32]. A linear time algorithm for checking satisfiability of propositional Horn clauses is given in [66].

Uncomputability of the predicate calculus (see Theorem 2.6.1) is due to Church and partial computability is due to Gödel. Theorem 2.6.3 was proved by Trakhtenbrot (see, e.g., [69]). LogSpace and thus PTime complexity of predicate calculus over finite domains was shown in [228]. The results on complexity of fixpoint logic over finite domains (Theorem 2.8.3) were proved in [35]. It was also considered, e.g., in [91, 228].

The high complexity of second-order logic is well-known. The so-called arithmetical and analytical hierarchies showing high complexity of the second-order logic were introduced and studied independently by Kleene and Mostowski. The theorem of NPTime-completeness of the existential fragment of second-order logic over finite structures (see a part of Theorem 2.7.2) is due to Fagin. In order to deal with the high complexity of second-order reasoning, one can apply second-order quantifier elimination techniques which appear quite powerful. The second-order quantifier elimination techniques have a long history (see, e.g., [2, 53, 75, 103, 141, 142, 180, 211, 212]). The fixpoint theorem 2.9.2, generalizing the Ackermann lemma of [2], is due to [142]. Two quantifier elimination algorithms based on these principles, extending the algorithm given in [211], are known as the Dls and the Dls* algorithms (see, e.g., [53, 54]).

Many-valued logics are studied in many sources. For a comprehensive treatment of the subject see, e.g., [22, 178, 224]. In this book we use the Kleene three-valued logic, which is also used as a semantics for Sql tables with unknown values.

3

Rough Sets

3.1 Introduction

The methodology we propose and develop in this book is founded on the concept of rough sets. In many AI applications one faces the problem of representing and processing incomplete, imprecise, and approximate data. Many of these applications require the use of approximate reasoning techniques. Before we introduce rough sets formally, let us begin with an intuitive example where representation of approximate data and reasoning with it is an essential component in the modeling process.

Example 3.1.1. Consider a UAV equipped with a sensor platform which includes a digital camera. Suppose that the UAV task is to recognize various situations on roads. It is assumed that the camera has a particular resolution. It follows that the precise shape of the road cannot be recognized if essential features of the road shape require a higher resolution then that provided by the camera. Figure 3.1 depicts a view from the UAV's camera, where a fragment of a road is shown together with three cars c1, c2, and c3.

Observe that due to the camera resolution there are collections of points that should be interpreted as being indiscernible with each other. The collections of indiscernible points are called *elementary sets*, using rough set terminology. In Figure 3.1, elementary sets are illustrated by dashed squares and correspond to pixels. Any point in a pixel is not discernible from any other point in the pixel from the perspective of the UAV. Elementary sets are then used to approximate objects that cannot be precisely represented by means of (unions of) elementary sets. For instance, in Figure 3.1, it can be observed that for some elementary sets one part falls within and the other outside the actual road boundaries (represented by curved lines).

Instead of a precise characterization of the road and cars, using rough set techniques, one can obtain approximate characterizations as depicted in Figure 3.2. Observe that the road sequence is characterized only in terms of

P. Doherty et al.: *Knowledge Representation Techniques*, Studfuzz **202**, 39–56 (2006)
www.springerlink.com

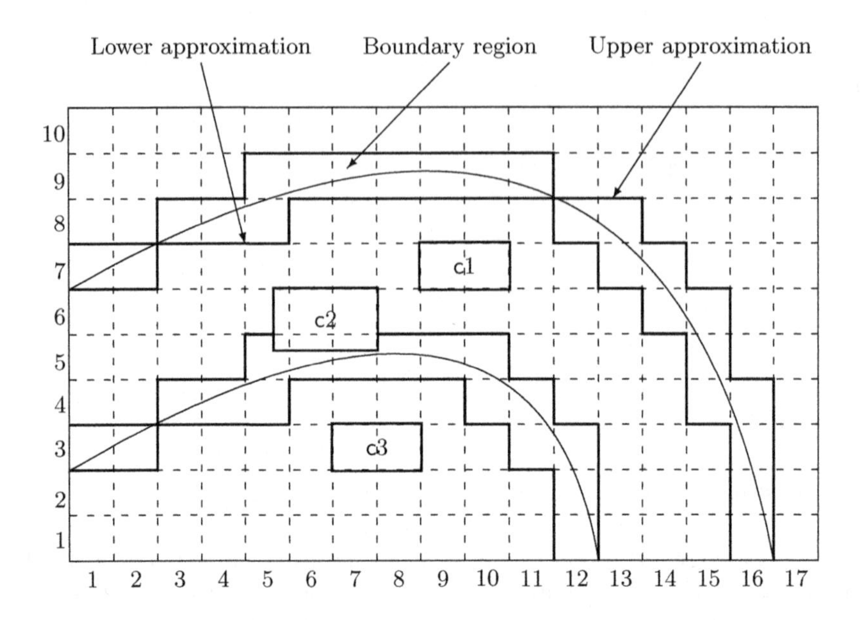

Fig. 3.1. Sensing a road considered in Example 3.1.1.

a lower and upper approximation of the actual road. A boundary region, containing points that are unknown to be inside or outside of the road's boundaries, is characterized by a collection of elementary sets marked with dots inside. Cars c1 and c3 are represented precisely, while car c2 is represented by its lower approximation (the thick box denoted by c2) and by its upper approximation (the lower approximation together with the region containing elementary sets marked by hollow dots inside). The region of elementary sets marked by hollow dots inside represents the boundary region of the car.

The approximations of the concepts are based on available information about points expressed by means of pixels to which the points belong. The lower approximation of a concept represents points that are known to be part of the concept, the boundary region represents points that might or might not be part of the concept, and the complement of the upper approximation represents points that are known not to be part of the concept. Consequently, car c1 is characterized as being completely on the road (inside the roads boundaries); it is unknown whether car c2 is completely on the road and car c3 is known to be outside, or off the road. □

As illustrated in Example 3.1.1, the rough set philosophy is founded on the assumption that we associate some information (data, knowledge) with every object of the universe of discourse. This information is often formulated in terms of attributes about objects. Objects characterized by the same infor-

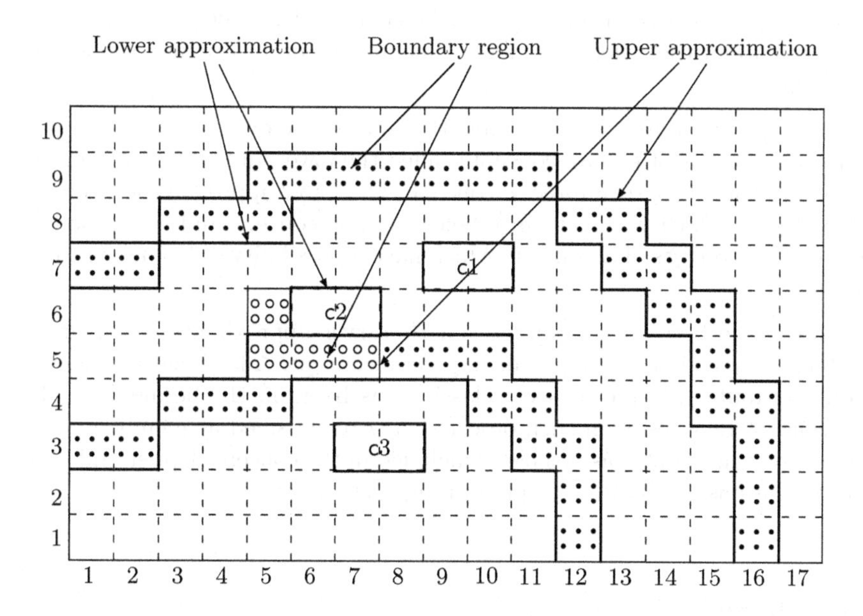

Fig. 3.2. The approximate view of the road considered in Example 3.1.1.

mation are interpreted as indiscernible (similar) in view of the available information about them. An indiscernibility relation, generated in this manner from the attribute/value pairs associated with objects, provides the mathematical basis of rough set theory.

Any set of all indiscernible (similar) objects is called an *elementary set*, and forms a basic granule (atom) of knowledge about the universe. Any union of some elementary sets in a universe is referred to as a *crisp set*; otherwise the set is referred to as being a *rough set*. In the latter case, two separate unions of elementary sets can be used to approximate the imprecise set, as we have seen in the example above. Since a relation is a set of tuples, a *rough relation* is defined to be a rough set of tuples.

Consequently, each rough set has what are called boundary-line cases, i.e., objects which cannot with certainty be classified either as members of the set or of its complement. This means that boundary-line cases cannot be properly classified by employing only the available information about objects. Obviously, crisp sets have no boundary-line elements at all.

The assumption that objects can be *observed* only through the information available about them leads to the view that knowledge about objects has a granular structure. Due to this granularity, some objects of interest cannot always be discerned given the information available, therefore the objects appear as the same (or similar). As a consequence, vague or imprecise concepts, in contrast to precise concepts, cannot be characterized solely in terms of information about their elements since elements are not always discernible from each other.

In the proposed approach, we assume that any vague or imprecise concept is replaced by a pair of precise concepts called the lower and the upper approximation of the vague or imprecise concept. The lower approximation consists of all objects which with certainty belong to the concept and the upper approximation consists of all objects which have a possibility of belonging to the concept.

The difference between the upper and the lower approximation constitutes the boundary region of a vague or imprecise concept. Additional information about attribute values of objects classified as being in the boundary region of a concept may result in such objects being re-classified as members of the lower approximation or as not being included in the concept. Upper and lower approximations are basic concepts in rough set theory.

3.2 Approximations

3.2.1 The Basic Concepts

We now introduce the concept of approximations more formally.

Let U be a set of objects. Any partition $\mathcal{E} = \{E_i \subseteq U : i \in I\}$ of U can be considered as a family of *elementary sets*. Of course, the choice of the family depends on a particular application.

Definition 3.2.1. *Let* U *be a set of objects,* $\mathcal{E} = \{E_i \subseteq U : i \in I\}$ *be a family of elementary sets and let* $X \subseteq U$. *The* lower approximation *and* upper approximation *of* X *w.r.t.* $\mathcal{E}$, *denoted by* $X_{\mathcal{E}^+}$ *and* $X_{\mathcal{E}^\oplus}$ *respectively, are defined by* $X_{\mathcal{E}^+} = \bigcup_{E_i \subseteq X} E_i$ *and* $X_{\mathcal{E}^\oplus} = \bigcup_{E_i \cap X \neq \emptyset} E_i$. □

Example 3.2.2. Let U be the set of non-negative reals. Elementary sets can be defined as intervals: $\mathcal{E} = \{[i, i+1) : i \in \omega\}$. Let $X = [2.4, 7.2]$. Then $X_{\mathcal{E}^+} = [3.0, 7.0)$ and $X_{\mathcal{E}^\oplus} = [2.0, 8.0)$. □

3.2.2 Representing Approximations in Logic

Approximations of sets and relations can be expressed in logic.

In order to construct a language for dealing with rough concepts, we introduce the following relation symbols for any rough relation R (see Figure 3.3):

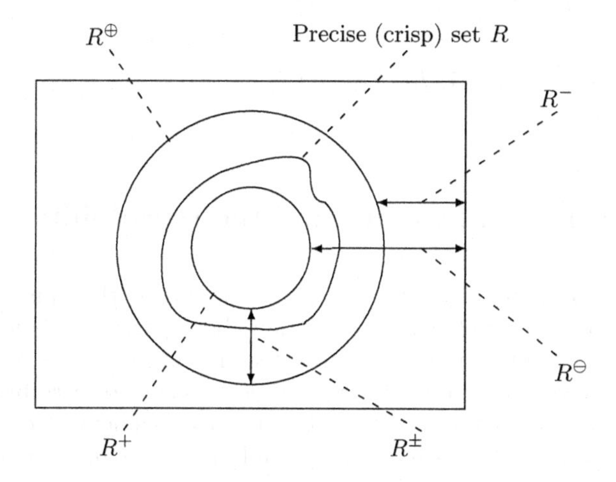

Fig. 3.3. Representation of a rough set in logic.

- R^+ – represents the positive facts known about the relation. R^+ corresponds to the lower approximation of R. R^+ is called the *positive region* (*part*) of R.
- R^- – represents the negative facts known about the relation. R^- corresponds to the complement of the upper approximation of R. R^- is called the *negative region* (*part*) of R.
- $R^{\pm}$ – represents the unknown facts about the relation. $R^{\pm}$ corresponds to the set difference between the upper and lower approximations of R. $R^{\pm}$ is called the *boundary region* (*part*) of R.
- $R^{\oplus}$ – represents the positive facts known about the relation together with the unknown facts. $R^{\oplus}$ corresponds to the upper approximation to R. $R^{\oplus}$ is called the *positive-boundary region* (*part*) of R.
- $R^{\ominus}$ – represents the negative facts known about the relation together with the unknown facts. $R^{\ominus}$ corresponds to the upper approximation of the complement of R. $R^{\ominus}$ is called the *negative-boundary region* (*part*) of R.

From the logical point of view, elementary sets can be represented by means of logical formulas or primitive relations, assuming their extensions form a partition of the universe.[1] Assume we are given elementary sets defined by formulas $\{\alpha_1(\bar{x}), \ldots, \alpha_n(\bar{x})\}$. Any relation can now be approximated as follows:

[1] In practice it is sufficient to require that the extensions form a covering of the universe - see Chapter 13.9.

$$R^{+}(\bar{x}) \stackrel{\text{def}}{\equiv} \bigvee_{1 \leq i \leq n} \{\alpha_i : \forall \bar{x}.[\alpha_i(\bar{x}) \rightarrow R(\bar{x})]\}$$

$$R^{\oplus}(\bar{x}) \stackrel{\text{def}}{\equiv} \bigvee_{1 \leq j \leq n} \{\alpha_j : \exists \bar{x}.[R(\bar{x}) \wedge \alpha_j(\bar{x})]\}.$$

3.3 Information Systems and Indiscernibility

One of the basic fundaments of rough set theory is the indiscernibility relation which is generated using information about particular objects of interest. Information about objects is represented in the form of a set of attributes and their associated values for each object. The indiscernibility relation is intended to express the fact that, due to lack of knowledge, we are unable to discern some objects from others simply by employing the available information about those objects. In general, this means that instead of dealing with each individual object we often have to consider clusters of indiscernible objects as fundamental concepts of our theories.

Let us now present this intuitive picture about rough set theory more formally.

Definition 3.3.1. *An* information system *is any pair* $\mathcal{A} = \langle U, A \rangle$, *where* U *is a non-empty finite set of* objects, *called the* universe, *and* A *is a non-empty finite set of functions, called* attributes, *such that* $a : U \rightarrow V_a$ *for every* $a \in A$. *The set* V_a *is called the* value set *of* a. *By the* information signature *of* $x \in U$ *w.r.t.* B, *where* $B \subseteq A$, *we understand* $\{\langle a, a(x) \rangle : a \in B\}$. □

Information systems are often represented in a form of tables with the first column containing objects and the remaining columns, separated by vertical double lines, containing values of attributes. Such tables are called *information tables.*

Table 3.1. Information table considered in Example 3.3.2.

Object	*Size*	*Color*
car1	large	red
car2	large	blue
car3	small	red

Example 3.3.2. Consider an information system $\mathcal{A} = \langle U, A \rangle$, where

- $U = \{\text{car1}, \text{car2}, \text{car3}\}$
- $A = \{Size, Color\}$; $Size(\text{car1}) = Size(\text{car2}) = \text{large}$, $Size(\text{car3}) = \text{small}$; $Color(\text{car1}) = Color(\text{car3}) = \text{red}$, $Color(\text{car2}) = \text{blue}$.

The information table corresponding to $\mathcal{A}$ is represented in Table 3.1. □

Note that in this definition, attributes are treated as functions on objects, where $a(x)$ denotes the value the object x has for the attribute a.

Any subset B of A determines a binary relation $\text{IND}_{\mathcal{A}}(B) \subseteq U \times U$, called an indiscernibility relation, defined as follows.

Definition 3.3.3. *Let $\mathcal{A} = \langle U, A \rangle$ be an information system and let $B \subseteq A$. By the* indiscernibility relation determined by B, *denoted by* $\text{IND}_{\mathcal{A}}(B)$, *we understand the equivalence relation*

$$\text{IND}_{\mathcal{A}}(B) = \{\langle x, x' \rangle \in U \times U : \forall a \in B.[a(x) = a(x')]\}.$$

If $\langle x, y \rangle \in \text{IND}_{\mathcal{A}}(B)$ we say that x and y are B-indiscernible.

Equivalence classes of the relation $\text{IND}_{\mathcal{A}}(B)$, denoted by $[x]_B$, are referred to as B-elementary sets. *The unions of B-elementary sets are called* B-definable sets. □

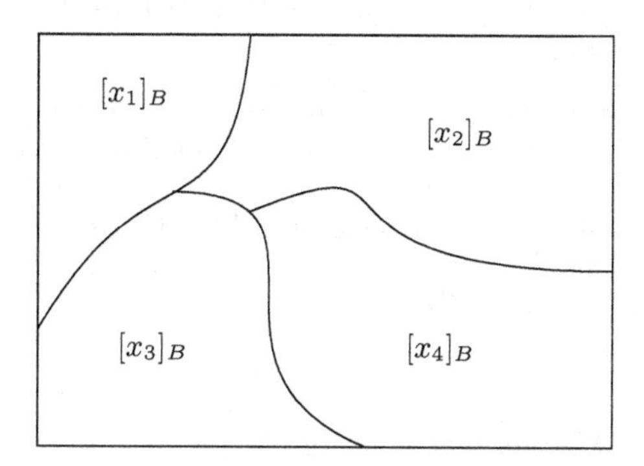

Fig. 3.4. A partition $\text{IND}_{\mathcal{A}}(B)$.

Of course, $\text{IND}_{\mathcal{A}}(B)$ is an equivalence relation. By U/B we denote the set of equivalence classes of relation $\text{IND}_{\mathcal{A}}(B)$. For example, in Figure 3.4, the partition of U defined by an indiscernibility relation $\text{IND}_{\mathcal{A}}(B)$ contains four equivalence classes, $[x_1]_B$, $[x_2]_B$, $[x_3]_B$ and $[x_4]_B$. An example of a B-definable set would be $[x_1]_B \cup [x_4]_B$, where $[x_1]_B$ and $[x_4]_B$ are B-elementary sets.

In Example 3.1.1 the indiscernibility relation is defined by a partition corresponding to pixels represented in Figures 3.1 and 3.2 by squares with dashed borders. Each square represents an elementary set. In the rough set approach the elementary sets are the basic building blocks (concepts) of our knowledge about reality.

The ability to discern between perceived objects is also important for constructing many entities like reducts, decision rules, or decision algorithms which are considered in later chapters (see Chapters 13 and 14).

The simplest *discernibility relation*, $\text{DIS}_{\mathcal{A}}(B)$, is defined as follows.

Definition 3.3.4. *Let $\mathcal{A} = \langle U, A \rangle$ be an information system and $B \subseteq A$. The* discernibility relation, *$\text{DIS}_{\mathcal{A}}(B) \subseteq U \times U$, is defined by $\langle x, y \rangle \in \text{DIS}_{\mathcal{A}}(B)$ if and only if $\langle x, y \rangle \notin \text{IND}_{\mathcal{A}}(B)$.* □

We now consider how to define sets of objects using formulas constructed from attribute/value pairs.

Definition 3.3.5. *An* elementary descriptor *(*descriptor*, for short) is any expression of the form $(a = v)$, where $a \in A$ and $v \in V_a$. A* generalized descriptor *is any formula of the form $\bigvee_{i=1}^{n}(a = v_i)$, where $a \in A$ and each $v_i \in V_a$. A* Boolean descriptor *is any Boolean combination of elementary descriptors. A* template *is a conjunction of elementary descriptors.* □

Strictly speaking, a descriptor $(a = v)$ should be written as a relational expression of the form $\lambda x.a(x) = v$, but we shall stick to the simplified form as in the above definition.

Let φ be a Boolean descriptor. The meaning of φ in $\mathcal{A}$, i.e., the set of all objects satisfying φ in $\mathcal{A}$, denoted $||\varphi||_{\mathcal{A}}$, is defined inductively as follows:

1. if φ is of the form $(a = v)$ then $||\varphi||_{\mathcal{A}} = \{x \in U \,:\, a(x) = v\}$
2. $||\varphi \wedge \varphi'||_{\mathcal{A}} = ||\varphi||_{\mathcal{A}} \cap ||\varphi'||_{\mathcal{A}}$
   $||\varphi \vee \varphi'||_{\mathcal{A}} = ||\varphi||_{\mathcal{A}} \cup ||\varphi'||_{\mathcal{A}}$
   $||\neg\varphi||_{\mathcal{A}} = U - ||\varphi||_{\mathcal{A}}$.

Definition 3.3.6. *We say that a set of objects $X \subseteq U$ is definable in $\mathcal{A}$ by some formula φ if and only if $X = ||\varphi||_{\mathcal{A}}$.*

Any $X \subseteq U$ definable in $\mathcal{A}$ is referred to as a crisp *(*precise*,* exact*) set; otherwise the set is referred to as a* rough *(*vague*,* imprecise*,* inexact*) set (relatively to $\mathcal{A}$).* □

3.4 Approximations in Information Systems

Let us now define approximations of sets in the context of information systems.

Definition 3.4.1. *Let $\mathcal{A} = \langle U, A \rangle$ be an information system, $B \subseteq A$ and $X \subseteq U$. The B-*lower approximation *and B-*upper approximation *of X, denoted by X_{B^+} and $X_{B^\oplus}$ respectively, are defined by $X_{B^+} = \{x : [x]_B \subseteq X\}$ and $X_{B^\oplus} = \{x : [x]_B \cap X \neq \emptyset\}$.* □

Definition 3.4.2. *The set consisting of objects in the* B-lower approximation X_{B+} *is also called the* B-positive region *of* X. *The set* $X_{B-} = U - X_{B\oplus}$ *is called the* B-negative region *of* X. *The set* $X_{B\pm} = X_{B\oplus} - X_{B+}$ *is called the* B-boundary region *of* X. □

Observe that the positive region of X consists of objects that can be classified with certainty as belonging to X using attributes from B. The negative region of X consists of those objects which can be classified with certainty as not belonging to X using attributes from B. The B-boundary region of X consists of those objects that cannot be classified either as belonging to X or as not belonging to X, using attributes from B.

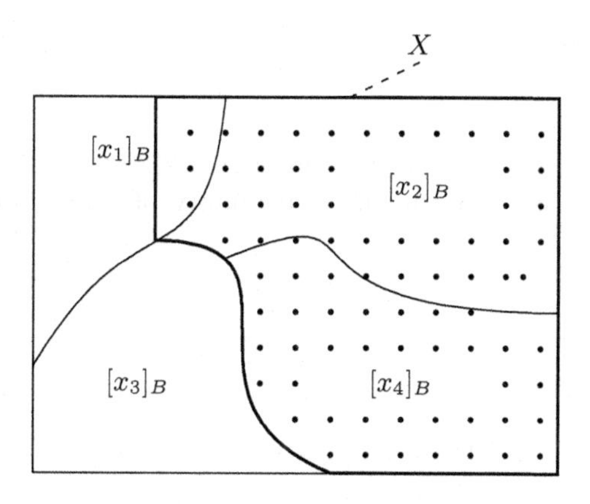

Fig. 3.5. A partition $\text{IND}_{\mathcal{A}}(B)$ and an imprecise set X.

For example, in Figure 3.5, The B-lower approximation of the set X (marked with dots), is $X_{B+} = [x_2]_B \cup [x_4]_B$. The B-upper approximation, is $X_{B\oplus} = [x_1]_B \cup [x_2]_B \cup [x_4]_B \equiv [x_1]_B \cup X_{B+}$. The B-boundary region, is $X_{B\pm} = [x_1]_B$. The B-negative region of X, is $X_{B-} = [x_3]_B \equiv U - X_{B\oplus}$.

The size of the boundary region of a set can be used as a measure of the quality of that set's approximation. One such measure is defined as follows.

Definition 3.4.3. *The* accuracy of approximation *of a finite nonempty set* X *is defined in terms of the following coefficient:*

$$\alpha_B(X) \stackrel{\text{def}}{=} \frac{|X_{B+}|}{|X_{B\oplus}|},$$

where $|X|$ *denotes the cardinality of* X. □

It is clear that $0 \leq \alpha_B(X) \leq 1$. If $\alpha_B(X) = 1$ then X is crisp with respect to B (X is precise with respect to B); otherwise, if $\alpha_B(X) < 1$ then X is rough with respect to B (X is vague or imprecise with respect to B).

3.5 Rough Sets and Membership Functions

In the context of rough set theory *rough membership functions* play an important rôle.

Definition 3.5.1. *Given a set of attributes B and a set X, a* rough membership function *μ_X^B is defined as*

$$\mu_X^B(x) \stackrel{\text{def}}{=} \frac{|X \cap [x]_B|}{|[x]_B|}.$$

□

It is clear that $0 \leq \mu_X^B(x) \leq 1$. The number $\mu_X^B(x)$ provides a ratio between how much of $[x]_B$ is in X and the cardinality of $[x]_B$. This is depicted in Figure 3.6.

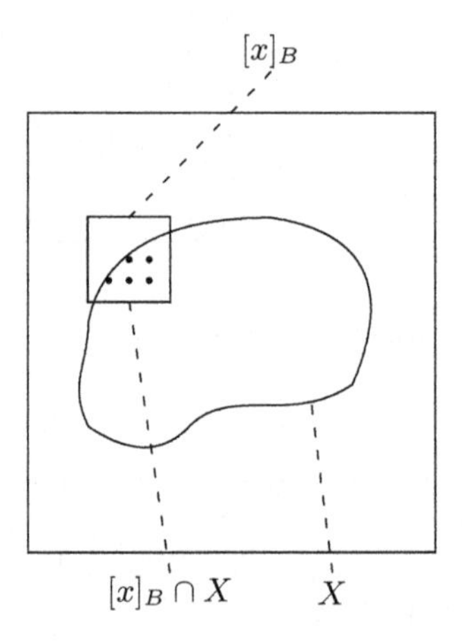

Fig. 3.6. Rough membership function.

The membership function $\mu_X(x)$ is similar to a conditional probability and its value can be interpreted as the degree of certainty to which x belongs to X. In other words, given X, the membership function indicates the degree of certainty to which x belongs to X.

Over finite domains, the rough membership function can be used to define approximations and the boundary region of a set as shown below:

$$X_{B+} = \{x \in U : \mu_X^B(x) = 1\}$$
$$X_{B\oplus} = \{x \in U : \mu_X^B(x) > 0\}$$
$$X_{B\pm} = \{x \in U : 0 < \mu_X^B(x) < 1\}.$$

In essence, classical rough sets are the weakest form of the quantitative idea above which abstracts away from the "degree to which" and simply uses "possibly in," instead.

3.6 Decision Systems and Decision Rules

Rough set techniques are often used as a basis for supervised machine learning using tables of data (see Chapter 14). In many cases the target of a classification task is represented by an additional attribute called a decision attribute. Information systems of this kind are called decision systems.

Definition 3.6.1. *Let $\langle U, A\rangle$ be an information system. A* decision system *is any system of the form $\mathcal{A} = \langle U, A, d\rangle$, where $d \notin A$ is the* decision attribute *and A is a set of* conditional attributes, *or simply* conditions. □

Let $\mathcal{A} = \langle U, A, d\rangle$ be given and let $V_d = \{v_1, \ldots, v_{r(d)}\}$. Decision d determines a partition $\{X_1, \ldots, X_{r(d)}\}$ of the universe U, where $X_k = \{x \in U : d(x) = v_k\}$ for $1 \leq k \leq r(d)$. The set X_i is called the i-th *decision class* of $\mathcal{A}$. By $X_{d(u)}$ we denote the decision class $\{x \in U : d(x) = d(u)\}$, for any $u \in U$.

Any object $x \in U$ belongs to a *decision class* $X_{d(x)}$ of $\mathcal{A}$. All decision classes of $\mathcal{A}$ create a partition of the universe U.

One can generalize the above definition to the case of decision systems of the form $\mathcal{A} = \langle U, A, D\rangle$ where the set $D = \{d_1, ...d_k\}$ of decision attributes and A are assumed to be disjoint. Formally this system can be treated as the decision system $\mathcal{A} = \langle U, C, d_D\rangle$ where $d_D(x) = \langle d_1(x), ..., d_k(x)\rangle$ for $x \in U$.

Similarly as in the case of information systems, decision systems can be naturally represented as *decision tables.* The first column of a decision table contains objects from U, the next columns contain values of conditional attributes and the remaining columns contain decision attributes. In order to separate conditional and decision attributes we use a double vertical line.

Example 3.6.2. Consider the situation described in Example 3.1.1. A similar representation of the road could be obtained using a more accurate camera by collecting data from the camera and creating an approximate description of the concept of road. In this example, a similar granularity of approximation dependent on the camera's resolution is assumed. Consider Table 3.2. The decision table on the lefthand side of the figure contains some sample data, where

- objects in the first column are points
- there are two attribute columns, *Pixel* and *Inside*.
  - *Pixel* refers to a pixel which is defined as an elementary set containing coordinates. For example, pixel $\langle i, j \rangle$ contains points $\{\langle x, y \rangle : i - 1 \leq x < i \text{ and } j - 1 \leq y < j\}$.
  - *Inside* represents information about whether a point is inside the boundaries of a road object.

The additional column "Approximation of Inside," provides us with the approximation obtained from the available data, where:

- "+" means that a given point is inside the road boundaries because it is contained in a pixel which is totally included in the road;
- "±" means that it is unknown whether the point is inside the road boundaries, because it is contained in a pixel, part of which is inside and part of which is outside of the road;
- "−" means that the point is outside the road boundaries because it is contained in a pixel which is totally outside of the road. □

Table 3.2. Decision table for determining *Inside* of Example 3.6.2

Point	*Pixel*	*Inside*	*Approximation of Inside*
$\langle 4.60, 5.50 \rangle$	$\langle 5, 6 \rangle$	TRUE	$+$
$\langle 4.50, 4.70 \rangle$	$\langle 5, 5 \rangle$	TRUE	$\pm$
$\langle 4.90, 4.01 \rangle$	$\langle 5, 5 \rangle$	FALSE	$\pm$
$\langle 4.50, 3.50 \rangle$	$\langle 5, 4 \rangle$	FALSE	$\pm$
$\langle 4.01, 3.99 \rangle$	$\langle 5, 4 \rangle$	TRUE	$\pm$
$\langle 5.50, 5.50 \rangle$	$\langle 6, 6 \rangle$	TRUE	$+$
$\langle 5.50, 4.50 \rangle$	$\langle 6, 5 \rangle$	TRUE	$\pm$
$\langle 5.50, 4.10 \rangle$	$\langle 6, 5 \rangle$	FALSE	$\pm$
$\langle 5.50, 3.40 \rangle$	$\langle 6, 4 \rangle$	FALSE	$-$
$\langle 6.40, 5.40 \rangle$	$\langle 7, 6 \rangle$	TRUE	$+$
$\langle 6.50, 4.80 \rangle$	$\langle 7, 5 \rangle$	TRUE	$\pm$
$\langle 6.50, 4.20 \rangle$	$\langle 7, 5 \rangle$	FALSE	$\pm$
$\langle 6.30, 3.40 \rangle$	$\langle 7, 4 \rangle$	FALSE	$-$

Let $\mathcal{A}$ be a decision system. A *condition template* of $\mathcal{A}$ is any conjunction of elementary (or generalized) descriptors only. A *decision rule* for $\mathcal{A}$ is any expression of the form $\varphi \Rightarrow \psi$, where φ is a condition template and ψ is a descriptor containing a decision attribute. Formulas φ and ψ are referred to as the *predecessor* and the *successor of the decision rule*. Decision rules are often called *if-then rules*.

A decision rule $\varphi \Rightarrow \psi$ is TRUE in $\mathcal{A}$ iff $\|\varphi\|_{\mathcal{A}} \subseteq \|\psi\|_{\mathcal{A}}$. Otherwise, one can measure its *truth degree* by introducing some inclusion measure of $\|\varphi\|_{\mathcal{A}}$ in $\|\psi\|_{\mathcal{A}}$. For example, one such measure which is widely used, is called a *confidence coefficient* and is defined as,

$$\begin{cases} \frac{|\|\varphi \wedge \psi\|_{\mathcal{A}}|}{|\|\varphi\|_{\mathcal{A}}|} & \text{for } |\|\varphi\|_{\mathcal{A}}| \neq 0 \\ 1 & \text{otherwise.} \end{cases}$$

Another measure of non-classical inclusion, called *support of the rule*, is defined as

$$\frac{|\|\varphi \wedge \psi\|_{\mathcal{A}}|}{|U|}.$$

An interesting class of decision rules consists of *minimal decision rules* in a given decision system $\mathcal{A}$, i.e., rules which are true in $\mathcal{A}$ but become not true after removing any conditional descriptor from them.

Each object x in a decision table determines a decision rule,

$$\bigwedge_{a \in C} (a = a(x)) \Rightarrow (d = d(x)),$$

where C is the set of conditional attributes and d is the decision attribute. Decision rules corresponding to some objects can have the same condition parts but different decision parts. Such rules are called *inconsistent* (*nondeterministic, conflicting*); otherwise the rules are referred to as *consistent* (*certain, deterministic, nonconflicting*) rules. Decision tables containing inconsistent decision rules are called *inconsistent* (*nondeterministic, conflicting*); otherwise the table is *consistent* (*deterministic, nonconflicting*).

Numerous methods based on the rough set approach combined with Boolean reasoning techniques have been developed for decision rule generation. When a set of rules have been induced from a decision table containing a set of training examples, they can be inspected to determine if they reveal any novel relationships between attributes that are worth pursuing further. In addition, the rules can be applied to a set of unseen cases in order to estimate their classification power.

For a systematic overview of rule generation and application methods the reader is referred to Chapter 14 in this book which covers the topic of machine learning and knowledge discovery.

3.7 Inducing Consistent Concept Descriptions

Up to now, we considered the approximations of raw data represented as decision tables. It is often the case that, for a particular application domain, additional qualitative knowledge about dependencies between concepts are often available in the form of expert knowledge. In this section, we consider methods for inducing concept approximations constrained by this additional knowledge.

Consider the traffic scenario domain used in previous examples. Given a set of facts represented by decision tables, an example of a dependency between concept approximations could be the rule,

> if a road is slippery and the speed of a car is high, then there is a high chance that an accident involving the car will occur.

An interesting issue arises as to how concept approximations can be induced using both the raw data in decision tables and qualitative knowledge associated with it. There are a number of approaches which immediately come to mind. For example,

- one can develop strategies for directly generating decision rules which preserve the qualitative dependencies between approximated concepts;
- one can tune decision rules generated from raw data relative to the qualitative dependencies.

Yet another approach, and one we will pursue in this book, involves using some ideas from the area of nonmonotonic reasoning together with the second approach above. For example, consider a case where our decision tables contain the conditional attributes c_1, c_2 and the decision attribute d. The following dependency might be provided by a domain expert:

> if $c_1 = \mathsf{high}$ is known and $c_2 = \mathsf{medium}$ is consistent with the available information by default[2] then $d = \mathsf{dangerous}$

Given a decision table containing these attributes, we would first generate the minimal decision rules representing the data.[3] We would then view the dependencies as representing constraints between the lower and upper approximations of the various decision classes. This can be done by interpreting the phrase "is known" as a lower approximation of its respective argument and the phrase "is consistent with the available information" as the upper approx-

[2] I.e., $(c_2 \neq \mathsf{medium})$ cannot be proved using the available information.

[3] For a technique of generating minimal rules see Chapter 14.

imation of its respective argument.[4] The minimal decision rules could then be tuned using the constraints between the lower and upper approximations of the decision classes in the qualitative dependency rules.

In the following, we provide an example of the main idea. The methodology illustrated in this example is fully elaborated in the second and third part of this book.

Example 3.7.1. Consider again the situation described in Example 3.1.1. Suppose now that one wants to determine whether a given car is completely within the boundaries of a road. The suitable information can be obtained on the basis of the pixels available at a particular level of resolution. Car c1 is completely within the road's boundaries, since all pixels covering c1 are completely within the road's boundaries; car c3 is outside of the road's boundaries, since all pixels covering c3 are outside of the road's boundaries; and it is unknown whether car c2 is completely within or outside of the road's boundaries, since some of the pixels covering c2 are in the boundary region of the road.

However, in certain cases, when additional knowledge is provided, one can classify c2 as being inside the road's boundaries. Namely, consider the following rule expressing knowledge about the road domain:

> if it is consistent with the available information that a car is completely within the boundaries of a road and its speed is known to be high then assume by default that the car is within the boundaries of the road.

The rule can be translated into the terminology of rough sets as follows:

> if a car is in the boundary region of a road and its speed is classified as high then assume that the car is completely within the boundaries of the road.

In the presence of additional information that car c2 is moving with high speed, one can conclude that it is completely within the boundaries of the road. □

3.8 Dependency of Attributes

An important issue in data analysis is to discover dependencies between attributes. Intuitively, a set of attributes D depends totally on a set of attributes C if the values of attributes from C uniquely determine the values of attributes

[4] Observe that rules of the form considered in the example are easily expressible in nonmonotonic formalisms, like default reasoning or circumscription (for more details see, e.g., Chapters 5 and 10).

from D. In other words, D depends totally on C, if there exists a functional dependency between values of C and D.

Formally, a dependency between attributes can be defined as follows.

Definition 3.8.1. *Let $\mathcal{A} = \langle U, A \rangle$ be an information system, and let D and C be subsets of A. By a* positive region of a partition U/D *with respect to C, denoted by* $\textsc{Pos}_C(D)$*, we mean the set:*[5]

$$\textsc{Pos}_C(D) = \bigcup_{X \in U/D} X_{C^+}.$$

The degree of the dependency $C \mapsto D$, $\gamma_{\mathcal{A}}(C, D)$, *is defined by*

$$\gamma_{\mathcal{A}}(C, D) \stackrel{\text{def}}{=} \frac{|\textsc{Pos}_C(D)|}{|U|}.$$

We say that D depends on C *to degree k, denoted by $C \mapsto_k D$, if $k = \gamma_{\mathcal{A}}(C, D)$.*

If $k = 1$ we say that D depends totally *on C, and if $k < 1$, we say that D* depends partially *(to* degree *k) on C.* □

Observe that, for any $C, D \subseteq A$, $0 \leq \gamma_{\mathcal{A}}(C, D) \leq 1$.

If D depends totally on C then $\textsc{Ind}_{\mathcal{A}}(C) \subseteq \textsc{Ind}_{\mathcal{A}}(D)$. This means that the partition generated by C is finer than the partition generated by D. Notice, that the concept of dependency discussed above corresponds to that considered in relational databases.

A geometrical interpretation of the positive region $\textsc{Pos}_C(D)$, is shown in Figure 3.7, where:

- dashed boxes represent a partition created by C
- the set of equivalence classes of $\textsc{Ind}_{\mathcal{A}}(D)$, $U/D = \{X_1, X_2, X_3\}$, where the three partitions are separated by the thick black line
- the union of the double-lined areas represents $\textsc{Pos}_C(D)$
- the dotted boxes represent the boundary region between classes.

In this example, it is clear that there is only a degree of dependency between C and D because some partitions in C do not uniquely map to a partition in U/D. For instance, if $x \in [x_k]_C$, we are unable to uniquely determine whether $x \in X_1$ or $x \in X_3$. On the other hand, if $x \in [x_l]_C$ then $x \in X_1$.

In summary, D is totally (partially) dependent on C, if all (some) elements of the universe U can be uniquely classified to blocks of the partition U/D, employing C.

[5] Recall that U/D denotes the set of equivalence classes of relation $\textsc{Ind}_{\mathcal{A}}(D)$.

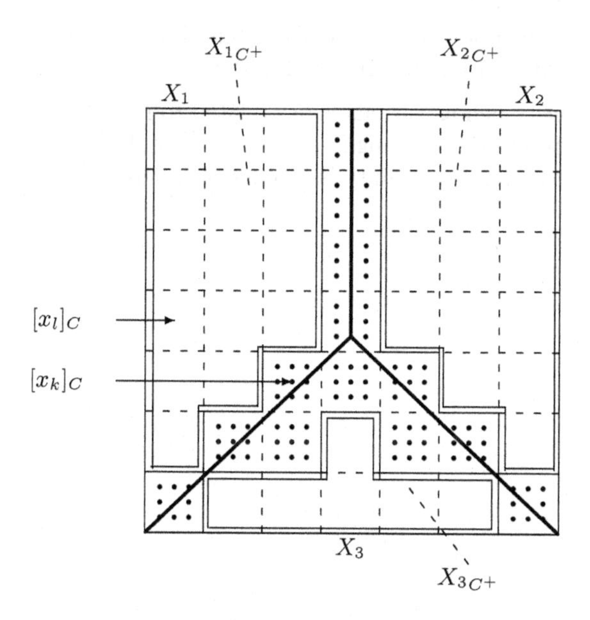

Fig. 3.7. Geometrical interpretation of positive region $\text{POS}_C(D)$.

3.9 Bibliographic Notes

In this chapter, we considered the basic concepts of rough set theory. The theory itself was originally proposed by Pawlak [148, 150].

Several generalizations of the classical rough set approach have been reported in the literature (for references see the papers and bibliography in [109, 146, 157, 158, 162, 163, 191, 203, 206, 204]). The quoted works also describe many case studies on rough set methods, as well as relationships among rough sets and other approaches to approximate reasoning.

A discussion of partial containment of sets and mereology has originated in [112, 113]. This approach has been generalized to the rough mereological approach, (see, e.g., [160, 164, 165, 189]).

Recently, it has been shown that the rough set approach can be used for synthesis of concept approximations in distributed environment of intelligent agents. The reader interested in intelligent agents and multiagent systems is referred, e.g., to [90, 67, 111]. In particular, the rough set methods are used for construction of interfaces between agents equipped with different sets of concepts [60, 61, 62, 145, 201] and for ontology approximation (see, e.g., [46, 138, 140, 185, 194]).

Readers interested in the above issues are also advised to consult the enclosed references (e.g., [145, 146, 158, 162, 163, 164, 201]).

Many important issues, such as various logics related to rough sets and many advanced algebraic properties of rough sets can be found in [143, 146, 162, 163, 164]. Reasoning under uncertainty is discussed in depth in [85].

4

Relational and Deductive Databases

4.1 Introduction

Relational and deductive databases provide basic tools for storing, querying and manipulating data. From the point of view of knowledge engineering, databases provide a fundamental layer on which other representation may be built. The choice of the underlying tools is then extremely important and seriously influences further use of the knowledge engineering techniques. In this chapter we sketch some possible choices concerning deductive database solutions. Let us start by introducing some basic definitions.

Definition 4.1.1. *A* relational database, *say B, is a relational structure*

$$B = \langle \text{DOM}, r_1^{a_1}, \ldots, r_k^{a_k} \rangle,$$

where

- DOM *is a finite non-empty set,*
- *for $1 \leq i \leq k$, $r_i^{a_i}$ is an a_i-argument relation, $r_i^{a_i} \subseteq \text{DOM}^{a_i}$.*

By a signature *or* vocabulary *of a database B, denoted by* SIG, *we mean a signature containing relation symbols $R_1^{a_1}, \ldots, R_k^{a_k}$ and constant symbols $C_1, \ldots, C_l$ representing all elements of the domain* DOM, *together with equality $=$. If the signature contains a binary relation which is a linear order on* DOM, *then we say that the database is* (linearly) ordered. *By the* size of the database *B, denoted by* SIZE(B), *we understand the number of elements in* DOM. □

According to the notational convention introduced in Section 2.6, we shall often use symbols $R_1^{a_1}, \ldots, R_k^{a_k}$ to denote both relations and the corresponding relation symbols.

P. Doherty et al.: *Knowledge Representation Techniques*, Studfuzz **202**, 57–76 (2006)
www.springerlink.com

In the field of relational databases the following conditions are usually assumed more or less explicitly:

- *domain closure axiom* (DCA): $\forall x. \bigvee_{i=1}^{\text{SIZE}(B)} \left(x = C_i\right)$, which states that all objects of the domain are named by constants
- *unique name assumption* (UNA): $\bigwedge_{1 \le i < j \le \text{SIZE}(B)} \left(C_i \neq C_j\right)$, which states that each object of the domain has a unique name
- *closed world assumption* (CWA): whenever a ground atom $p(\bar{t})$ is not entailed by the database, then assume that $\neg p(\bar{t})$ holds.[1]

In some applications it is necessary to drop the CWA assumption. An alternative to CWA is the *open world assumption* (OWA), where positive and negative facts are represented in the database and all facts not explicitly listed in the database are assumed to be unknown. The CWA and the OWA represent two ontological extremes. In Chapter 6 we discuss situations which permit the application of the CWA *locally* in a particular context.

A deductive database consists of two parts: an extensional and intensional part. The extensional database is usually equivalent to a traditional relational database and the intensional database contains a set of definitions of relations that are not explicitly stored in the database. Intuitively, intensional relations correspond to views known from relational databases, however the deductive approach offers a much more expressive formalism for defining the contents of views. Accordingly, we have the following definition.

Definition 4.1.2. *By a* deductive database *we understand a relational database augmented with an additional set of formulas (sometimes called* rules*) defining fresh relations in terms of a chosen logic. The relational database is called an* extensional database *(*EDB*), and the set of formulas is called an* intensional database *(*IDB*). We say that a relation (relation symbol) is* intensional *in a database if it appears in the intensional database only, otherwise it is called* extensional. □

Accordingly, we divide the set of relation symbols REL into two parts: EXTREL and INTREL, of *extensional* and *intensional relation symbols*, respectively.

Definition 4.1.3. *A* database query *is any polynomially bounded mapping*

[1] Observe that whenever the CWA is in force, there is no need to represent all relevant information. Only positive information, i.e., facts stating what is true, should be kept in the data base. The negative facts, specifying what is false, are inferred implicitly. Since generally negative facts vastly outnumber positive facts, using the CWA greatly simplifies the resulting representation.

$$Q : \text{STRUC}[\text{SIG}] \longrightarrow \text{STRUC}[\text{SIG}']$$

from finite relational structures of vocabulary SIG *to structures of vocabulary* SIG$'$. *That is, there is a polynomial* p *such that for any* $B \in$ STRUC[SIG], *the size of* $Q(B)$, SIZE($Q(B)$), *is not greater than* p(SIZE(B)). □

Observe that we require queries to be deterministic, i.e., returning a unique answer for each input database. The requirement as to the polynomial bound reflects the tractability demand.

It is also worth noting here that the concept of queries is independent of any particular query language. In order to make the connection between queries and expressions of query languages, we accept the following definition.

Definition 4.1.4. *We say that an expression* E *of a query language* defines a database query Q *(or that* Q *is expressed by* E*) if, for any database* B*, the value of* E *in* B *is defined and is equal to the output of query* Q *evaluated on database* B. □

In what follows, when a query language is fixed, we often do not distinguish queries from expressions defining the queries.

Using a standard relational database with SQL as its querying mechanism is inadequate from the knowledge representation perspective. The SQL designers tried to keep a good balance between the expressiveness of the language and the computational complexity of the underlying querying machinery. Unfortunately, there are even simple, but still efficiently computable queries that are not expressible in pure SQL. For instance, given a genealogy parent-child relation, one cannot express a query that computes all antecedents of a given family member.[2] On the other hand, many such queries are still efficiently computable. In order to compute them using SQL, one has to use a host programming language, such as, e.g., C, C++ or JAVA, and encode the queries. Such a hard-coding of queries is far from the declarative style one would like to maintain for knowledge representation. Another approach to querying databases incorporates various logical formalisms to represent queries. This is much more natural from the point of view of knowledge engineering. Logical queries can then be asked to deductive databases directly or, what is a more common in practice, to use existing relational database management systems (RDBMS) as "low level" tools and extend them with a deductive front-end mechanism. The architecture of such databases is shown in Figure 4.1.

When one analyzes various querying mechanisms, complexity issues are of great importance. Complexity is basically measured w.r.t. the size of the database domain (*data complexity*) and w.r.t. the size of the query (*expression*

[2] In fact, new versions of SQL allow use of a restricted form of recursion, but this does not solve the expressiveness problems in their full generality.

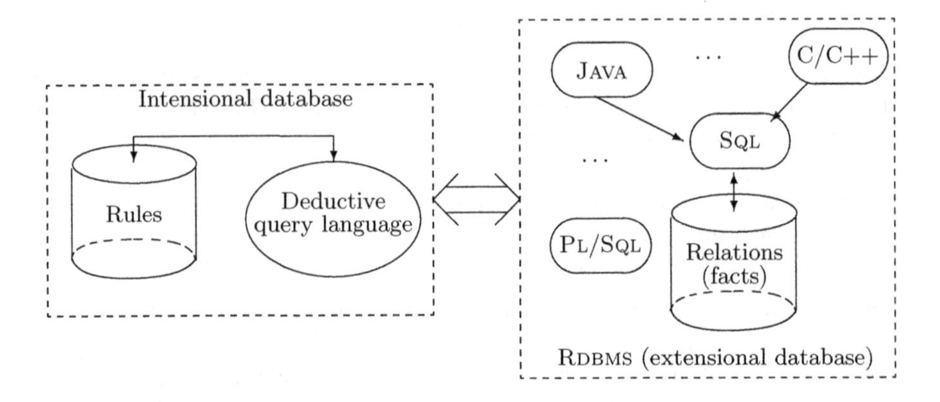

Fig. 4.1. Deductive front-end to a relational database systems.

complexity). In this book we accept the common assumption that queries are considered to be fixed, i.e., to have a constant length and thus we concentrate primarily on the data complexity. We also assume that *space data complexity* refers to the auxiliary memory needed to calculate a relation. In general, we shall deal with the following database querying problem (for details see also Sections 4.2 – 4.6).

Problem 4.1.5 (Database Querying Problem). Let Q be a fixed query expressed in a given query language. The *database querying problem* is defined by its input-output relation:

INPUT: database instance B
OUTPUT: the least relation defined by Q and entailed by B provided that such a relation exists or information that the relation does not exist relative to B. □

4.2 Predicate Calculus as a Query Language

One of the possible approaches for representing intensional databases is to use predicate calculus as the language for representing rules and queries.

Let $B = \langle \text{DOM}, R_1, \ldots, R_k \rangle$, where $R_1, \ldots, R_n \in \text{EXTREL}$ are extensional relation symbols. Consider formula $A(x_1, \ldots, x_n)$ over the signature of B, where $x_1, \ldots, x_n$ are all free variables of A. Such a formula, called a *first-order query*, defines the following relation, say R:[3]

[3] Having defined such relations, we can use them in further definitions.

$$R(a_1, \ldots, a_n) \text{ holds provided that} \qquad (4.1)$$
$$a_1, \ldots, a_n \in \textsc{Dom} \text{ and } A(a_1, \ldots, a_n) \text{ is valid in } B.$$

In particular, first-order formulas without free variables return TRUE or FALSE as value and thus correspond to "yes-or-no" queries.

Let us emphasize that no form of recursion is allowed, i.e., no intensional relation is allowed to refer to itself directly or indirectly.

Example 4.2.1. Given a unary relation $Person(x)$, denoting that x is a person, and a binary relation $Child(x, y)$ denoting that x is a child of y, the following first-order query defines the unary relation $Parent(x)$:

$$Parent(x) \equiv [Person(x) \wedge \exists y.Child(y, x)]$$

denoting that person x is a parent. □

Such a query language is quite efficient due to the following theorem.

Theorem 4.2.2. *The problem of computing first-order queries is in* PTIME *and in* LOGSPACE. □

Assume that extensional relations are implemented. The following algorithm checks whether a given tuple belongs to the output relation for a given first-order query Q in the prenex normal form, with free variables $x_1, \ldots, x_k$, and provides us with a proof of Theorem 4.2.2.

Algorithm 4.2.3.

INPUT: database instance B and tuple $\bar{a} = \langle a_1, \ldots, a_k \rangle$
OUTPUT: TRUE *if $\bar{a}$ is in the relation defined by Q, entailed by B,* FALSE *in the opposite case.*

1. *The case when Q has no quantifiers: one has to determine the truth value of each atomic formula occurring in Q for the given input tuple $\bar{a}$. This is done by checking whether a relevant portion of $\bar{a}$ appears in the relation corresponding to the atom. When the truth values of atoms are established, one evaluates the resulting Boolean formula and returns the obtained value.*
2. *The case when Q is of the form $\exists x.A$: all possible values of x are tried. If some value satisfying A is found, the answer is* TRUE, *otherwise it is* FALSE.
3. *The case when Q is of the form $\forall x.A$: all possible values of x are tried. If some value falsifying A is found, the answer is* FALSE, *otherwise it is* TRUE. □

Time and space data complexity of Algorithm 4.2.3 can easily be evaluated by first observing that the input query Q is fixed. Thus its length is treated as a constant, so the depth of the recursion is constant, too, as the recursive calls are invoked for each quantifier appearing in Q. Next assume there are n elements in the database domain. When applying a binary encoding, $\log(n)$ space is required to encode each element. Traversing the database and checking truth values of atoms requires a constant space. Each level of recursion, corresponding to the evaluation of a quantifier, requires storing the current value assigned to the quantified variable, i.e., requires a space of the size $\log(n)$. Consequently, the algorithm works in logarithmic space, and so also in polynomial time.

In order to see that Theorem 4.2.2 holds, we have to compute the answer to a first-order query using at most logarithmic auxiliary space. In order to do this, it is sufficient to generate all possible tuples in some order and use, for each tuple, Algorithm 4.2.3 to accept the tuple as one belonging to the output relation or not. Logarithmic space is then sufficient.

Note that standard SQL has essentially the same querying power as the predicate calculus.

4.3 Fixpoint Calculus as a Query Language

Similarly to the predicate calculus, the fixpoint calculus can be used as a query language. By a *fixpoint query* we thus understand any query expressed by means of a formula of the fixpoint calculus. The whole idea is very similar to that presented in Section 4.2, except that now one deals with a much more expressive formalism. In fact, the following theorem applies.

Theorem 4.3.1. *Computing fixpoint queries is in* PTIME *and* PSPACE. *Moreover, all queries computable in polynomial time are expressible in the fixpoint calculus, provided that the database is linearly ordered.* □

Thus fixpoint calculus can be considered as the most powerful tractable querying mechanism, as it expresses all tractable queries. On the other hand, fixpoint formulas are not as straightforward to use for expressing queries. We therefore tend to treat the calculus as a low level language and lean towards the Semi-Horn Query Language (SHQL), which is fully declarative and more amenable to knowledge engineering purposes (see Section 4.6).

Fixpoint formulas have a very nice computational characterization. Namely, given an extensional database B, we have the following definitions of the least fixpoint and the greatest fixpoint.

$$\text{LFP}\, R(\bar{x}).A(R) = \bigcup_{k>1} A^k(R := \text{FALSE}) \tag{4.2}$$

$$\text{GFP}\, R(\bar{x}).A(R) = \bigcap_{k>1} A^k(R := \text{TRUE}) \tag{4.3}$$

where $A^k(R := \text{FALSE})$ is an abbreviation for $\underbrace{A(\ldots(A}_{k-\text{times}}(R := \text{FALSE}))\ldots)$.
Both the least and the greatest fixpoint can now be computed using variations of the following algorithm.

Algorithm 4.3.2 (Naîve algorithm for computing fixpoint relations).

INPUT: formula of the form LFP $P.A(P)$ *or* GFP $P.A(P)$
OUTPUT: relation Result defined by the input formula

Result:= Init;
while Result changes do Result:= A(Result)

where

$$Init = \begin{cases} \text{FALSE } \textit{when input is defined by } \text{LFP}\, P.A(P) \\ \text{TRUE } \textit{when input is defined by } \text{GFP}\, P.A(P). \end{cases}$$

□

It is easily observed that the naîve algorithm operates in time and space polynomial in the size of the database. However, one can save a considerable amount of computation by applying better algorithms and known optimization techniques.[4]

Algorithm 4.3.2 can easily be generalized to deal with simultaneous fixpoints by assuming that *Result* and *Init* are tuples of relations rather than a single relation.

4.4 Datalog

DATALOG is a simple deductive database language based on the logic programming paradigm. However, unlike logic programming, DATALOG does not allow the use of complex terms and works on finite domains only.

DATALOG is based on the language of the predicate calculus. The DATALOG syntax is defined by the following rule:

⟨DATALOG RULE⟩ ::=
  ⟨ATOMIC FORMULA⟩ ← [⟨ATOMIC FORMULA⟩
                {, ⟨ATOMIC FORMULA⟩}].

[4] See Section 4.7 for pointers to the relevant literature.

The atomic formula to the left is called the *head of the rule* and the list of atomic formulas to the right is called the *body of the rule.* It is usually assumed that any variable occurring in the rule's head also occurs in the rule's body (such rules are called *safe*).

Remark 4.4.1. In the book we will not require a variable occurring in the rule's head to occur also in the rule's body. In cases violating the safety condition, we assume that the total relation referring to variables that are in the rule's head and not in its body are added to the rule's body. More precisely, consider the rule

$$R(\bar{x}) \leftarrow R_1(\bar{x}_1), \ldots, R_k(\bar{x}_k). \tag{4.4}$$

Let $\bar{z} \stackrel{\text{def}}{=} \bar{x} - (\bar{x}_1 \cup \ldots \cup \bar{x}_k)$ be nonempty.[5] Let $T(\bar{z})$ be a new relation symbol, representing the total relation, i.e., for all $\bar{z}$, $T(\bar{z})$ holds. Rule

$$R(\bar{x}) \leftarrow R_1(\bar{x}_1), \ldots, R_k(\bar{x}_k), T(\bar{z}),$$

equivalent to (4.4), is safe. Note that $T(\bar{z})$ is finite, since the database domain is finite.

The semantics of unsafe rules, as accepted in this book, is given by assuming that the total relation binding the "unsafe" variables is implicitly added to each unsafe rule. □

A rule with the empty body is called a *fact.*[6]

Any finite set of DATALOG rules is called a DATALOG *program.*

Observe that recursion is allowed in DATALOG. However, no explicit existential quantifiers are allowed and negation appears neither in the head nor in the body of DATALOG rules. In Sections 4.5, 4.6 and further parts of the book, we shall discuss some languages extending DATALOG in various ways, relaxing the restrictions as to the use of quantifiers and negation.

The meaning of a rule is that the conjunction of atoms of the body implies the head of the body, assuming that all variables are universally quantified. More precisely, if $\bar{t}, \bar{t}_1, \ldots, \bar{t}_k$ are tuples of terms then a rule of the form:

$$R(\bar{t}) \leftarrow R_1(\bar{t}_1), \ldots, R_k(\bar{t}_k)$$

represents the following first-order formula:

$$\forall \bar{x}.[(R_1(\bar{t}_1) \land \ldots \land R_k(\bar{t}_k)) \rightarrow R(\bar{t})],$$

[5] By $\bar{x} \cup \bar{y}$ we shall mean all variables that are in $\bar{x}$ or $\bar{y}$ (after removing duplicates), and by $\bar{x} - \bar{y}$ we shall always mean all variables that are in $\bar{x}$ and not in $\bar{y}$.

[6] The empty body is considered to be TRUE.

where $\bar{x}$ are all the variables appearing in the rule. Observe that the above formula is equivalent to:

$$\forall \bar{y}.[\exists \bar{z}.(R_1(\bar{t}_1) \wedge \ldots \wedge R_k(\bar{t}_k)) \rightarrow R(\bar{t})],$$

where $\bar{y}$ are all variables appearing in the rule's head and $\bar{z} = \bar{x} - \bar{y}$.

Assume that a DATALOG program consists of a number of rules of the following form:

$$\begin{aligned} &R(\bar{x}) \leftarrow A_1(\bar{y}_1) \\ &\ldots \\ &R(\bar{x}) \leftarrow A_n(\bar{y}_n). \end{aligned} \tag{4.5}$$

The set of rules (4.5) can be represented equivalently as the following formula:

$$[\exists \bar{z}_1.A_1(\bar{y}_1) \vee \ldots \vee \exists \bar{z}_n.A_n(\bar{y}_n)] \rightarrow R(\bar{x}), \tag{4.6}$$

where, for $k = 1, \ldots, n$, $\bar{z}_k = \bar{y}_k - \bar{x}$.

There may be many relations R satisfying formula (4.6). The semantics attached to DATALOG states that the least such relation is the desired result. The explicit definition of R is expressed by using the fixpoint calculus as follows:

$$R(\bar{x}) = \text{LFP}\, R(\bar{x}).[\exists \bar{z}_1.A_1(\bar{y}_1) \vee \ldots \vee \exists \bar{z}_n.A_n(\bar{y}_n)]. \tag{4.7}$$

In the case where a DATALOG program defines many relations, the definition of the underlying semantics is provided by the least simultaneous fixpoint over all relations occurring in the program.

Example 4.4.2. Consider a domain consisting of regions and locations, where each location is directly included in exactly one region.

Assume that the inclusion of a location in region is defined by an extensional relation $DirectRegion(x, y)$, denoting that location x is directly included in region y. Assume further that regions form a hierarchy w.r.t. inclusion and that we are given an extensional relation $Includes(x, y)$, meaning that region x includes region y. Consider the following DATALOG rules:

$$\begin{aligned} &InRegion(x, y) \leftarrow DirectRegion(x, y) \\ &InRegion(x, y) \leftarrow Includes(y, z), InRegion(x, z). \end{aligned}$$

The rules define relation $InRegion(x, y)$ denoting that location x is directly or indirectly included in region y.

The semantics of the above program is given by the following simultaneous fixpoint:

$$\begin{array}{l}\text{LFP}\, \mathit{DirectRegion}(x,y), \mathit{Includes}(y,z), \mathit{InRegion}(x,y).\\ \quad [\text{EDB}(\mathit{DirectRegion}),\\ \quad \text{EDB}(\mathit{Includes}),\\ \quad \mathit{DirectRegion}(x,y) \vee \exists z.(\mathit{Includes}(y,z) \wedge \mathit{InRegion}(x,z))],\end{array}$$

where EDB(R) abbreviates the conjunction of facts about relation R included in the extensional database. □

4.5 Datalog¬

DATALOG¬ extends DATALOG by allowing negation in the body of rules. Such an extension raises many problems concerning the semantics and complexity of the resulting language. In fact, there are many extensions of DATALOG that allow one to use negation in rules, but most of them either lose the intuitive semantics based on predicate calculus or lose their tractability. In this section we discuss two of the most widely accepted variants of DATALOG¬, the *stratified* and *well-founded* DATALOG¬.

The DATALOG¬ syntax is defined by the following rule:

⟨DATALOG¬ RULE⟩ ::=
⟨ATOMIC FORMULA⟩ ← [[¬]⟨ATOMIC FORMULA⟩
{, [¬]⟨ATOMIC FORMULA⟩}].

As usual, any finite set of DATALOG¬ rules is called a DATALOG¬ program.

4.5.1 Stratified Programs

One of the most interesting classes of DATALOG¬ programs, appearing in practical implementations, consists of stratified programs. Intuitively, a program is stratified if it consists of layers (called strata) such that each relation is fully defined in a single strata and each negative literal occurring in a rule of a given layer refers to a relation defined in one of the previous layers. The formal definition of stratification follows.

Definition 4.5.1. *By a* stratification *of a* DATALOG¬ *program P we mean a sequence of* DATALOG¬ *programs $P^1, \ldots, P^n$, with P^1 possibly empty, such that:*

- $P^1 \cup \ldots \cup P^n = P$ *and for all* $1 \leq i \neq j \leq n$, $P^i \cap P^j = \emptyset$
- *for any relation R of P and* $1 \leq i \leq n$*:*

- *if R occurs in P^i then all the rules with R in their heads are in $P^1 \cup \ldots \cup P^i$*
- *if R occurs in P^i under negation then all the rules with R in their heads are in $P^1 \cup \ldots \cup P^{i-1}$.*

Given a stratification $P^1, \ldots, P^n$ of P, each P^i is called a stratum *of the stratification. A program is called* stratified *if it has a stratification.* □

Example 4.5.2. The following DATALOG$^\neg$ program:

$$R(x) \leftarrow S(x,y), R(y) \tag{4.8}$$
$$R(x) \leftarrow \neg S(x,y), R(x) \tag{4.9}$$
$$S(x,y) \leftarrow S(y,x) \tag{4.10}$$

is stratified and has a stratification $\{(4.10)\}, \{(4.8), (4.9)\}$.

Program:

$$T(x) \leftarrow U(x)$$
$$U(x) \leftarrow \neg T(x)$$

cannot be stratified. □

There is a simple and efficient test allowing one to check whether a given DATALOG$^\neg$ program is stratified. Let P be a DATALOG$^\neg$ program. By the *dependency graph* P we shall mean a graph with vertices labelled by intensional relations of P and containing two types of edges:

- there is a *positive edge* $\langle Q, R\rangle$ in the graph if and only if there is a rule in P in which Q appears positively in the rule's body and R appears in the rule's head
- there is a *negative edge* $\langle Q, R\rangle$ in the graph if and only if there is a rule in P in which Q appears negatively in the rule's body and R appears in the rule's head.

The program P is stratified if no cycle in its dependency graph contains a negative edge.

Example 4.5.3. Consider the dependency graphs corresponding to programs of Example 4.5.2 illustrated in Figure 4.2.

The first graph contains no cycle with a negative edge and consequently, the first program is stratified. The second graph contains a cycle with a negative edge, thus the second program is indeed not stratified. □

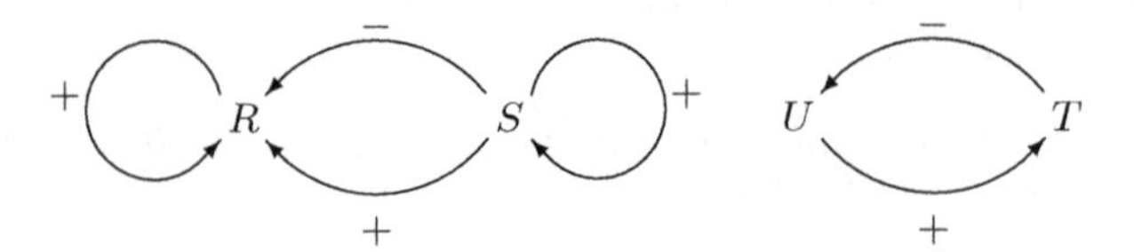

Fig. 4.2. Dependency graphs corresponding to programs of Example 4.5.2.

A program may have various different stratifications, but its semantics is independent of any particular stratification. Given a program P and its stratification $\{P^1, \ldots, P^n\}$, we attach a semantics to P by applying, in order, programs $P^1, \ldots, P^n$ and accepting the closed world assumption. Let $\bar{R}_i$ be the set of all intensional relations of P^i. Then one first computes $\text{LFP}\,\bar{R}_1.P^1$, then having computed $\bar{R}_1$ one computes $\text{LFP}\,\bar{R}_2.P^2$ and so on until all strata are applied.

4.5.2 Well-founded Semantics

Although stratified semantics provides a simple and natural approach to DATALOG$^\neg$ programs, it has one major limitation. Namely, it cannot be applied to all DATALOG$^\neg$ programs.

We now present another semantics for DATALOG$^\neg$ programs which is called *well-founded.* The new semantics can be applied to all DATALOG$^\neg$ programs. Moreover, well-founded semantics can be viewed as a natural extension of stratified semantics because both agree on stratified programs.

The main conceptual distinction between stratified and well-founded semantics lies in different assumptions concerning the answer to a DATALOG$^\neg$ query. The former assumes that each fact is either true or false. The latter admits that the truth value of a fact may be unknown.

We start with some preliminary notions.

Definition 4.5.4. *Let P be a* DATALOG$^\neg$ *program over a signature with the set of constants C. A* ground instance of a rule $r \in P$ *is any rule obtainable from r by uniformly substituting all free variables occurring in r by constant symbols from C. A set of all ground instances of all rules from P is called the* ground instance of a program P. *The set of all ground atoms occurring in the ground instance of a* DATALOG$^\neg$ *program P is denoted by* $\text{ATM}(P)$. □

Definition 4.5.5. *A* three-valued interpretation of a DATALOG$^\neg$ program P *is a mapping from* $\text{ATM}(P)$ *into the set* $\{\text{TRUE}, \text{FALSE}, \text{UNKNOWN}\}$. □

Let I be a three-valued interpretation of a DATALOG$^\neg$ program P. We denote by I^0 (respectively I^1) the set of all atoms from $\text{ATM}(P)$ whose truth value w.r.t. I is FALSE (respectively TRUE).

It is often convenient to represent a three-valued interpretation of a DATALOG$^\neg$ program by listing the positive and negative facts and omitting the unknown ones. For instance, the interpretation I, where $I(R(a)) = \text{TRUE}$, $I(R(b)) = \text{UNKNOWN}$, $I(Q(b)) = \text{UNKNOWN}$ and $I(Q(a)) = \text{FALSE}$ can be represented as $I = \{R(a), \neg Q(a)\}$.

Let $\leq$ be the truth ordering, as defined in Section 2.10. The least (w.r.t. $\leq$) interpretation, denoted by $I_\perp$, is an interpretation where all atoms are FALSE.

Given a three-valued interpretation I, we can extend it in a standard way to Boolean combinations of atoms, as done in Section 2.10 (see formula (2.10)).

Definition 4.5.6. *Given a* DATALOG$^\neg$ *program P and an interpretation I of P, the* positivized version *of P w.r.t. I is the program obtained from the ground instance of P by replacing each negative literal $\neg A$ by $I(\neg A)$ (i.e., by* FALSE, TRUE *or* UNKNOWN*). Such programs are called* 3-DATALOG$^\neg$ programs.[7] □

Example 4.5.7. Consider a DATALOG$^\neg$ program P given by

$$W(a) \leftarrow \neg R(a)$$
$$Q(a) \leftarrow \neg R(a), W(a)$$
$$S(a) \leftarrow \neg T(a)$$
$$T(a) \leftarrow Q(a), \neg S(a)$$
$$U(a) \leftarrow \neg T(a), W(a), S(a)$$

and the interpretation $I_\perp = \{\neg W(a), \neg R(a), \neg Q(a), \neg S(a), \neg T(a), \neg U(a)\}$. The positivized version of P w.r.t. $I_\perp$, P_+, is given by

$$W(a) \leftarrow \text{TRUE}$$
$$Q(a) \leftarrow \text{TRUE}, W(a)$$
$$S(a) \leftarrow \text{TRUE}$$
$$T(a) \leftarrow Q(a), \text{TRUE}$$
$$U(a) \leftarrow \text{TRUE}, W(a), S(a).$$

□

Definition 4.5.8. *Given a 3-*DATALOG$^\neg$ *program P, the* three-valued immediate consequence operator, *T_P, on an interpretation I of P, is a mapping defined as follows, where $A \in \text{ATM}(P)$:*

[7] Note that 3-DATALOG$^\neg$ programs do not contain negation symbols.

$$T_P(A) \stackrel{\text{def}}{=} \begin{cases} \text{TRUE} & \textit{if } A \textit{ is a fact or there is a rule of the form} \\ & A \leftarrow R_1, \ldots, R_k \textit{ such that} \\ & I(R_1 \wedge \ldots \wedge R_k) = \text{TRUE} \\ \text{FALSE} & \textit{if there is no rule with } A \textit{ in its head or,} \\ & \textit{for each rule of the form } A \leftarrow R_1, \ldots, R_k, \\ & I(R_1 \wedge \ldots \wedge R_k) = \text{FALSE} \\ \text{UNKNOWN} & \textit{otherwise.} \end{cases}$$

□

The immediate consequence operator T_P has the following property.

Theorem 4.5.9. *Let P be a 3-DATALOG$^\neg$ program. Then the sequence $\{T_P^i(I_\bot)\}_{i>0}$ is non-decreasing w.r.t. the truth ordering and converges to the least fixpoint of T_P.* □

Let P be a DATALOG$^\neg$ program and let I be a three-valued interpretation of P. We write $C_P(I)$ to denote the least fixpoint of T_{P_+}, where P_+ is the positivized version of the program P w.r.t. I.

Example 4.5.10 (Example 4.5.7 continued). We compute $C_P(I_\bot)$:

$$(T_{P_+})^1(I_\bot) = \{W(a), \neg R(a), \neg Q(a), S(a), \neg T(a), \neg U(a)\}$$
$$(T_{P_+})^2(I_\bot) = \{W(a), \neg R(a), Q(a), S(a), \neg T(a), U(a)\}$$
$$(T_{P_+})^3(I_\bot) = \{W(a), \neg R(a), Q(a), S(a), T(a), U(a)\}$$
$$(T_{P_+})^4(I_\bot) = (T_{P_+})^3(I_\bot).$$

Hence, $C_P(I_\bot) = \{W(a), \neg R(a), Q(a), S(a), T(a), U(a)\}$. □

Definition 4.5.11. *Let P be a DATALOG$^\neg$ program. A three-valued interpretation I of ATM(P) is called a* stable model *of P iff $C_P(I) = I$.* □

Definition 4.5.12. *Let P be a DATALOG$^\neg$ program. The* well-founded semantics *assigns to P the three-valued interpretation of ATM(P) consisting of all positive and negative facts belonging to all stable models of P. We call this interpretation the* well-founded model of *P.* □

The above definition does not provide an efficient method for computing well-founded models of DATALOG$^\neg$ programs. Below we present a tractable method to compute well-founded models.

Let P be a DATALOG$^\neg$ program. We start by defining the following sequence of interpretations:

$$I_0 \stackrel{\text{def}}{=} I_\bot$$
$$I_{i+1} \stackrel{\text{def}}{=} C_P(I_i).$$

It can be shown that, for all $i > 0$,

$$I_0 \leq I_2 \ldots \leq I_{2i} \leq I_{2i+2} \leq \ldots \leq \; I_{2i+1} \leq I_{2i-1} \leq \ldots \leq I_1.$$

Thus, the even subsequence is non-decreasing and the odd one is non-increasing. Since there are finitely many three-valued interpretations for a given program, each of those subsequences becomes constant at some point. Let I_* denote the limit of the subsequence $\{I_{2i}\}_{i \geq 0}$ and let I^* denote the limit of the subsequence $\{I_{2i+1}\}_{i \geq 0}$. We define a three-valued interpretation of P, denoted by I_*^*, by

$$I_*^*(A) = \begin{cases} \text{TRUE} & \text{if } I_*(A) = I^*(A) = \text{TRUE} \\ \text{FALSE} & \text{if } I_*(A) = I^*(A) = \text{FALSE} \\ \text{UNKNOWN} & \text{otherwise.} \end{cases}$$

The following result holds.

Theorem 4.5.13. *Let P be a* DATALOG$^\neg$ *program and let I_*^* be as specified above. Then I_*^* is the well-founded model of P.* □

Example 4.5.14 (Example 4.5.10 continued). We now compute the well-founded model of program P. We start with $I_0 = I_\perp$. Applying the C_P operator we get the following sequence of three-valued interpretations of P:

$$I_1 = \{W(a), Q(a), \neg R(a), S(a), T(a), U(a)\}$$
$$I_2 = \{W(a), Q(a), \neg R(a), \neg S(a), \neg T(a), \neg U(a)\}$$
$$I_3 = I_1$$
$$I_4 = I_2.$$

Therefore,

$$I_* = \{W(a), Q(a), \neg R(a), \neg S(a), \neg T(a), \neg U(a)\}$$
$$I^* = \{W(a), Q(a), \neg R(a), S(a), T(a), U(a)\}.$$

In consequence, $I_*^* = \{W(a), Q(a), \neg R(a)\}$. □

4.6 Semi-Horn Query Language (SHQL)

The *semi-Horn Query Language*, SHQL, is a purely declarative query language. Use of negation in a query is interpreted as classical negation, a class of mixed quantifiers is allowed in queries, and both intentional and extensional predicates may occur anywhere in the query. In SHQL a query is expressed as a theory consisting of semi-Horn formulas.[8]

SHQL is used as follows. Given the task of computing a definition of an intensional predicate Q (or asking whether a tuple is an instance of Q) relative to a relational database B consisting of the relations $R_1, \ldots, R_n$, we first provide an implicit definition of Q in terms of a semi-Horn theory, $Th(Q)$. The theory $Th(Q)$ is only constrained by the fact that it must be semi-Horn. All quantifiers and logical connectives are interpreted classically. The explicit definition of Q is computed in time and space polynomial in the size of the database, where by explicit definition we mean a formula which is equivalent to Q but does not refer to Q.

The computation process can be described in two stages.

In the first stage, we provide a PTIME compilation process which returns an explicit definition characterizing the intensional predicate. If the query is defined by means of nonrecursive semi-Horn rules, then the output is a first-order formula expressing an explicit definition of Q. The output relation is then computed in PTIME. If the query is recursive, than the output is a logically equivalent fixpoint formula expressing an explicit definition of Q. In this case, the output relation is also computed in PTIME. This is done using quantifier elimination techniques.[9]

In the second stage, we use the explicit definition of Q computed in the first stage to compute a suitable relation in the relational database that satisfies Q. After computing the output relation, we check whether such a relation is consistent (or, in other words, *coherent*), with the database contents, i.e., $B \cup Th(Q) \not\models$ FALSE. Assuming that this is the case, we know that the output relation exists and can now accept the answer computed previously. Both checking that the query is coherent and computing the output relation can be done efficiently because calculating fixpoint queries and fixpoint satisfiability checking over finite domains are both in PTIME.[10]

Given a query $Th(Q)$, the combined problem of checking that the query is coherent, finding an explicit definition of Q and explicitly computing the answer is in general NPTIME-complete. However, the syntactic restriction to semi-Horn theories makes the problem solvable in polynomial time. Most importantly, SHQL is a highly expressive language which covers all PTIME queries and is at the same time purely declarative. Querying with SHQL is as natural as querying with classical logic and the compilation step is completely

[8] In fact, here we consider a subset of the original SHQL, sufficient for our purposes and still retaining both the ability of expressing all PTIME queries and tractability.

[9] Note that this technique can be used for theories outside the semi-Horn class, also applying techniques of second-order quantifier elimination, but neither the complexity results nor a successful reduction are guaranteed.

[10] Recall that in the case of first-order queries it is even in LOGSPACE.

transparent to the user. The following example illustrates how SHQL may be applied.

Example 4.6.1. This example demonstrates how intensional and extensional predicates may be used anywhere in the SHQL query. In particular, in comparison with rule-based queries such as logic programming or DATALOG, both intensional and extensional predicates may occur in both the head and body of any implication.

Assume we have a database B, containing information about whether persons are "rich", "smart" or "experienced", denoted by the unary extensional predicates, *Rich*, *Smart*, and *Experienced*, respectively. Suppose we are interested in selecting all rich persons and perhaps some others and we only want to consider those who are smart and experienced. Let Q denote the unary intensional predicate that describes the required relation. The first condition is then expressed by the formula

$$\forall x.[Rich(x) \to Q(x)],$$

while the second condition is expressed by the formula

$$\forall x.[Q(x) \to (Smart(x) \land Experienced(x))].$$

The implicit query is then defined as the conjunction of the above formulas and we are interested in obtaining both the least and the greatest relation Q satisfying the above conditions. Since the above formulas are semi-Horn, part of SHQL, explicit definitions of those relations can be computed automatically. The least relation is defined by the formula

$$\forall x.[Q(x) \equiv Rich(x)]$$

and the greatest relation is defined by the formula

$$\forall x.[Q(x) \equiv (Smart(x) \land Experienced(x))].$$

Observe that our query has a side-effect due to the transitivity of implication. Namely the following coherence condition should also hold:[11]

$$\forall x.[Rich(x) \to (Smart(x) \land Experienced(x))].$$

Thus, for instance, if a database contains an element e such that $Rich(e)$ and $\neg Smart(e)$, then the query is inconsistent with the database. □

[11] Such conditions are computed automatically. In fact, when applying the SHQL methodology one automatically generates a condition which is both sufficient and necessary for assuring that the query is consistent with the database.

The SHQL methodology is founded on second-order quantifier elimination techniques and basically uses variants of the following theorem, where the notation is explained in Section 2.9.[12]

Theorem 4.6.2. *Let $A(\bar{x}, \bar{z}, Q)$ and $C(Q)$ be arbitrary first-order formulas positive w.r.t. Q and let $B(Q)$ be an arbitrary first-order formula negative w.r.t. Q. Then:*

- *for any formula $T(Q)$ of the form $\forall \bar{x}.[A(\bar{x}, \bar{z}, Q) \to Q(\bar{x})] \wedge B(Q)$ we have:*
  - *the explicit definition of the least Q satisfying $T(Q)$ is given by* $Q(\bar{x}) \equiv \text{LFP}\, Q(\bar{x}).A(\bar{x}, \bar{z}, Q)$
  - *the coherence condition for $T(Q)$ is defined by the following formula:* $B(Q(\bar{t}) := \text{LFP}\, Q(\bar{x}).A(\bar{x}, \bar{z}, Q)[\bar{t}])$.
- *for any formula $S(Q)$ of the form $\forall \bar{x}.[Q(\bar{x}) \to A(\bar{x}, \bar{z}, Q))] \wedge C(Q)$ we have:*
  - *the explicit definition of the greatest Q satisfying $S(Q)$ is given by* $Q(\bar{x}) \equiv \text{GFP}\, Q(\bar{x}).A(\bar{x}, \bar{z}, Q)$
  - *the coherence condition for $S(Q)$ is defined by the following formula:* $C(Q(\bar{t}) := \text{GFP}\, Q(\bar{x}).A(\bar{x}, \bar{z}, Q)[\bar{t}])$. □

Observe that in the case where formula A in Theorem 4.6.2 is not recursive, i.e., does not contain Q, the resulting definition for Q, as well as the suitable coherence condition, can easily be formulated in terms of first-order logic. This follows from the fact that in such a case both $\text{LFP}\, Q(\bar{x}).A(\bar{x}, \overline{z_i})$ and $\text{GFP}\, Q(\bar{x}).A(\bar{x}, \overline{z_i})$ are equivalent to $A(\bar{x}, \overline{z_i})$. We also have the following theorem.

Theorem 4.6.3. *Let B be a relational database.*

- *Any implicit query $A(Q)$ to B, where $A(Q)$ is a semi Horn formula w.r.t. Q, is computable in polynomial time in the size of the database. If $A(Q)$ is a nonrecursive formula, then Q is computable in logarithmic space.*
- *Any* PTIME *query can be expressed by means of semi-Horn formulas provided that B is linearly ordered.* □

[12] We do not discuss the whole underlying methodology here, since we provide necessary theorems in further parts of the book, adopting them to simpler situations. The interested reader will find pointers to SHQL in Section 4.7.

4.7 Bibliographic Notes

There are many good books concerning database systems. For a comprehensive presentation of the subject and the relevant bibliography the reader is referred, e.g., to [1, 220, 222, 223].

The relational model of database systems is defined in [37]. A survey of relational databases is given, e.g., in [40] and of the relational database theory in [96]. Applications of logics in databases are outlined in [126]. Practical query languages, including SQL, are discussed, e.g., in [41, 220, 222]. An introductory textbook on deductive databases is [38]. Further readings on deductive databases can be found in [125, 173]. Many useful mathematical results applicable to deductive databases are also presented in [68]. For pointers to first-order logic as a query language, in particular concerning the complexity results, see also Chapter 2.

The notion of closed world databases has been introduced by Reiter in [175]. One of the problems with Reiter's original CWA logic is that it does not deal properly with disjunctive data bases. Accordingly, several other approaches to the CWA have been proposed. Most of them are discussed in detail in [116].

DATALOG is a query language based on the logic programming paradigm. Many results concerning the semantics of DATALOG are then obtained by adapting the corresponding logic programming results. Particularly important are results on fixpoint semantics, worked out in the context of logic programming in [225].

Fixpoint logic as a database language has been considered by many authors. Monotone fixpoint queries were originally defined in [35]. Other fixpoint languages, including among others *inflationary queries*, were studied in [83, 84]. The characterization of fixpoints given by (4.2) and (4.3) is based on the Knaster and Tarski fixpoint theorem for monotone functionals defined over partial orders (see [213]). Theorem 4.3.1 has been independently proved in [91] and [228]. A comprehensive presentation of complexity related issues is given in [92].

Methods for evaluating recursive and fixpoint queries, together with query optimization techniques, are discussed, e.g., in [1, 7, 80, 223]. In the book we concentrated on the *bottom-up approach*. There are many papers on *top-down* approaches, originating from logic programming. One of the most general approaches to top-down evaluation, studied in [221], uses techniques similar to those developed for logic programs - see, e.g., [101]. For a comprehensive discussion of related topics see also [76].

A great deal of attention has been devoted to the use of negation in DATALOG-like languages. Stratified DATALOG$^\neg$ has been independently proposed by many authors [4, 36, 105, 226]. Well-founded and three-valued semantics have been studied in [18, 73, 170, 227]. The fact that the stratified semantics agrees with well-founded semantics on stratified programs has been proved in [227]. For a comprehensive presentation of the subject see [1]. A survey of approaches, in the context of logic programming, is also given in [5].

The SHQL language together with its underlying methodology is introduced in [55], where, among others, Theorems 4.6.2 and 4.6.3 were provided. The methodology is based on the second-order quantifier elimination technique introduced in [142] and the DLS algorithm for second-order quantifier elimination given in [53]. For some further results concerning the methodology see also [49]. Another application of the DLS related techniques in the context of databases is presented in [93].

5

Non-Monotonic Reasoning

5.1 Introduction

Traditional logics are *monotonic*, i.e., adding new premises (axioms) will never invalidate previously inferred conclusions (theorems), or, equivalently, the set of conclusions non-decreases monotonically with the set of premises. Formally, a logic is monotonic if and only if it satisfies the condition that for any sets of premises S and S',

$$S \subseteq S' \text{ implies } Cn(S) \subseteq Cn(S'),$$

where Cn denotes the semantic consequence operator of a given logic, i.e., the operator which to each set of formulas S assigns the set $Cn(S)$ of all formulas entailed from S in the logic.

In the last two decades, there has been a great deal of interest in logical systems that relax the property of monotonicity. These have been studied primarily in the context of commonsense reasoning. What typifies this kind of inference is its ability to deal with incomplete information. In everyday life we are constantly faced with situations in which relevant facts used to reason about various aspects of the world are not known to hold with complete certainty. Yet, in order to act, we must be able to draw conclusions based on such facts even if available evidence is insufficient to assure their correctness. Clearly, such conclusions are risky and may be invalidated when new information becomes available. The basic understanding is that such conclusions more often than not do actually hold and it is an exception when this is not the case, thus the risk in jumping to such conclusions is minimal. A classical example of such a situation is the following.

> Assume that I am planning to make a trip by car. To begin with, I must decide where my car actually is. Given no evidence to the

P. Doherty et al.: *Knowledge Representation Techniques*, Studfuzz **202**, 77–99 (2006)
www.springerlink.com

contrary, it is reasonable to conclude that my car is located where I last parked it.

Of course, the above conclusion may turn out to be incorrect. The car may have been towed away or stolen, but such instances would be uncommon. In order to plan and act, it is essential that such weak conclusions can be drawn.

There are a number of points of view as to how nonmonotonic formalisms should be classified according to type. One viewpoint advocated in the literature is that there are two basic types of non-monotonic reasoning: *default reasoning* and *autoepistemic reasoning*.

By the former we mean the drawing of a rational conclusion, from less than conclusive information, in the absence of evidence leading to the contrary. What typifies default reasoning is its defeasibility: any conclusion derived by default can be invalidated by providing new information. A classical example of default reasoning is the rule stating: "In the absence of evidence to the contrary, assume that a bird flies."

By autoepistemic reasoning, we understand reaching a conclusion, from an incomplete representation of complete information, under the assumption that if the conclusion were false, its negation would be explicitly represented. A typical example of autoepistemic reasoning is the rule stating: "If your name is not on a list of winners, assume that you are a loser." The motivation for using this rule stems from the observation that the number of losers is usually much greater than the number of winners, so explicitly keeping all the losers would be impractical. It should be observed that autoepistemic reasoning is not defeasible: its conclusions cannot be invalidated by providing new evidence, since we assume complete information about the considered aspect of the world. However, this type of reasoning is non-monotonic because conclusions derivable by employing autoepistemic rules change non-monotonically with respect to the particular context in which the rules are embedded.[1]

A number of non-monotonic formalisms have been studied in the literature. In this chapter we provide a brief introduction to the most prominent of them, namely, *default logic* and *circumscription*. Both of these formalisms can be used to model default and autoepistemic reasoning. Another reason for making this choice is that both default logic and circumscription will be used throughout the book.

---

[1] If the list of winners consists of John and Mary, our rule allows us to infer that Bill is a loser. However, if the list of winners consists of John, Mary and Bill, this conclusion will no longer be derivable.

5.2 Default Logic

5.2.1 Foundations

As its name suggests, default logic provides a formal framework for default reasoning.

The basic construct used in default logic is that of a *default* rule. This is an expression of the form

$$\frac{A(\bar{x}) : B(\bar{x})}{C(\bar{x})} \tag{5.1}$$

where $A(\bar{x})$, $B(\bar{x})$ and $C(\bar{x})$ are first-order formulas whose free variables are among those of $\bar{x} = \langle x_1, \ldots, x_n \rangle$. $A(\bar{x})$ is called the *prerequisite*, $B(\bar{x})$ the *justification*, and $C(\bar{x})$ the *consequent of the default.*[2]

The default (5.1) has the following interpretation: For all individuals represented by $\bar{x} = \langle x_1, \ldots, x_n \rangle$, if $A(\bar{x})$ is believed and $B(\bar{x})$ is consistent with what is believed, then $C(\bar{x})$ is to be believed.[3]

As an example, consider a prototypical fact that birds usually fly. This can be represented as a rule stating that: "In the absence of evidence to the contrary, assume about any particular bird that it flies." In default logic this can be represented by the default rule

$$\frac{Bird(x) : Flies(x)}{Flies(x)}.$$

In default logic, the primary objects of interest are *default theories.*

Definition 5.2.1. *A* default theory *is a pair $T = \langle W, D \rangle$, where W is a set of first-order sentences, axioms of T, and D is a set of defaults.* □

Intuitively, a default theory is viewed as a representation of one or more aspects of the world under consideration. The axioms in a theory are intended to contain all information known to be true. The default rules extend this information by supporting plausible conclusions. A set of beliefs about the world represented by a theory T is called an *extension* of T. To define this notion, we need some preliminary terminology.

A default $A(\bar{x}) : B(\bar{x}) / C(\bar{x})$ is called *open* if and only if at least one of $A(\bar{x})$, $B(\bar{x})$, $C(\bar{x})$ contains a free variable; otherwise, it is called *closed.* A default

[2] In plain text, the default (5.1) is often written in the form $A(\bar{x}) : B(\bar{x}) / C(\bar{x})$. If $A(\bar{x})$ = TRUE, the default is called *prerequisite-free* and it is written as $: B(\bar{x}) / C(\bar{x})$.

[3] In the AI literature, theorems derivable using non-monotonic formalisms are usually referred to as *beliefs*.

theory is said to be *open* if and only if it contains at least one open default; otherwise, it is said to be *closed.*

It is important to note that free variables in a default rule are viewed as implicitly universally quantified and with a scope covering the whole default rule. Consequently, an open default represents a general schema that can be applied to various tuples of individuals.

An *instance of an open default* is the result of uniformly replacing all free occurrences of variables by ground terms. More specifically, an instance of an open default $A(\bar{x}) : B(\bar{x})/C(\bar{x})$, where $\bar{x} = \langle x_1, \ldots, x_n \rangle$, is any closed default of the form $A(\bar{t}) : B(\bar{t})/C(\bar{t})$, where $\bar{t} = \langle t_1, \ldots, t_n \rangle$ is an n-tuple of ground terms uniformly replacing all free occurrences of variables $\bar{x}$. For example, the closed defaults $Bird(\mathsf{Tweety}) : Flies(\mathsf{Tweety})/Flies(\mathsf{Tweety})$ and $Bird(\mathsf{Joe}) : Flies(\mathsf{Joe})/Flies(\mathsf{Joe})$ are both instances of open default $Bird(x) : Flies(x)/Flies(x)$.

Since open defaults represent general inference schemata which can be applied to various tuples of individuals, it is natural to identify such a default with the set of all its instances. Adopting this solution, we shall be able to eliminate open defaults and, in consequence, to limit our attention to closed default theories only.

To formalize this idea, we define a mapping which to every default theory T assigns a closed theory $\text{CLOSED}(T)$, obtainable from T by replacing all T's open defaults by sets of their instances. To make the specification of the above transformation as simple as possible, we put two restrictions on the theories under consideration. Firstly, we assume that both the axioms of a theory and consequents of its defaults are universal formulas. Secondly, we limit ourselves to languages without function symbols.[4]

Definition 5.2.2. *Let $T = \langle W, D \rangle$ be a default theory over a language $\mathcal{L}$. a* closure *of T, written $\text{CLOSED}(T)$, is a closed default theory obtained from T by replacing all open defaults from D by sets of their instances constructible using individual constants of $\mathcal{L}$.* □

The notion of an extension of a default theory is explicitly defined for closed theories only. If T is an open theory, its extensions are identified with those of $\text{CLOSED}(T)$.

The definition of an extension of a closed default theory uses a fixpoint construction.

Definition 5.2.3. *Let $T = \langle W, D \rangle$ be a closed default theory. For any set of sentences S, let $\Gamma(S)$ be the smallest set of sentences satisfying the following properties:*

[4] Although the default theories we consider are quite restrictive, they are sufficient for many practical applications.

1. $\Gamma(S) = Cn(\Gamma(S))$
2. $W \subseteq \Gamma(S)$
3. If $\frac{A:B}{C} \in D$, $A \in \Gamma(S)$ and $\neg B \notin S$, then $C \in \Gamma(S)$.

A set of sentences E is an extension *of T if and only if $E = \Gamma(E)$, i.e., if and only if E is a fixed point of Γ.* □

Example 5.2.4. Let $T = \langle W, D \rangle$, where

$$W = \{Bird(\mathsf{Tweety})\}, \quad D = \left\{ \frac{Bird(x) : Flies(x)}{Flies(x)} \right\}.$$

$$\text{Closed}(T) = \langle W, D' \rangle, \text{ where } D' = \left\{ \frac{Bird(\mathsf{Tweety}) : Flies(\mathsf{Tweety})}{Flies(\mathsf{Tweety})} \right\}.$$

Closed(T), and hence T, has one extension given by

$$E = Cn(\{Bird(\mathsf{Tweety}), Flies(\mathsf{Tweety})\}).$$

□

Example 5.2.5. Let $T = \langle W, D \rangle$, where

$$W = \{Republican(\mathsf{Nixon}) \wedge Quaker(\mathsf{Nixon})\}$$

$$D = \left\{ \frac{Republican(x) : \neg Pacifist(x)}{\neg Pacifist(x)}, \frac{Quaker(x) : Pacifist(x)}{Pacifist(x)} \right\}.$$

Closed(T) $= \langle W, D' \rangle$, where

$$D' = \left\{ \frac{Republican(\mathsf{Nixon}) : \neg Pacifist(\mathsf{Nixon})}{\neg Pacifist(\mathsf{Nixon})}, \frac{Quaker(\mathsf{Nixon}) : Pacifist(\mathsf{Nixon})}{Pacifist(\mathsf{Nixon})} \right\}.$$

Closed(T), and hence T, has two extensions:

$$E_1 = Cn(\{Republican(\mathsf{Nixon}) \wedge Quaker(\mathsf{Nixon}), \neg Pacifist(\mathsf{Nixon})\})$$
$$E_2 = Cn(\{Republican(\mathsf{Nixon}) \wedge Quaker(\mathsf{Nixon}), Pacifist(\mathsf{Nixon})\}).$$

□

If a default theory has more than one extension, each of them is considered as an alternative set of beliefs about the world represented by the theory. Observe that there are theories that lack any extensions, as shown in the example below.

Example 5.2.6. Let $T = \langle W, D \rangle$, where $W = \emptyset$ and $D = \left\{ \frac{: P(\mathsf{a})}{\neg P(\mathsf{a})} \right\}$. This theory lacks an extension. This is because the consequence of the default denies its justification. If the default is not applied, then there is no way to derive $\neg P(\mathsf{a})$, so that we are forced to apply it. However, if we do this, $\neg P(\mathsf{a})$ will obtain the status of a belief and the default will become inapplicable.

To show formally that T lacks an extension, observe that the only candidates are $E_1 = Cn(\{\neg P(\mathsf{a})\})$ and $E_2 = Cn(\emptyset)$. Since $\Gamma(E_1) = Cn(\emptyset) \neq E_1$ and $\Gamma(E_2) = Cn(\{\neg P(\mathsf{a})\}) \neq E_2$, neither E_1 nor E_2 are extensions of T. □

5.2.2 Basic Properties of Default Theories

The notion of an extension can be given the following pseudo-iterative specification.[5]

Theorem 5.2.7. *If $T = \langle W, D \rangle$ is a closed default theory, then a set E of sentences is an extension of T if and only if $E = \bigcup_{i=0}^{\infty} E_i$, where $E_0 = W$ and, for $i \geq 0$, $E_{i+1} = Cn(E_i) \cup \{C \mid (A{:}B/C) \in D,\ A \in E_i,\ \neg B \notin E\}$.* □

Theorem 5.2.7 gives rise to a number of corollaries.

Corollary 5.2.8. *A closed default theory $T = \langle W, D \rangle$ has an inconsistent extension if and only if W is inconsistent.* □

Corollary 5.2.9. *If a closed default theory has an inconsistent extension, then this is its only extension.* □

Corollary 5.2.10. *If E and F are extensions of a closed default theory and $E \subseteq F$, then $E = F$.* □

Definition 5.2.11. *Let $T = \langle W, D \rangle$ be a closed default theory and suppose that E is an extension of T. The* set of generating defaults *for E w.r.t. T, written $\mathrm{GD}(E,T)$, is defined by*

$$\mathrm{GD}(E,T) = \{(A{:}B/C) \in D \mid A \in E \text{ and } \neg B \notin E\}.$$

□

Let D be any set of closed defaults. By $\mathrm{Cons}(D)$ we denote the set of consequents of the defaults from D.

The following result holds.

[5] Note that the specification below is not strictly iterative since the definition of E_i ($i \geq 1$) refers to E.

Theorem 5.2.12. *If E is an extension of a closed default theory $T = \langle W, D \rangle$, then $E = Cn(W \cup \text{Cons}(\text{Gd}(E, T)))$.* □

5.2.3 Normal Default Theories

As has been observed, there are default theories which lack extensions. It is then reasonable to investigate whether there are restricted classes of theories for which extensions are guaranteed to exist.

Definition 5.2.13. *Any default of the form* $\frac{A(\bar{x}) : B(\bar{x})}{B(\bar{x})}$ *is said to be* normal. *A theory $\langle W, D \rangle$ is said to be* normal *if and only if every default in D is normal.* □

Normal default theories are sufficient for modeling many practically occurring situations. An important property associated with normal default theories is the existence of extensions.

Theorem 5.2.14. *Every closed (and hence also open) normal default theory has an extension.* □

Another important property of normal default theories is that they are monotonic with respect to the addition of new defaults.

Theorem 5.2.15. *Let D_1 and D_2 be sets of closed normal defaults such that $D_1 \subseteq D_2$. Let E_1 be an extension of $T_1 = \langle W, D_1 \rangle$ and let $T_2 = \langle W, D_2 \rangle$. Then T_2 has an extension E_2 such that $E_1 \subseteq E_2$.* □

It should be emphasized that Theorem 5.2.15 need not hold for non-normal default theories. Consider, for instance, the theory $T = \langle W, D \rangle$, where $W = \emptyset$ and $D = \left\{ \frac{: P(\mathsf{a})}{Q(\mathsf{a})} \right\}$. T has only one extension, namely, $E = Cn(\{Q(\mathsf{a})\})$. Augmenting T by a new default, $\frac{: \neg P(\mathsf{a})}{\neg P(\mathsf{a})}$, we obtain a new theory with one extension given by $F = Cn(\{\neg P(\mathsf{a})\})$. Clearly, $E \not\subseteq F$.

The next theorem shows that combining beliefs from distinct extensions can lead to inconsistency.

Theorem 5.2.16. *If a closed normal default theory has distinct extensions E and F, then $E \cup F$ is inconsistent.* □

It is worth observing that Theorem 5.2.16 may fail for non-normal default theories. Nevertheless, even in such a case, beliefs from distinct extensions should not be kept together. To illustrate this, consider the theory $T = \langle W, D\rangle$, where $W = \emptyset$ and $D = \left\{\frac{: P(\mathsf{a})}{\neg Q(\mathsf{a})}, \frac{: Q(\mathsf{a})}{\neg P(\mathsf{a})}\right\}$. T has two extensions $E = Cn(\{\neg Q(\mathsf{a})\})$ and $F = Cn(\{\neg P(\mathsf{a})\})$. Clearly, $E \cup F$ is consistent. However, since $\neg Q(\mathsf{a})$ is justified by the consistency of $P(\mathsf{a})$ and $\neg P(\mathsf{a})$ is justified by the consistency of $Q(\mathsf{a})$, $\neg Q(\mathsf{a})$ and $\neg P(\mathsf{a})$ can not be regarded as coexisting beliefs.

5.2.4 Semi-Normal Default Theories

In many naturally occurring situations, the most common non-monotonic rules are those of the form: "If A, believe B unless you believe otherwise." Since any such rule translates into the normal default $A\!:\!B/B$, it is not surprising that normal defaults are extremely common in practice. This raises the question as to whether non-normal defaults are required in modeling. Unfortunately, the answer is positive. The problem is that many non-monotonic rules, which in isolation are naturally represented by normal defaults, must be re-represented when considered in a wider context. The following is a classical example.

Example 5.2.17. Suppose we are given the following:

Bill is a high school dropout
Typically, high school dropouts are adults
Typically, adults are employed.

These commonsense facts are naturally represented by the following normal default theory T:

$$W = \{Dropout(\mathsf{Bill})\}$$

$$D = \left\{\frac{Dropout(x)\!:\!Adult(x)}{Adult(x)}, \frac{Adult(x) : Employed(x)}{Employed(x)}\right\}.$$

Although intuition dictates that one should remain agnostic about the employment status of Bill, T forces us to believe that he is employed. The problem arises because dropouts are atypical adults as regards the state of employment, so that the transitivity from "dropout" via "adult" to "employed" is intuitively unjustified. In order to block this transitivity, the second default can be replaced by the non-normal default,

$$\frac{Adult(x)\!:\!Employed(x) \wedge \neg Dropout(x)}{Employed(x)},$$

which is inapplicable to known dropouts. □

This example describes the following situation which occurs quite often in the modeling process. A non-monotonic rule is provided which, in isolation, is naturally represented by a normal default $A(\bar{x}) : B(\bar{x})/B(\bar{x})$. However, when the rule is embedded in a context, some exceptional circumstances occur, say $E(\bar{x})$, in which the application of the rule is unacceptable. To avoid counter-intuitive inferences, the rule has to be modified to rule out such exceptions. The most natural solution is to replace the existing rule by a new rule,

$$\frac{A(\bar{x}) : B(\bar{x}) \wedge \neg E(\bar{x})}{B(\bar{x})},$$

which is inapplicable for tuples of objects that are known to satisfy E.

Definition 5.2.18. *Any default of the form* $\dfrac{A(\bar{x}) : B(\bar{x}) \wedge C(\bar{x})}{B(\bar{x})}$ *is said to be* semi-normal. *a default theory is said to be* semi-normal *if and only if all of its defaults are semi-normal.*[6] □

Although semi-normal default theories are very convenient from a practical point of view, their behavior is not as regular as that of normal default theories. In particular, they lack (in general) two important properties of normal default theories: the existence of extensions and semi-monotonicity. This is illustrated by the following examples.

Example 5.2.19. Consider the theory T consisting of the empty set of axioms and the following three semi-normal default rules:

$$d_1 = \frac{: P(\mathsf{a}) \wedge \neg Q(\mathsf{a})}{\neg Q(\mathsf{a})} \qquad d_2 = \frac{: Q(\mathsf{a}) \wedge \neg R(\mathsf{a})}{\neg R(\mathsf{a})} \qquad d_3 = \frac{: R(\mathsf{a}) \wedge \neg P(\mathsf{a})}{\neg P(\mathsf{a})}.$$

This theory has no extension. The reason is that applying any one default forces the application of one of the other two, but applying the latter contradicts the justification of the rule previously applied. If one begins by applying d_1, one is then forced to apply d_3, but the application of d_3 denies the justification of d_1. If one begins with d_2, one is forced to apply d_1, but the application of d_1 contradicts the justification of d_2. Finally, beginning with d_3 leaves d_2 applicable, but applying d_2 denies the justification of d_3. □

Example 5.2.20. Let $T = \left\langle \emptyset, \left\{ \dfrac{: P(\mathsf{a}) \wedge Q(\mathsf{a})}{Q(\mathsf{a})} \right\} \right\rangle$. T has one extension, namely, $E = Cn(\{Q(\mathsf{a})\})$. Adding the default $: \neg P(\mathsf{a})/\neg P(\mathsf{a})$, a new theory is obtained whose single extension is $F = Cn(\{\neg P(\mathsf{a})\})$. Clearly $E \not\subseteq F$. □

[6] Note that any normal default $A : B/B$ may be identified with the semi-normal default $A : B \wedge \text{TRUE}/B$. It follows, therefore, that the class of semi-normal defaults includes normal defaults.

5.2.5 Complexity Results for Default Logic

In this section, a number of complexity results for default logic are provided. Attention is limited to propositional theories.[7]

Theorem 5.2.21. *The problem of deciding whether a finite default theory has an extension is Σ_2^P-complete. The problem remains Σ_2^P-complete even when restricted to semi-normal default theories.* □

Theorem 5.2.22.

1. *The problem of determining whether a formula A is a member of some extension of a finite default theory T is Σ_2^P-complete. The problem remains Σ_2^P-complete even if T is a normal default theory.*
2. *The problem of determining whether a formula A is a member of all extensions of a finite default theory T is Π_2^P-complete. The problem remains Π_2^P-complete even if T is a normal default theory.* □

The above results are not very encouraging. They can be slightly improved if we restrict ourselves to restricted classes of default theories. However, even for very simple classes, the above problems are generally NPTIME-complete or CO-NPTIME-complete (see Section 5.4 for references).

5.2.6 Prioritized Default Logic

Prioriterizing the set of default rules in a default theory is a useful technique. Consider the following example.

Example 5.2.23. Suppose we are given the following:

John is an adult full-time student.
Typically, full-time students are not employed.
Typically, adults are employed.

This is naturally represented by

$$W = \{Adult(\mathsf{John}) \wedge FullTimeStudent(\mathsf{John})\}$$

$$D = \left\{ \frac{FullTimeStudent(x) : \neg Employed(x)}{\neg Employed(x)}, \frac{Adult(x) : Employed(x)}{Employed(x)} \right\}.$$

[7] A default theory is said to be propositional if both its axioms and the prerequisites, the justifications and the consequents of its defaults are formulas of propositional logic.

This theory has two extensions which differ in the employment status of John. One of them, namely that containing *Employed*(John), is intuitively unacceptable. Given that John is an adult full-time student, we clearly prefer to conclude that he is not employed. Stated another way, we would like to give the first default priority over the second one.

One alternative which can be used to express priorities between defaults is to use semi-normal default rules. In the example above, it is sufficient to replace the second default by

$$\frac{Adult(x) : Employed(x) \wedge \neg FullTimeStudent(x)}{Employed(x)}.$$

□

One obvious drawback to introducing semi-normal defaults is that they lack the nice properties of normal defaults. Another alternative then is to use normal default rules and specify priorities between defaults explicitly. We call this a *prioritized default logic.*

Definition 5.2.24. *A* prioritized default theory *is a triple $\langle D, W, < \rangle$, where $\langle D, W \rangle$ is an ordinary default theory and $<$ is a strict partial order on D. Intuitively, $d_1 < d_2$ means that d_2 has higher priority than d_1.* □

Let D be a set of closed defaults and suppose that $D_1 \subseteq D$ and $D_2 \subseteq D$. Assume further that $<$ is a strict partial order on D. We say that D_1 is *preferred* to D_2 w.r.t $<$, written $D_2 < D_1$, if and only if

- For each $d, d' \in D$ such that $d < d'$, $\langle d, d' \rangle \notin [D_1 - D_2] \times [D_2 - D_1]$.
- There exist $d, d' \in D$ such that $d < d'$ and $\langle d', d \rangle \in [D_1 - D_2] \times [D_2 - D_1]$.

Intuitively, the first item states that if d is a member of D_1, but not a member of D_2, and d' is a member of D_2, but not a member of D_1, then it must not be the case that d' is preferred to d. Similarly, the second item states that there is at least one d which is in D_1, but not in D_2, and at least one d' which is in D_2, but not in D_1, such that d' is preferred to d.

Example 5.2.25. Let $D_1 = \{d_1, d_2, d_4\}$, $D_2 = \{d_2, d_3, d_5\}$. D_1 is preferred to D_2 w.r.t. $<$ given that $d_3 < d_1$. On the other hand, neither D_1 is preferred to D_2 nor D_2 is preferred to D_1 if $<$ is given by $d_3 < d_1$ and $d_4 < d_5$. □

Let D be a set of closed defaults and suppose that $\mathcal{D} = \{D_1, \ldots, D_n\}$, where each $D_i \subseteq D$. Let $<$ be a strict partial order on D, interpreted as usual. We say that D_i is $<$-maximal in the class $\mathcal{D}$ if and only if there is no $D_j \in \mathcal{D}$ such that D_j is preferred to D_i.

Definition 5.2.26. *Let* $\langle W, D, < \rangle$ *be a closed prioritized default theory and suppose that* $E_1, \ldots, E_n$ $(n \geq 0)$ *are all extensions of* $T = \langle W, D \rangle$. E_i *is said to be an extension of* $\langle W, D, < \rangle$ *if and only if* $\text{GD}(E_i, T)$ *is* $<$*-maximal in the class* $\{\text{GD}(E_1, T), \ldots, \text{GD}(E_n, T)\}$.[8] □

Let us reconsider Example 5.2.23. Normal default rules can be used to represent the facts provided that the first default, d_1, is given higher priority than the second one, d_2. In other words, the relation $<$ is specified such that $d_2 < d_1$. The original theory, T, has two extensions:

$$E_1 = Cn(\{Adult(\mathsf{John}) \wedge FullTimeStudent(\mathsf{John}), \neg Employed(\mathsf{John})\}$$
$$E_2 = Cn(\{Adult(\mathsf{John}) \wedge FullTimeStudent(\mathsf{John}), Employed(\mathsf{John})\}.$$

Obviously, $\text{GD}(E_1, T)$, but not $\text{GD}(E_2, T)$, is $<$-maximal in the class $\{\text{GD}(E_1, T), \text{GD}(E_2, T)\}$.

5.3 Circumscription

5.3.1 Introduction

Circumscription is a powerful non-monotonic formalism centered around the following idea:

> the objects (tuples of objects) that can be shown to satisfy a certain relation or property are conjectured to be *all* the objects (tuples of objects) satisfying that relation or property.

For instance, to circumscribe the relation of "being red" is to assume that any object that cannot be proved to be red is not red. Another way to view circumscription is as an expressive generalization of the closed-world assumption.

Unlike default logic, where non-monotonicity is modelled by the means of special expressions, namely default rules, circumscription operates completely in the framework of classical logic. For example, to express a non-monotonic rule, one usually introduces a special relation constant, *Ab*, standing for "abnormal," with the intention that abnormal objects are those violating the rule. Given this intuition, the rule characterizing the conjecture that birds normally fly can be represented by the sentence

$$\forall x.[(Bird(x) \wedge \neg Ab(x)) \rightarrow Flies(x)]. \tag{5.2}$$

[8] Recall that $\text{GD}(E, T)$ is the set of generating defaults for E w.r.t. T (see Definition 5.2.11).

It should be emphasized that applying the standard inference mechanism associated with classical logic to the theory consisting of the sentence (5.2) and an additional axiom $Bird(\mathsf{Tweety})$ is inadequate for inferring that Tweety flies. The reason is that classical logic offers no way to conclude that the bird Tweety is normal, i.e., $\neg Ab(\mathsf{Tweety})$ (in the absence of information about Ab). Circumscription supplies the additional inferential power. This is done by circumscribing the relation of being abnormal. By circumscribing this property, one can infer that any bird that cannot be proved abnormal is normal and, consequently, that it flies. Circumscribing relations is achieved by adding an additional second-order axiom to the original theory.

Circumscription can be viewed as a form of minimization – to circumscribe a relation is to minimize its extension.[9] Observe also that circumscription can be used for maximizing relations, since maximizing a relation corresponds to minimizing its negation.

Circumscription has the following specific features:

1. it is always the task of the user to specify the relations whose extensions are to be minimized. The user can also specify relations whose extensions may vary during the minimization process. This is important for generating positive facts. Circumscription provides a general mechanism which can be applied to arbitrarily chosen minimized and varied relations in a specific theory
2. unlike the case for default logic, all individuals in the domain, not only those denoted by individual constants, are influenced by the process of minimization
3. circumscription is based on syntactic manipulations. Given a theory T, a list $P_1, \ldots, P_n$ of relation symbols to be minimized and a list $Q_1, \ldots, Q_m$ of relation symbols that are allowed to vary, the circumscription of $P_1, \ldots, P_n$ in T with variable $Q_1, \ldots, Q_m$ amounts to implicitly adding to T a special second-order sentence, called a *circumscription axiom*, which captures the desired minimization.

5.3.2 Definition of Circumscription

Let $\mathcal{L}$ be a fixed first-order language with equality. The objects under consideration, referred to as *circumscriptive theories* are finite sets of sentences stated in $\mathcal{L}$. Since each such set is equivalent to the conjunction of its members, a circumscriptive theory may always be viewed as a single sentence. In

[9] An extension of a relation P is the set of objects (tuples of objects) satisfying P. Note that the term "extension" is used here in a quite different sense than in the section devoted to default logic.

the sequel, we will not distinguish between a theory T and the sentence denoting the conjunction of all members of T. We write $T(P_1, \ldots, P_n)$ to indicate that some (but not necessarily all) of the relation symbols occurring in T are among $P_1, \ldots, P_n$.

An n-ary *relation expression* is an expression of the form

$$\lambda x_1 \ldots x_n.\, A(x_1, \ldots, x_n) \qquad (n \geq 0)$$

where $x_1, \ldots, x_n$ are individual variables and $A(x_1, \ldots, x_n)$ is any formula of first- or second-order logic. We identify an n-ary relation symbol P with the relation expression $\lambda x_1 \ldots x_n.\, P(x_1, \ldots, x_n)$. Similarly, an n-ary relation variable X is identified with the relation expression $\lambda x_1 \ldots, x_n. X(x_1, \ldots, x_n)$.

The following are relation expressions (below X is a relation variable):

$$\lambda x.\, P(x); \qquad \lambda xy.\, (X(x) \vee P(y)); \qquad \lambda x.\, \text{FALSE}.$$

An n-ary relation expression U is intended to represent an n-ary relation which is usually referred to as the *extension* of U.

In the sequel, an n-ary relation expression $\lambda x_1 \ldots, x_n. A(x_1, \ldots, x_n)$ will often be written as $\lambda \bar{x}.\, A(\bar{x})$, where $\bar{x}$ stands for a tuple $\langle x_1, \ldots, x_n \rangle$.

Let U be a relation expression of the form $\lambda \bar{x}.\, A(\bar{x})$, where $\bar{x} = \langle x_1, \ldots, x_n \rangle$, and suppose that $\bar{t} = \langle t_1, \ldots, t_n \rangle$ is an n-tuple of terms. The *application* of U to $\bar{t}$, written $U(\bar{t})$, is the formula $A(\bar{t})$. For instance, the application of $\lambda xy.\, (P(x) \rightarrow Q(y))$ to $\bar{t} = \langle a, z \rangle$ is the formula $P(a) \rightarrow Q(z)$.

If U and V are relation expressions of the same arity, then $U \leq V$ stands for $\forall \bar{x}.\, (U(\bar{x}) \rightarrow V(\bar{x}))$.[10] Similarly, if $\bar{U} = \langle U_1, \ldots, U_n \rangle$ and $\bar{V} = \langle V_1, \ldots, V_n \rangle$ are similar tuples of relation expressions, i.e., for $1 \leq i \leq n$, U_i and V_i are of the same arity, then $\bar{U} \leq \bar{V}$ is an abbreviation for $\bigwedge_{i=1}^{n} [U_i \leq V_i]$. We write $\bar{U} = \bar{V}$ for $(\bar{U} \leq \bar{V}) \wedge (\bar{V} \leq \bar{U})$, and $\bar{U} < \bar{V}$ for $(\bar{U} \leq \bar{V}) \wedge \neg(\bar{V} \leq \bar{U})$.

Definition 5.3.1. *Let $\bar{P} = \langle P_1 \ldots, P_n \rangle$ be a tuple of distinct relation symbols, $\bar{S} = \langle S_1, \ldots, S_m \rangle$ be a tuple of distinct relation symbols disjoint with $\bar{P}$, and let $T(\bar{P}, \bar{S})$ be a theory. The* circumscription *of $\bar{P}$ in $T(\bar{P}, \bar{S})$ with varied $\bar{S}$, written* $\text{CIRC}(T; \bar{P}; \bar{S})$, *is the sentence*

$$T(\bar{P}, \bar{S}) \wedge \forall \bar{X} \forall \bar{Y}.\, \neg[T(\bar{X}, \bar{Y}) \wedge \bar{X} < \bar{P}] \tag{5.3}$$

where $\bar{X} = \langle X_1 \ldots, X_n \rangle$ and $\bar{Y} = \langle Y_1, \ldots, Y_m \rangle$ are tuples of relation variables similar to $\bar{P}$ and $\bar{S}$, respectively.[11] □

[10] Note that $U \leq V$ means that the extension of U is a subset of the extension of V.

[11] $T(\bar{X}, \bar{Y})$ is the sentence obtained from $T(\bar{P}, \bar{S})$ by replacing all occurrences of $P_1 \ldots, P_n$ by $X_1 \ldots, X_n$, respectively, and all occurrences of $S_1 \ldots, S_m$ by $Y_1 \ldots, Y_m$, respectively.

Observe that (5.3) can be rewritten as

$$T(\bar{P},\bar{S}) \wedge \forall \bar{X} \forall \bar{Y}. \{[T(\bar{X},\bar{Y}) \wedge [\bar{X} \leq \bar{P}]] \rightarrow [\bar{P} \leq \bar{X}]\}$$

which, in turn, is an abbreviation for

$$T(\bar{P},\bar{S}) \wedge$$
$$\forall \bar{X} \forall \bar{Y}. \left\{ \left[T(\bar{X},\bar{Y}) \wedge \bigwedge_{i=1}^{n} \forall \bar{x}.(X_i(\bar{x}) \rightarrow P_i(\bar{x})) \right] \rightarrow \bigwedge_{i=1}^{n} \forall \bar{x}.(P_i(\bar{x}) \rightarrow X_i(\bar{x})) \right\}.$$

Definition 5.3.2. *A formula A is said to be a* consequence of the circumscription *of $\bar{P}$ in $T(\bar{P},\bar{S})$ with variable $\bar{S}$ if and only if* $\textsc{Circ}(T;\bar{P};\bar{S}) \models A$.[12]

□

Example 5.3.3. Let T consist of the following formulas:

$Bird(\mathsf{Tweety})$
$\forall x.[(Bird(x) \wedge \neg Ab(x)) \rightarrow Flies(x)]$.

Let $\bar{P} = \langle Ab \rangle$ and $\bar{S} = \langle Flies \rangle$.

$$\begin{aligned} \textsc{Circ}(T;\bar{P};\bar{S}) = {} & T(\bar{P},\bar{S}) \wedge \\ & \forall X \forall Y. \{ [Bird(\mathsf{Tweety}) \wedge \forall x.[(Bird(x) \wedge \neg X(x)) \rightarrow Y(x)] \wedge \\ & \qquad \forall x.[X(x) \rightarrow Ab(x)]] \rightarrow \forall x.[Ab(x) \rightarrow X(x)] \}. \end{aligned}$$

In its basic form, the idea is to find relational expressions for X and Y that when substituted into the theory T, will result in strengthening the theory so additional inferences can be made. For example, substituting $\lambda x.\textsc{False}$ for X and $\lambda x.Bird(x)$ for Y, one can conclude that

$$\textsc{Circ}(T;\bar{P};\bar{S}) \models T \wedge A \tag{5.4}$$

where A is

$$\{\forall x.[(Bird(x) \wedge \neg\textsc{False}) \rightarrow Bird(x)] \wedge \forall x.[\textsc{False} \rightarrow Ab(x)]\} \rightarrow \forall x.[Ab(x) \rightarrow \textsc{False}].$$

Since A can be simplified to the logically equivalent sentence

$$\forall x.[Ab(x) \rightarrow \textsc{False}],$$

which in turn is equivalent to $\forall x.\neg Ab(x)$, one can infer by (5.4) that

$$\textsc{Circ}(T;\bar{P};\bar{S}) \models Flies(\mathsf{Tweety}).$$

□

[12] Here $\models$ denotes the entailment relation of the second-order logic.

5.3.3 Semantics of Circumscription

Given a relational structure $M = \langle \textsc{Dom}, \{f_i^{\textsc{Dom}} : i \in I\}, \{R_j^{\textsc{Dom}} : j \in J\}\rangle$, we write $\textsc{Dom}(M)$ to denote the domain of M. If R_j is a relation symbol, then $M(R_j)$ stands for $R_j^{\textsc{Dom}}$. The class of all models of a theory T will be denoted by $\textsc{Mod}(T)$.

The semantics of circumscription is based on the concept of sub-models.

Definition 5.3.4. *Let $\bar{P}$, $\bar{S}$ and $T(\bar{P}, \bar{S})$ be as in Definition 5.3.1. Let M and N be models of T. We say that M is a $(\bar{P}, \bar{S})$-*submodel *of N, written $M \leq^{(\bar{P},\bar{S})} N$, if and only if*

1. $\textsc{Dom}(M) = \textsc{Dom}(N)$
2. *$M(R) = N(R)$, for any relation symbol R not in $\bar{P} \cup \bar{S}$*
3. *$M(R) \subseteq N(R)$, for any relation symbol R in $\bar{P}$.* □

We write $M <^{(\bar{P},\bar{S})} N$ if and only if $M \leq^{(\bar{P},\bar{S})} N$, but not $N \leq^{(\bar{P},\bar{S})} M$. A model M of T is $(\bar{P}, \bar{S})$-*minimal* if and only if T has no model N such that $N <^{(\bar{P},\bar{S})} M$.

We write $\textsc{Mod}^{(\bar{P},\bar{S})}(T)$ to denote the class of all $(\bar{P}, \bar{S})$- minimal models of T.

Theorem 5.3.5. *For any T, $\bar{P}$ and $\bar{S}$,*

$$\textsc{Mod}(\textsc{Circ}(T; \bar{P}; \bar{S})) = \textsc{Mod}^{(\bar{P},\bar{S})}(T).$$

□

Example 5.3.6. Let us reconsider the theory T from Example 5.3.3:

$Bird(\mathsf{Tweety})$
$\forall x.[(Bird(x) \land \neg Ab(x)) \to Flies(x)]$.

It should be emphasized that the sentence $Flies(\mathsf{Tweety})$ is not derivable using circumscription if the relation symbol $Flies$ is not varied. That is

$$\textsc{Circ}(T; Ab; \langle\rangle) \not\models Flies(\mathsf{Tweety}).$$

To understand the reason why, consider a relational structure M such that $\textsc{Dom}(M) = \{\mathsf{Tweety}\}$, $M(Bird) = M(Ab) = \{\mathsf{Tweety}\}$ and $M(Flies) = \emptyset$. It is clear that M is a model of T. However, M is also an $(Ab, \langle\rangle)$-minimal model of T (we cannot make $M(Ab)$ smaller while preserving the truth of T and $M(Flies)$). Thus, since $Flies(\mathsf{Tweety})$ is false in M, this formula cannot be derived by circumscribing Ab in T without varying relations (in view of Theorem 5.3.5). □

Example 5.3.7. Let T consist of the following:

$$R(\mathsf{Nixon}) \wedge Q(\mathsf{Nixon})$$
$$\forall x.[(R(x) \wedge \neg Ab_1(x)) \to \neg P(x)]$$
$$\forall x.[(Q(x) \wedge \neg Ab_2(x)) \to P(x)].$$

This is the standard "Nixon diamond" theory (see Example 5.2.5) with R, P, Q standing for *Republican*, *Pacifist* and *Quaker*, respectively.[13] Let M and N be models of T such that

$$\text{DOM}(M) = \text{DOM}(N) = \{\mathsf{Nixon}\},$$
$$M(R) = N(R) = \{\mathsf{Nixon}\}, \quad M(Q) = N(Q) = \{\mathsf{Nixon}\}$$

and

$$\begin{array}{ll} M(P) = \{\mathsf{Nixon}\} & N(P) = \emptyset \\ M(Ab_1) = \{\mathsf{Nixon}\} & N(Ab_1) = \emptyset \\ M(Ab_2) = \emptyset & N(Ab_2) = \{\mathsf{Nixon}\}. \end{array}$$

It is easily observed that for any $\bar{S}$, both M and N are $(\overline{Ab}, \bar{S})$-minimal models of T, where $\overline{Ab} = \langle Ab_1, Ab_2 \rangle$. Furthermore, $M \models P(\mathsf{Nixon})$ and $N \models \neg P(\mathsf{Nixon})$. It follows, therefore, that when $\overline{Ab}$ is circumscribed in T, it can not be inferred whether Nixon is a pacifist or not. The best that can be obtained is the disjunction $\neg Ab_1(\mathsf{Nixon}) \vee \neg Ab_2(\mathsf{Nixon})$, stating that Nixon is either normal as a republican or as a quaker. □

5.3.4 Properties of Circumscription

It is reasonable to ask whether circumscription preserves satisfiability if the original first-order theory which we circumscribe over is satisfiable. As the following example shows, this is not always the case.

Example 5.3.8. Let T consist of the following:

$$\exists x.P(x) \wedge [\forall y.P(y) \to x = s(y)]$$
$$\forall x.P(x) \to P(s(x))$$
$$\forall x \forall y.s(x) = s(y) \to x = y.$$

Recall that ω denotes the set of natural numbers. Consider the relational structure M, where $\text{DOM}(M) = \omega$, $M(P) = \omega$ and s is the successor function.

[13] Observe that we use two abnormality relations here, namely Ab_1 and Ab_2. This is because being abnormal with respect to pacifism as a quaker is a different notion than being abnormal with respect to pacifism as a republican.

It is easily observed that M is a model of T, so that T is satisfiable. On the other hand, $\text{CIRC}(T; P; \langle\rangle)$ is unsatisfiable. We leave it to the reader to show that there are no $(P, \langle\rangle)$-minimal models of T. □

In view of Example 5.3.8, it is natural to seek classes of theories for which circumscription preserves satisfiability.

Definition 5.3.9. *A theory T is* well-founded *w.r.t.* $\langle\bar{P}, \bar{S}\rangle$ *if and only if , for every* $M \in \text{MOD}(T)$*, there is an* $N \in \text{MOD}^{(\bar{P},\bar{S})}(T)$ *such that* $N \leq^{(\bar{P},\bar{S})} M$. □

Theorem 5.3.10. *If T is satisfiable and well-founded w.r.t.* $\langle\bar{P}, \bar{S}\rangle$*, then also* $\text{CIRC}(T; \bar{P}; \bar{S})$ *is satisfiable.* □

An important class of well-founded theories are the universal theories.

Theorem 5.3.11. *For any disjoint tuples $\bar{P}$ and $\bar{S}$ of relation symbols, every universal theory is well-founded w.r.t.* $\langle\bar{P}, \bar{S}\rangle$. □

Corollary 5.3.12. *For any disjoint tuples $\bar{P}$ and $\bar{S}$ of relation symbols and any satisfiable universal theory T,* $\text{CIRC}(T; \bar{P}; \bar{S})$ *is satisfiable.* □

We now consider the expressive power of circumscription with respect to well-founded theories.

Since circumscription minimizes circumscribed relations, one may expect that it never yields new positive instances of such a relation. For well-founded theories this is indeed the case.

Theorem 5.3.13. *If T is well-founded w.r.t.* $\langle\bar{P}, \bar{S}\rangle$*,* $P \in \bar{P}$ *is an n-ary relation symbol and $\bar{t}$ is an n-tuple of ground terms, then*

$\text{CIRC}(T; \bar{P}; \bar{S}) \models P(\bar{t})$ *if and only if* $T \models P(\bar{t})$. □

Theorem 5.3.14. *If T is well-founded w.r.t.* $\langle\bar{P}, \bar{S}\rangle$*, R is an n-ary relation symbol not in $\bar{P}, \bar{S}$ and $\bar{t}$ is an n-tuple of ground terms, then*

1. $\text{CIRC}(T; \bar{P}; \bar{S}) \models R(\bar{t})$ *if and only if* $T \models R(\bar{t})$
2. $\text{CIRC}(T; \bar{P}; \bar{S}) \models \neg R(\bar{t})$ *if and only if* $T \models \neg R(\bar{t})$. □

Theorems 5.3.13 and 5.3.14 state that the only new ground literals that one may hope to obtain from well-founded theories by circumscription are negative instances of circumscribed relations, and both positive and negative instances of variable relations.

5.3.5 Prioritized Circumscription

It is often useful to prioritize the minimization of relations in circumscriptive theories. This section describes an extension to circumscription for doing this. Let us start with the following example.

Example 5.3.15. Suppose we are given a theory T consisting of the following:[14]

$$Adult(\mathsf{John}) \wedge FullTimeStudent(\mathsf{John})$$
$$\forall x.[(Adult(x) \wedge \neg Ab_1(x)) \rightarrow Employed(x)]$$
$$\forall x.([FullTimeStudent(x) \wedge \neg Ab_2(x)) \rightarrow \neg Employed(x)].$$

Given these axioms, John may be considered a normal full-time student ($\neg Ab_2(\mathsf{John})$) or a normal adult ($\neg Ab_1(\mathsf{John})$), but not both. This is because there are two conflicting minimizations of (Ab_1, Ab_2) with *Employed* varied, leading to $\neg Ab_1(\mathsf{John}) \wedge Employed(\mathsf{John})$ and $\neg Ab_2(\mathsf{John}) \wedge \neg Employed(\mathsf{John})$, respectively. Consequently, all that can be inferred is the disjunction

$$\neg Ab_1(\mathsf{John}) \vee \neg Ab_2(\mathsf{John}). \tag{5.5}$$

Clearly, the inference (5.5) is too weak to capture our intuitions concerning this example. As has already been noted, given that John is a full-time student, it is not reasonable to remain agnostic about his employment status. Rather, one is prepared to assume that he is unemployed.

One means of avoiding unintuitive minimizations is to add a new piece of information. In our example, it is sufficient to supply the theory T with

$$\forall x.FullTimeStudent(x) \rightarrow Ab_1(x).$$

□

The obvious drawback to the solution outlined in Example 5.3.15 is that in practical applications there may be a great number of unwanted minimizations, so that their elimination will require many additional axioms.

There is another way to avoid undesirable minimizations. The circumscriptive technique can be generalized by introducing a priority ordering on relations to be minimized. The idea is that any conflict arising from the minimization of two relations is resolved in favor of the relation with the higher priority. This generalization of circumscription is referred to as *prioritized* circumscription.

Suppose one wants to minimize the extensions of a pair of relation symbols, say P_1 and P_2, in a theory T. Assume for simplicity that no additional relations

[14] This is a circumscriptive version of the default theory considered in Example 5.2.23.

can be varied during the minimization. In standard circumscription this is expressed by the sentence

$$T(P_1, P_2) \wedge \forall X_1 X_2. \{[T(X_1, X_2) \wedge (\langle X_1, X_2\rangle \leq \langle P_1, P_2\rangle)] \rightarrow [\langle P_1, P_2\rangle \leq \langle X_1, X_2\rangle]\}$$

where, for arbitrary pairs of similar relation expressions, $\langle U_1, U_2\rangle$ and $\langle V_1, V_2\rangle$, $\langle U_1, U_2\rangle \leq \langle V_1, V_2\rangle$ is defined by $(U_1 \leq V_1) \wedge (U_2 \leq V_2)$. Assume now that one wants to minimize P_1 and P_2 in T, the former at higher priority than the latter. To express this, one may retain the original circumscription definition with one exception: instead of minimizing $\langle P_1, P_2\rangle$ with respect to $\leq$, one employs the relation $\preceq$ specified by

$$\langle U_1, U_2\rangle \preceq \langle V_1, V_2\rangle \text{ if and only if } (U_1 \leq V_1) \wedge (U_1 = V_1 \rightarrow U_2 \leq V_2).$$

Notice that $\langle U_1, U_2\rangle \preceq \langle V_1, V_2\rangle$ does not imply $U_2 \leq V_2$. If $U_1 < V_1$, then $\langle U_1, U_2\rangle \preceq \langle V_1, V_2\rangle$, for any U_2, V_2.

The relation $\preceq$ can be naturally extended into similar tuples of relation expressions:

$$\langle U_1, \ldots, U_n\rangle \preceq \langle V_1, \ldots, V_n\rangle \stackrel{\text{def}}{\equiv} U_1 \leq V_1 \wedge \bigwedge_{i=2}^{n} \left[\left(\bigwedge_{j=1}^{i-1} U_j = V_j\right) \rightarrow U_i \leq V_i\right]$$

and further, into similar lists of tuples consisting of relation expressions:

$$\langle \bar{U}_1, \ldots, \bar{U}_n\rangle \preceq \langle \bar{V}_1, \ldots, \bar{V}_n\rangle \stackrel{\text{def}}{\equiv} \bar{U}_1 \leq \bar{V}_1 \wedge \bigwedge_{i=2}^{n} \left[\left(\bigwedge_{j=1}^{i-1} \bar{U}_j = \bar{V}_j\right) \rightarrow \bar{U}_i \leq \bar{V}_i\right].$$

$\langle \bar{U}_1, \ldots, \bar{U}_n\rangle \prec \langle \bar{V}_1, \ldots, \bar{V}_n\rangle$ is written as an abbreviation for

$$[\langle \bar{U}_1, \ldots, \bar{U}_n\rangle \preceq \langle \bar{V}_1, \ldots, \bar{V}_n\rangle] \wedge \neg[\langle \bar{V}_1, \ldots, \bar{V}_n\rangle \preceq \langle \bar{U}_1, \ldots, \bar{U}_n\rangle].$$

Given these preliminaries, a formal definition of prioritized circumscription can now be provided. To minimize a tuple $\bar{P}$ of relation symbols, one starts by partitioning $\bar{P}$ into disjoint sublists $\overline{P^1}, \ldots, \overline{P^n}$. The intention is that the elements of $\overline{P^1}$ are minimized at the highest priority, those of $\overline{P^2}$ at the next highest priority, etc. This is expressed by writing $\overline{P^1} > \cdots > \overline{P^n}$.

Definition 5.3.16. *Let $\bar{P}$, $\bar{S}$ and $T(\bar{P}, \bar{S})$ be as in Definition 5.3.1. The* prioritized circumscription *of $\bar{P}$ in $T(\bar{P}, \bar{S})$ with variable $\bar{S}$ with respect to priorities $\overline{P^1} > \cdots > \overline{P^n}$, written*

$$\text{Circ}(T; \overline{P^1} > \cdots > \overline{P^n}; \bar{S}),$$

is the sentence

$$T(\bar{P}, \bar{S}) \wedge \forall \bar{X} \forall \bar{Y}. \neg[T(\bar{X}, \bar{Y}) \wedge \bar{X} \prec \bar{P}]. \tag{5.6}$$

Here $\bar{X} \prec \bar{P}$ *is* $\langle \overline{X^1}, \ldots, \overline{X^n} \rangle \prec \langle \overline{P^1}, \ldots, \overline{P^n} \rangle$, *where each* $\overline{X^i}$ *is a tuple of relation variables similar to* $\overline{P^i}$. □

Notice that (5.6) may be equivalently written as

$$T(\bar{P}, \bar{S}) \wedge \forall \bar{X} \forall \bar{Y}. [T(\bar{X}, \bar{Y}) \wedge [\bar{X} \preceq \bar{P}] \rightarrow [\bar{P} \preceq \bar{X}]].$$

As usual, the set of consequences of prioritized circumscription is identified with the set of all formulas entailed by $\text{Circ}(T; \overline{P^1} > \cdots > \overline{P^n}; \bar{S})$.

Example 5.3.17. Consider the theory of Example 5.3.15, with A, FTS, E and j standing for *Adult*, *FullTimeStudent*, *Employed* and John, respectively. We circumscribe Ab_1, Ab_2 in T with variable E. Since the minimization of Ab_2 is to be preferred over that of Ab_1, Ab_2 is given a higher priority than Ab_1.

$$\begin{aligned}
&\text{Circ}(T; Ab_2 > Ab_1; E) \equiv T \wedge \\
&\quad \forall X_1, X_2, Y. \Big\{ \Big[A(\mathsf{j}) \wedge FTS(\mathsf{j}) \wedge \forall x.[(A(x) \wedge \neg X_1(x)) \rightarrow Y(x)] \wedge \\
&\qquad\qquad \forall x.[(FTS(x) \wedge \neg X_2(x)) \rightarrow \neg Y(x)] \wedge \\
&\qquad\qquad \forall x.[X_2(x) \rightarrow Ab_2(x)] \wedge \\
&\qquad\qquad \forall x.[(X_2(x) \equiv Ab_2(x))) \rightarrow \forall x.(X_1(x) \rightarrow Ab_1(x))] \Big] \rightarrow \\
&\qquad\qquad \Big[\forall x.[Ab_2(x) \rightarrow X_2(x)] \wedge \\
&\qquad\qquad \forall x.[(Ab_2(x) \equiv X_2(x)) \rightarrow \forall x.(Ab_1(x) \rightarrow X_1(x))] \Big] \Big\}.
\end{aligned}$$

Substituting $\lambda x.\, FTS(x)$ for X_1, $\lambda x.\, \text{False}$ for X_2, $\lambda x.\, A(x) \wedge \neg FTS(x)$ for Y, and doing the appropriate calculations, one obtains

$$\text{Circ}(T; Ab_2 > Ab_1; E) \models A_1 \wedge A_2 \wedge A_3,$$

where

$$A_1 = \forall x. \neg Ab_2(x)$$
$$A_2 = \forall x.\, Ab_1(x) \equiv D(x)$$
$$A_3 = \neg E(\mathsf{j}).$$

□

5.4 Bibliographic Notes

The version of default logic presented in this chapter is in large part due to Reiter, who introduced this formalism in [176]. Semi-normal default theories and their importance from the standpoint of practical applications were first discussed in [177].

Reiter's presentation of default logic is purely syntactical. The first attempt at developing semantic foundations for this formalism is presented in [115] where a semantic characterization for normal default theories is provided. The general idea is to consider each default as a mapping from classes of models into classes of models such that the range of the mapping restricts its input class into those models in which the consequent of the default is true. This semantics has been generalized into arbitrary default theories in [70]. In [20] another semantics for default logic has been presented. It is based on Kripke structures.

The general complexity results for default logic presented here are provided in [79]. The reader interested in complexity results for various restricted classes of propositional default theories should consult [98, 208].

Prioritized default logic, for normal default theories, was introduced in [6, 28]. The formulation here is slightly different, but the resulting formalism is equivalent, for normal theories, to one of formalisms presented in [28].

Reiter [176] provides a proof theory for normal default theories.

There have been several other formulations for default logic. In contrast to Reiter's original approach, all of them enjoy the property of existence of extensions. In particular, three of them, *justified default logic* [115], *cumulative default logic* [26] and *constrained default logic* [43] are worth noting in particular.

There are several books covering the subject of default logic. The most detailed exposition of Reiter's approach, restricted to the propositional case, can be found in [118]. The books [19, 27, 116] go well beyond classical default logic. The first two provide, in addition, a detailed presentation of justified default logic, whereas the third contains a brief introduction to cumulative default logic. The reader interested in computational aspects of default logic should consult [179].

Circumscription was introduced by McCarthy. In [119, 120] he proposed two circumscriptive logics, known in the AI literature as *predicate circumscription* and *formula circumscription*, respectively. The second of these formalisms has been slightly extended and reformulated in [103]. It is this version of circumscription, usually referred to as *second-order circumscription*, that has been presented here. All the results stated in Section 5.3 are due to Lifschitz [103, 104]. A good survey of second-order circumscription is given in [107].

Circumscription has one major advantage and one major weakness when compared with other non-monotonic formalisms. On the positive side, circumscription is embedded in a standard logical framework. Once a circumscription axiom is constructed, all deductions proceed in classical monotonic logic. Unfortunately, the circumscription axiom is expressed in second-order logic which makes the technique problematic to use from a computational point of view. To alleviate this problem, several researchers have studied the possibility of reducing certain classes of circumscriptive theories into first-order logic. The first attempt in this direction was made in [103], where various classes of circumscriptive theories are isolated where they can be reduced to first-order equivalents. In [75] and [53] two general algorithms reducing second-order quantifiers have been presented. The former, called the SCAN algorithm is based on second-order resolution, whereas the latter, the DLS algorithm, uses a lemma, originally provided in 1935 by Ackermann [2]. Both of these algorithms can be used to reduce a reasonably wide class of circumscriptive theories. Their implementations can be found on the Internet (see `http://www.mpi-sb.mpg.de/units/ag2/projects/scan/scan` for SCAN, and `http://www.ida.liu.se/labs/kplab/projects/dls`, for DLS).

A number of other circumscriptive logics have been proposed in the AI literature. Of these, the most interesting is *pointwise circumscription* [106]. This formalism is very powerful and can be used to model various complex inference patterns which are difficult to express in the framework of other circumscriptive logics. The book [116] provides a detailed survey of various circumscriptive formalisms.

Two other non-monotonic formalisms have been studied extensively. These are *autoepistemic logic* and a family of logics modelling the *closed world assumption* (CWA–logics).

Autoepistemic logic was introduced by Moore [129, 130] to model autoepistemic reasoning. The books [116, 118] provide a detailed exposition of autoepistemic logic.

The closed world assumption is one of the earliest contributions to non-monotonic reasoning. It was proposed by Reiter in [175] in the context of deductive databases.[15] Limitations of Reiter's original formulation led several researchers to formalize more sophisticated logics dealing with the closed world assumption. Some of them are discussed in [116].

---

[15] See also Chapter 4.

Part II

From Relations to Knowledge Representation

6

Rough Knowledge Databases

6.1 Introduction

Consider an autonomous system such as a ground robot or an unmanned aerial vehicle operating in a highly complex and dynamic environment. For systems of this sort to function in an intelligent and robust manner, it is useful to have both deliberative and reactive capabilities. Such systems combine the use of reactive and deliberative capabilities in achieving task goals. Reactive capabilities are necessary so the system can react to contingencies which arise unexpectedly and demand immediate response with little room for deliberation as to what the best response should be. Deliberative capabilities are useful in the sense that internal representations of aspects of the system's operational environment can be used to predict the course of events in the near or intermediate future. These predictions can then be used to determine more selective actions or better responses in the present which potentially save the system time, effort and resources in the course of achieving task goals.

Due to the complexity of the operational environments in which robotics systems such as these generally operate and the inaccuracy of sensor data about the environment acquired through different combinations of sensors, these systems can not assume to have complete information or models about their surrounding environment, or the effects of their actions on these environments. On the other hand, the deliberative component is dependent on the synthesis, management, update and use of incomplete qualitative models of the operational environment represented internally in the system architecture. These internal models are used for reasoning about the system's environment and the effects of its actions on the environment while attempting to achieve task goals. In spite of the lack of complete information, such systems quite often have or can acquire additional information which can be used in certain contexts to assume additional knowledge about the incomplete parts of the specification. This information may be of a normative or default nature, may

P. Doherty et al.: *Knowledge Representation Techniques*, Studfuzz **202**, 103–127 (2006)
www.springerlink.com

include rules-of-thumb particular to the operational domain in question, or may include knowledge implicit in the result of executing a sensing action.

One potentially useful approach that can be pursued in the development of on-line reasoning capabilities and representation of qualitative models of aspects of an autonomous system's operational environment is the use of traditional database technology combined with techniques originating from artificial intelligence research with knowledge-based systems. There are a number of different compositions of technologies that may be pursued ranging from more homogeneous logic programming based deductive database systems to heterogeneous systems which combine the use of traditional relational database technology with specialized front-end reasoning engines.

The latter approach is pursued in this chapter, but with a number of modifications to the standard deductive database framework. These modifications are made necessary by the requirement of representing and reasoning about incomplete qualitative models of the operational environments in which autonomous systems are embedded. A number of fundamental generalizations of standard semantic concepts used in the traditional deductive database approach are advocated:

- the extensional database (EDB) which represents and stores base relations and properties (facts) about the external environment, or the system's internal environment, is given a formal semantics based on the use of rough sets. The extension of a database relation or property contains explicit positive and negative information in addition to implicitly represented boundary information which is defined as the difference between upper and lower approximations of the individual relations and properties
- the intensional database (IDB) contains two rule sets generating implicit positive and negative information, respectively, via application of the rule sets in the context of the facts in the EDB. The closed-world assumption is *not* applied to the resulting information generated from the EDB/IDB pair
- the open-world assumption is to be applied to the extensional and intensional database pair which can be locally closed in a dynamic manner via the use of *contextually closed queries* (CCQs). A CCQ consists of the query itself, a context represented as a set of integrity constraints and a local closure policy specified in terms of the minimization/maximization of selected relations. The contextually closed query layer (CCQ layer) represents the closure mechanism and is used to answer individual CCQs.

In effect, the CCQ layer permits the representation of additional normative, default or closure information associated with the operational environment at hand and the particular view of the environment currently used by the querying agent. Together with the rough set semantics for relations, a rough set knowledge database in this context represents an incompletely specified world model with dynamic policies which permit the local closure of parts of the world model when querying it for information.

The combination of the EDB, IDB, and CCQ layer is called a *rough knowledge database* (RKDB). The computational basis for the inference engine used to query the RKDB is based on the use of circumscription, quantifier elimination, and the ability to automatically generate syntactic characterizations of the upper and lower bounds of rough relations in the RKDB.

6.1.1 Open- and Closed-World Reasoning

What is meant intuitively by open- and closed-world reasoning? In traditional databases, reasoning is often based on the assumption that information stored in a specific database contains a complete specification of the application environment at hand. If a tuple is not in a base relational table, it is assumed not to have that specific property. In the case of deductive databases, if the tuple is not in a base relational table or any intensional relational tables generated implicitly by the application of intensional rules, it is again assumed not to have these properties. Under this assumption, an efficient means of representing negative information about the world depends on applying the closed-world assumption (CWA). Recall that in this case, atomic information about the world, absent in a world model (represented as a database), is assumed to be false.

On the other hand, for many applications such as the autonomous systems applications already mentioned, the assumption of complete information is not feasible nor realistic and the CWA can not be used. In such cases an Open-World Assumption (OWA), where information not known by an agent is assumed to be unknown, is often accepted, but this complicates both the representational and implementational aspects associated with inference mechanisms and the use of negative information. The CWA and the OWA represent two ontological extremes. Quite often, a reasoning agent does have or acquires additional information which permits the application of the CWA *locally* in a particular context. In addition, if it does have knowledge of what it does not know, this information is valuable because it can be used in plan generation to acquire additional information through use of sensors.

In such a context various forms of *Local Closed World* (LCW) assumptions have been found to be useful. Such assumptions provide a compromise between CWA and OWA, allowing one to close the world locally. In the current chapter we provide a semantics and a methodology for LCW reasoning which provides an intuitive and general framework for integrating LCW reasoning in knowledge databases used by intelligent agents.

The approach exhibits the following features:

- it is applicable to deductive databases

- integrity constraints take on an important role in characterizing LCW assumptions in a principled manner. In most knowledge databases the relationships between pieces of information are expressed by means of integrity constraints (e.g., defined by means of classical first-order formulas). In the case of applying LCW policies locally to particular relations, one minimizes those relations. However, in such cases the integrity constraints have to be preserved. This can result in implicit changes to some additional relations. However, the integrity constraints are still preserved thus the knowledge structure represented continues to satisfy the desired properties
- the use of integrity constraints and local closure policies are decoupled from the knowledge database itself and associated dynamically with individual agent queries. The agent's themselves possess local views and preferences about the world model that may or may not be shared by other agents or even the same agent using a different query or context
- the approach permits selected fixing, varying and minimizing of specific relations in integrity constraints. This provides the user with a flexibility in defining LCW constraints and brings the approach close to the methodology used in circumscription-based knowledge representation. It should be emphasized that the implementation is not always dependent on the use of circumscription
- at the semantic level we use rough sets to represent database information. Namely, rough sets contain information about tuples known to be in a relation (the lower approximation of the relation), tuples that are or might be in the relation (the upper approximation of the relation), tuples known not to be in the relation (the complement of the upper approximation of the relation) and tuples for which it is unknown whether they belong to the relation (the difference between the upper and lower approximation of the relation).

6.1.2 The Architecture of RKDBs

Let us now discuss the architecture of RKDBs as understood in the book. Such databases constitute a kernel in the architecture of knowledge-based systems. The architecture of a RKDB is illustrated in Figure 6.1.

The most fundamental layer of the database is the extensional database. We assume that the extensional database contains positive and negative facts. The facts that are not explicitly listed in the extensional database are assumed to be unknown in this layer of the database. Thus in the extensional database layer we apply the open-world assumption. The intensional database layer provides rules that define some new relations, but also rules allowing one to extend the positive and negative parts of the extensional relations. The outermost, most advanced layer, which we call the *contextual closure query*

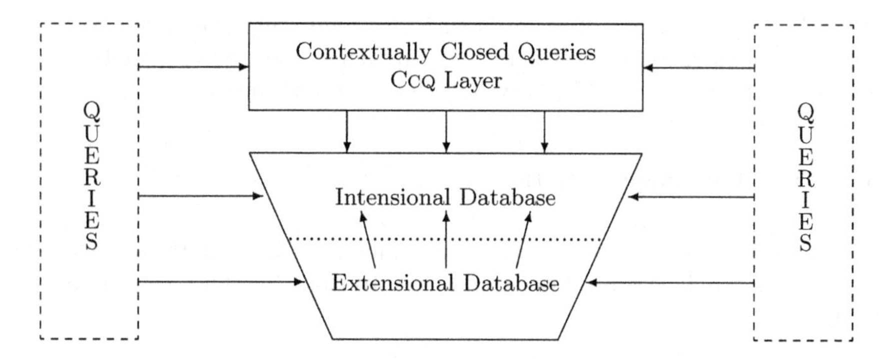

Fig. 6.1. The Architecture of Knowledge Databases.

layer (CCQ layer), consists of the CCQ inference mechanism which includes the query/answer mechanism used by individual CCQs applied to the two lower layers of the RKDB.

The extensional database consists of rough relations. According to the methodology we apply, the rules of the intensional database function as rough set transducers,[1] transforming combinations of rough extensional and intensional relations into new relations which satisfy the constraints of the intensional rules. As in the extensional database, in the intensional layer the open-world assumption is applied. Local closure context policies (LCC policies) allow us to minimize chosen relations (or their complements), while at the same time preserving the imposed integrity constraints. Queries are asked via the outermost layer, but in some applications it might be useful to submit queries to the intensional or even extensional layer.

6.2 The Languages of RKDBs

6.2.1 The Language of EDBs

The extensional database consists of sets of positive and negative facts. We thus assume that the language of the extensional database is a set of literals, i.e., formulas of the form $R(\bar{c})$ or $\neg R(\bar{c})$, where R is a relation symbol and $\bar{c}$ is a tuple of constant symbols. It is assumed that the extensional database is consistent, i.e., it does not contain both $R(\bar{c})$ and $\neg R(\bar{c})$, for some relation R and tuple $\bar{c}$.

Observe that an extensional database can be a standard relational database or its extension by some algorithms or agents providing additional data. For

[1] Rough set transducers are explained in Chapter 7.

example, in practice one often deals with classifiers supplying databases with interpretations of lower level data, e.g., inducing from sensor signals higher level concepts represented or viewed as database relations.

6.2.2 The Language of IDBs

The intensional database is intended to infer new facts, both positive and negative via application of intensional rules to the EDB. The rules are of the form

$$\pm P(\bar{x}) \leftarrow \pm P_1(\bar{x}_1), \ldots \pm P_k(\bar{x}_k)$$

where $\pm$ is either the empty string or the negation symbol $\neg$.[2]

The rules can be divided into two layers, the first for inferring positive and the second for inferring negative facts. The first layer of rules (called the *positive* IDB *rule layer*), used for inferring positive facts, is of the form

$$P(\bar{x}) \leftarrow \pm P_1(\bar{x}_1), \ldots \pm P_k(\bar{x}_k) \tag{6.1}$$

while the second layer of rules (called the *negative* IDB *rule layer*), used for inferring negative facts, is of the following form

$$\neg P(\bar{x}) \leftarrow \pm P_1(\bar{x}_1), \ldots \pm P_k(\bar{x}_k). \tag{6.2}$$

Observe that intensional rules can be provided by an expert or even be induced from data using machine learning techniques (see Chapter 14).

6.2.3 The Language of Integrity Constraints and LCC Policies

Integrity constraints are expressed as formulas of classical first-order logic. Intuitively, they can be considered as implicit definitions of intentional relations, which are minimized or maximized by the LCC (Local Contextual Closure) assumptions in a specific LCC policy. In the following sections, in order to obtain tractable instances of the general algorithm, we will impose some syntactic restrictions on the syntactic form of integrity constraints, together with the LCC assumptions in a specific LCC policy (see Section 6.5).

LCC *policies* are expressions of the form

$$\text{LCC}[L_1, \ldots, L_p; K_1, \ldots, K_r] : I, \tag{6.3}$$

[2] We do not require the safety condition for rules which is assumed in DATALOG. We deal with unsafe rules as discussed in Remark 4.4.1.

where $L_1, \ldots, L_p$ are (positive or negative) literals, $K_1, \ldots, K_r$ are relation symbols not appearing in L_i's and I is a set of integrity constraints. Literals $L_1, \ldots, L_p$ are minimized assuming relations $K_1, \ldots, K_r$ can vary.[3] By an LCC *assumption* we mean a minimization or maximization of a single literal from $L_1, \ldots, L_p$ in (6.3).

We sometimes omit the part ":I" of (6.3) if the corresponding integrity constraints are known from the context.

6.3 The Semantics of RKDB's

6.3.1 Notational Conventions

Let us denote the facts in the extensional database by EDB and the facts in the intensional database by IDB. Let $R_1, \ldots, R_n$ be all relations in the RKDB. For a specific relation R in the RKDB, we denote the positive ground literals of R in the EDB by $\text{EDB}^+(R)$ and the negative ground literals of R in the EDB by $\text{EDB}^-(R)$. We shall use the following notation:

- EDB^+, to denote the positive part of the EDB which is $\bigcup_{i=1}^{n} \text{EDB}^+(R_i)$
- EDB^-, to denote the negative part of the EDB which is $\bigcup_{i=1}^{n} \text{EDB}^-(R_i)$.

The EDB is then equivalent to $\text{EDB}^+ \cup \text{EDB}^-$, $\text{EDB} \equiv \text{EDB}^+ \cup \text{EDB}^-$.

For a specific relation R in the RKDB, we denote the positive literals of R in the IDB generated by the positive intensional rules of form (6.1) by $\text{IDB}^+(R)$ and the negative literals of R in the IDB generated by the negative intensional rules of form (6.2) by $\text{IDB}^-(R)$. Assume also that:

- by IDB^+, we denote the positive part of the IDB which is $\bigcup_{i=1}^{n} \text{IDB}^+(R_i)$
- by IDB^-, we denote the negative part of the IDB which is $\bigcup_{i=1}^{n} \text{IDB}^-(R_i)$.

The IDB is then equivalent to $\text{IDB}^+ \cup \text{IDB}^- \cup \text{EDB}^+ \cup \text{EDB}^-$,

[3] Thus LCC corresponds to circumscription of $I \cup \text{EDB} \cup \text{IDB}$ with relations $L_1, \ldots, L_p$ minimized (positive literals) or maximized (negative literals) and $K_1, \ldots, K_r$ allowed to vary.

$$\text{IDB} \equiv \text{IDB}^+ \cup \text{IDB}^- \cup \text{EDB}^+ \cup \text{EDB}^-.$$

Let DOM be a finite set (a domain of the database). The semantics of constant symbols and variables is given by a valuation:

$$v : \text{CONST} \cup V_{\mathsf{I}} \longrightarrow \text{DOM}.$$

The valuation v is then extended to the case of vectors of constants and variables in the usual way. We also assume that the unique names assumption (UNA) holds.

In the semantics defined in the following sections, all relations are interpreted as rough sets of tuples.

The symbol $\Vdash$ denotes the RKDB *entailment relation* and the symbol $\models$ denotes the classical two-valued entailment relation.

By indexing relations with EDB, IDB and LCC we indicate that they are considered in the particular context as relations of the extensional, intensional and CCQ layer of the RKDB, respectively.

6.3.2 The Semantics of EDBs

The semantics of the extensional database is given by rough sets of tuples. Let R be a relational symbol appearing in the extensional database. Then R is interpreted as the rough set whose positive part contains all tuples $v(\bar{c})$ for which the literal $R(\bar{c})$ is in the database and the negative part contains all tuples $v(\bar{c})$ for which the literal $\neg R(\bar{c})$ is in the database. All other tuples are in the boundary region of R.

$$\text{EDB} \Vdash R(\bar{c}) \text{ if and only if } R(\bar{c}) \in \text{EDB}^+(R)$$
$$\text{EDB} \Vdash \neg R(\bar{c}) \text{ if and only if } R(\bar{c}) \in \text{EDB}^-(R),$$

where R is a relation of the EDB and $\bar{c}$ is a tuple of constants.

Rough relations for the EDB are then defined as follows:

$$R^+_{\text{EDB}} = \{v(\bar{c}) : \text{EDB} \Vdash R(\bar{c})\}$$
$$R^-_{\text{EDB}} = \{v(\bar{c}) : \text{EDB} \Vdash \neg R(\bar{c})\}$$
$$R^\pm_{\text{EDB}} = \{v(\bar{c}) : \text{EDB} \not\Vdash R(\bar{c}) \text{ and } \text{EDB} \not\Vdash \neg R(\bar{c})\}.$$

6.3.3 The Semantics of IDBs

The semantics of the intensional database is given by rough sets of tuples after application of the intensional rules to the extensional database.

In order to provide the semantics of IDB, we require use of the definition of the Feferman-Gilmore translation, used in translating three-valued logic formulas into the classical two-valued logic.

Definition 6.3.1. *By a* Feferman-Gilmore translation of formula α, *denoted by* $\text{FG}(\alpha)$, *we shall mean the formula obtained from* α *by replacing all negative literals of the form* $\neg R(\bar{y})$ *by* $R^-(\bar{y})$ *and all positive literals of the form* $R(\bar{y})$ *by* $R^+(\bar{y})$. □

Let $\bar{S} = \langle S_1, \ldots, S_p \rangle$ contain all relation symbols of the form R^+ and R^-, where R is a relation symbol occurring in an IDB rule. For any relation S_i, all rules with S_i^+ (respectively S_i^-) in their heads should be gathered into a single formula of the form

$$\forall \bar{y}_i.[S_i^{\pm}(\bar{y}_i) \leftarrow \alpha_i(\bar{y}_i)],$$

where

- $\alpha_i(\bar{y}_i) \equiv \bigvee_j \exists \bar{z}_j.\beta_{ij}(\bar{y}_i, \bar{z}_j)$
- each of $\beta_{ij}(\bar{y}_i, \bar{z}_j)$ denotes the body of the appropriate rule
- $\bar{z}_j$ are variables appearing in the rule's body and not in the rule's head
- $\pm$ stands for $+$ or $-$, respectively.

Define

$$\bar{S}_{\text{IDB}} \stackrel{\text{def}}{\equiv} \text{LFP}\, \bar{S}.[\text{FG}(\alpha_1), \ldots, \text{FG}(\alpha_p)]. \tag{6.4}$$

In some cases the IDB may be inconsistent. This happens when there is a relation R such that $R^+ \cap R^- \neq \emptyset$. In what follows we require that the IDB is consistent, i.e., for all IDB relations R we have, $R^+ \cap R^- = \emptyset$. This consistency criterion can be verified in time polynomial in the size of the database.

The semantics of IDB rules are then defined as follows:

$\text{IDB} \Vdash R(\bar{c})$ if and only if $R(\bar{c}) \in \text{EDB}^+(R) \cup \text{IDB}^+(R)$,
$\text{IDB} \Vdash \neg R(\bar{c})$ if and only if $R(\bar{c}) \in \text{EDB}^-(R) \cup \text{IDB}^-(R)$,

where R is a relation in the EDB or in the head of an intensional rule, $\bar{c}$ is a tuple of constants, and $\text{IDB}^+(R)$ and $\text{IDB}^-(R)$ are computed from the simultaneous fixpoint definition, $\bar{S}_{\text{IDB}}$, defined by formula (6.4).

Rough relations for the IDB are then defined as follows:

$$R^+_{\text{IDB}} = \{v(\bar{c}) : \text{IDB} \Vdash R(\bar{c})\}$$
$$R^-_{\text{IDB}} = \{v(\bar{c}) : \text{IDB} \Vdash \neg R(\bar{c})\}$$
$$R^{\pm}_{\text{IDB}} = \{v(\bar{c}) : \text{IDB} \nVdash R(\bar{c}) \text{ and } \text{IDB} \nVdash \neg R(\bar{c})\}.$$

Observe that,

$$\text{EDB} \Vdash R(\bar{c}) \text{ implies IDB} \Vdash R(\bar{c})$$
$$\text{EDB} \Vdash \neg R(\bar{c}) \text{ implies IDB} \Vdash \neg R(\bar{c}).$$

Remark 6.3.2. In the case that one wants to distinguish between facts entailed solely by application of intensional rules, this can be done in a straightforward manner, but as a rule, one is interested in querying both EDB and IDB together, thus the choice of RKDB entailment from the IDB. □

6.3.4 The Semantics of the CCQ Layer and LCC Policies

The inference mechanism associated with the CCQ layer is intended to provide a form of *contextual closure* relative to part of the EDB and IDB when querying the RKDB. A *contextually closed query* consists of

- the *query* itself, which can be any fixpoint or first-order query
- the *context* represented as a set of one or more integrity constraints
- a *local closure policy* representing the closure context and consisting of a minimization policy representing the local closure.

Any LCC *policy* consists of a context and a local closure policy. LCC policies may also be viewed as approximations with rough relations in the EDB and IDB as input, a transducer consisting of one or more integrity constraints and a minimization policy, and modified rough relations in the RKDB as output.

Let the EDB and IDB be defined as before, I denote a finite set of integrity constraints, and let RKDB:LCC$[\bar{L}; \bar{K}]{:}I$ denote querying the three layers of the RKDB with a specific LCC policy LCC$[\bar{L}; \bar{K}]{:}I$. Then,

$$\text{RKDB: LCC}[\bar{L}; \bar{K}]{:}I \Vdash R(\bar{c}) \text{ if and only if}$$
$$\text{CIRC}(I \cup \text{IDB} \cup \text{EDB}; \bar{L}; \bar{K}) \models R(\bar{c}),$$
$$\text{RKDB: LCC}[\bar{L}; \bar{K}]{:}I \Vdash \neg R(\bar{c}) \text{ if and only if}$$
$$\text{CIRC}(I \cup \text{IDB} \cup \text{EDB}; \bar{L}; \bar{K}) \models \neg R(\bar{c}),$$

where the notation is as in Section 6.3.1, under the assumption that the circumscriptive theory is consistent.

Thus the CCQ layer results in dynamically redefining some relations in EDB and IDB in order to satisfy integrity constraints in a particular query. Of course, the EDB and IDB are not permanently modified, they are simply temporarily modified relative to the contextual query in question.

A relation R which is minimized, maximized or allowed to vary is defined as the following rough relation:

$$R^{+}_{\text{LCC}} = \{v(\bar{c}) : \text{RKDB: LCC}[\bar{L};\bar{K}]{:}I \Vdash R(\bar{c})\}$$
$$R^{-}_{\text{LCC}} = \{v(\bar{c}) : \text{RKDB: LCC}[\bar{L};\bar{K}]{:}I \Vdash \neg R(\bar{c})\}$$
$$R^{\pm}_{\text{LCC}} = \{v(\bar{c}) : \text{RKDB: LCC}[\bar{L};\bar{K}]{:}I \nVdash R(\bar{c}) \text{ and } \text{RKDB: LCC}[\bar{L};\bar{K}]{:}I \nVdash \neg R(\bar{c})\}.$$

Intuitively, this means that the positive part of R contains tuples present in all extensions of R satisfying the integrity constraints I, the boundary part contains tuples present in some extensions of R satisfying I, but not in all of them, and the negative part of R contains tuples not present in any extension of R satisfying I.

The relations that are not minimized, maximized or allowed to vary are not changed, thus their semantics is that given by the EDB and IDB layers of the RKDB.

6.4 The Computation Method

6.4.1 The Pragmatics of Computing Contextual Queries

A contextual query in its simplest form involves the (implicit) generation of the extension of a relation R in the context of a set of integrity constraints and a minimization policy, and asking whether one or more tuples is a member of that relation relative to a background RKDB consisting of EDB $\cup$ IDB. Essentially, we are required to implicitly compute R^{+}_{LCC}, R^{-}_{LCC}, and $R^{\pm}_{\text{LCC}}$ and determine whether the tuple or tuples are in any of the resulting rough set partitions of R. Based on this specification, one can show that in some cases, where the LCC policy and integrity constraints associated with the query is restricted appropriately, querying the relation R can be done very efficiently. One of the more important results of the considered technique is that one can automatically generate syntactic characterizations of each of the partitions of a rough set relation without actually generating their explicit extensions. The syntactic characterizations can then be used to efficiently query the RKDB consisting of the EDB/IDB pair.

Since integrity constraints are not associated with the EDB/IDB pair, but with an agent asking a query, the integrity constraints associated with an agent are not necessarily satisfied together with the EDB/IDB. Checking satisfiability is tractable in this context due to the first-order or fixpoint nature of the integrity constraints and the finiteness of the database. Under additional syntactic restrictions, the satisfiability of the circumscriptive theory can also be guaranteed. In the case of inconsistency, this would imply the need for the specification and computation of specific update policies.

6.4.2 The Algorithm

The algorithm presented below applies to the general case, i.e., to the problem which is co-NPTIME complete. However, in Section 6.5 we show specializations of the algorithm to some cases, where PTIME complexity is guaranteed. The inputs to the algorithm are:

- an extensional database EDB
- an intensional database IDB
- a set of integrity constraints I
- an LCC policy LCC$[\bar{L}; \bar{K}]$:I
- a relation symbol R.[4]

As output, the algorithm returns the definition of the relation R obtained by applying the LCC policy and preserving integrity constrains in I, according to the semantics defined in section 6.3. Recall that CIRC$(T; \bar{L}; \bar{K})$ used below stands for the circumscription of theory T with $\bar{L}$ minimized and $\bar{K}$ allowed to vary (see Definition 5.3.1).

1. Construct the formula $C \equiv$ CIRC$(I \cup$ IDB $\cup$ EDB$; \bar{L}; \bar{K})$ representing the given LCC policy applied to the IDB and EDB
2. eliminate second-order quantifiers from the formula obtained in step 1. In general, the elimination may fail and the result is the initial second-order formula C. However, if certain restrictions concerning the form of I are assumed, the elimination of second-order quantifiers is guaranteed (see Section 6.5)
3. calculate the intersection of all extensions of R satisfying formula C. If there is not any relation R satisfying C, terminate and return the answer "unsatisfiable," meaning that either the EDB and IDB pair is inconsistent, or the integrity constraints can not be satisfied
4. calculate the union of all extensions of R satisfying formula C
5. for any tuple $\bar{c}$:
   - if $v(\bar{c})$ is in the intersection calculated in step 3, add $v(\bar{c})$ to R^+
   - if $v(\bar{c})$ is not in the union calculated in step 4, add $v(\bar{c})$ to R^-
   - if none of the above two cases applies, then $v(\bar{c})$ is in $R^{\pm}$.

In practice, one uses particular second-order quantifier elimination algorithms, which may fail. Since second-order formulas are useless as results, it is reason-

[4] The relation symbol R can be viewed as part of the query which consists of a number of relations that are required to compute the full query.

able to return the answer "unknown" when the elimination algorithm used in Step 2 fails.[5]

Observe also, that in practice it is often better to calculate syntactic definitions of new relations rather than calculating their extensions as is done in the above inefficient algorithm.

6.4.3 Expressiveness of the Approach

An interesting question arises as to whether the current approach allows one to express all tractable LCC policies, where by a *tractable* LCC *policy* we mean any LCC policy such that all minimized, maximized and varied relations are PTIME-computable w.r.t. the size of the underlying databases.

The following characterization shows that the method presented is strong enough to express all tractable LCC policies. In other words, any tractable LCC policy can always be reformulated in the form used in Lemma 6.4.1 below. In Section 6.5, we provide additional syntactic characterizations of LCC policies which guarantee tractability.

Lemma 6.4.1. Assume that the database is linearly ordered. Then all tractable LCC policies can be expressed as policies of the form

$$\text{LCC}[\bar{L}; \bar{K}] : \{\beta_i(\bar{x}) \to L_i(\bar{x}) : L_i \in \bar{L}\},$$

where each $\beta_i(\bar{x})$ is a first-order formula positive w.r.t. L_i. □

Lemma 6.4.1 follows from the observation that any relation computable in PTIME can be expressed by means of the least fixpoint of a formula of the form $\beta_i(\bar{x})$ such that $\forall \bar{x}.\beta_i(\bar{x}) \to L_i(\bar{x})$ holds, provided that the database domain is ordered (for references see Section 6.7). Since all minimized, maximized and varied relations are assumed to be tractable, they can be expressed by the least fixpoints of formulas of the form $\text{LFP}\, L_i(\bar{x}).\beta_i(\bar{x})$, thus also by the policy

$$\text{LCC}[\bar{L}; \bar{K}] : \{\beta_i(\bar{x}) \to L_i(\bar{x}) : L_i \in \bar{L}\}.$$

6.5 The Case of Universal LCC Policies

In general, the problem of querying the database in the presence of unrestricted integrity constraints is CO-NPTIME complete. On the other hand,

[5] In this case the algorithm is only sound, but not complete relative to the semantics provided in Section 6.3.4.

some classes of LCC policies for which the computation mechanism is in PTIME can be isolated. In this section, we consider a restriction on integrity constraints which allows us to compute explicit definitions of the new relations as first-order and fixpoint formulas. In such cases computing contextually closed queries is in PTIME.

Assume that the integrity constraints have the following form:

$$\forall \bar{x}.[\alpha(\bar{x}) \rightarrow \beta(\bar{x})], \tag{6.5}$$

where α and β are arbitrary first-order formulas.

Observe that, in general, some LCC policies might contain conflicting requirements that the same relation is to be minimized and maximized at the same time. In order to exclude such situations, we introduce a notion of marking defined below. Intuitively, we mark relations with symbol 'min' to indicate that a given relation is to be minimized, and with symbol 'max' to indicate that the relation is to be maximized.

Definition 6.5.1. By a *marking of relation symbols for policy* LCC$[\bar{L};\bar{K}]{:}I$ we understand a mapping assigning to any relation symbol, both in the local closure policy LCC$[\bar{L};\bar{K}]$ and I in the LCC policy, the least subset of $\{\text{min}, \text{max}\}$ that is closed under the following rules:

1. for any relation symbol S appearing in $\bar{L}$ positively, 'min' is in the set of marks of S
2. for any relation symbol S appearing in $\bar{L}$ negatively, 'max' is in the set of marks of S
3. if $\alpha(R) \rightarrow \beta(S)$ is in I, $R, S \in \bar{L} \cup \bar{K}$ and S occurs in β positively and is marked by 'min', or S occurs in β negatively and is marked by 'max' then:
   - if R occurs positively in α, then 'min' is in the set of marks of R
   - if R occurs negatively in α, then 'max' is in the set of marks of R
4. if $\alpha(R) \rightarrow \beta(S)$ is in I, $R, S \in \bar{L} \cup \bar{K}$ and α contains a positive occurrence of R and R is marked by 'max', or α contains a negative occurrence of R and R is marked by 'min' then:
   - if S occurs positively in β, then 'max' is in the set of marks of S
   - if S occurs negatively in β, then 'min' is in the set of marks of S.

An LCC$[\bar{L};\bar{K}]{:}I$ policy is called *uniform* if no relation symbol is marked by both 'max' and 'min'. □

Example 6.5.2. Let us consider the following integrity constraint:

$$[Car(x) \wedge Red(x)] \rightarrow RedCar(x). \tag{6.6}$$

The marking for the policy

$$\text{LCC}\,[\{RedCar(x), Car(x)\}; \{Red(x)\}] : (6.6)$$

assigns the mark 'min' to all the relation symbols. Thus the policy is uniform. On the other hand, the marking for the policy

$$\text{LCC}\,[\{RedCar(x), \neg Car(x)\}; \{Red(x)\}] : (6.6)$$

assigns the mark 'min' to *Red* and the marks {'min','max'} to *Car* and *RedCar*. Thus the later policy is not uniform. □

Let us now introduce the notion of universal LCC policies useful in many applications.

Definition 6.5.3. By a *universal* LCC *policy* we understand any uniform policy $\text{LCC}\,[\bar{L}; \bar{K}]{:}I$ in which I is a set of constraints of the form

$$\forall \bar{y}.[(\pm P_1(\bar{x}_1) \wedge \ldots \wedge \pm P_k(\bar{x}_k)) \rightarrow \pm P(\bar{x})] \tag{6.7}$$

where $\pm$ stands for $\neg$ or the empty symbol, $P_1, \ldots, P_k, P$ are relation symbols and $\bar{y}$ is a vector of all variables occurring in $\bar{x}_1, \ldots, \bar{x}_k, \bar{x}$. □

In the case of universal LCC policies we have a computation method much more efficient than that described in Section 6.4.2. In the rest of this section we will consider only universal LCC policies $\text{LCC}\,[\bar{L}; \bar{K}]{:}I$, for given tuples of literals $\bar{L}$,$\bar{K}$ and a set I of integrity constraints.

In the computation method for universal policies, we first construct minimal rough relations satisfying the EDB, IDB and the integrity constraints, where minimality is defined w.r.t. the information ordering defined below.

Definition 6.5.4. *Let R and S be rough relations. We define the* information ordering on rough relations, *denoted by $R \sqsubseteq S$, as follows:*

$$R \sqsubseteq S \stackrel{\text{def}}{\equiv} R^+ \subseteq S^+ \text{ and } R^- \subseteq S^-.$$

□

In order to find minimal w.r.t. $\sqsubseteq$ rough relations satisfying I, EDB and IDB we will use the following tautologies of first-order logic:

$$\forall \bar{x}.[\alpha(\bar{R}) \rightarrow (\beta(\bar{R}) \vee M(\bar{y}))] \equiv \forall \bar{x}.[(\alpha(\bar{R}) \wedge \neg M(\bar{y})) \rightarrow \beta(\bar{R})] \tag{6.8}$$
$$\forall \bar{x}.[(\alpha(\bar{R}) \wedge M(\bar{y})) \rightarrow \beta(\bar{R})] \equiv \forall \bar{x}.[\alpha(\bar{R}) \rightarrow (\beta(\bar{R}) \vee \neg M(\bar{y}))],$$

where it is assumed that all double negations $\neg\neg$ are removed.

Next we define the notion of expansion. The intuition behind this notion is that any implication $(A_1 \wedge \ldots \wedge A_k) \rightarrow B$ is logically equivalent to

$$(\neg B \wedge A_1 \wedge \ldots \wedge A_{i-1} \wedge A_{i+1} \wedge \ldots \wedge A_k) \rightarrow \neg A_i.$$

Thus, viewed as a rule, it contributes to the generation not only facts about B, but also facts about all $\neg A_i$, for $1 \leq i \leq k$.

Definition 6.5.5. *Let I be an integrity constraint of the form:*

$$\forall \bar{x}.[(\pm R_1(\bar{y}_1) \wedge \ldots \wedge \pm R_m(\bar{y}_m)) \rightarrow \pm S(\bar{z})], \tag{6.9}$$

Let $\mathcal{P} = \text{Lcc}[\bar{L}; \bar{K}]{:}I$ be an Lcc *policy. By the* expansion of I w.r.t. $\mathcal{P}$, *denoted by $Exp^{\mathcal{P}}(I)$, we understand the least set of constraints of the form,*

$$\forall \bar{x}.[(\bigwedge_k L_k(\bar{x}_k)) \rightarrow \pm L(\bar{x}_S)],$$

obtained from (6.9) by applying the tautologies (6.8), such that any instance of a (possibly negated) literal of (6.9) containing a relation symbol occurring in $\bar{L}; \bar{K}$, is a consequent of exactly one constraint. □

Example 6.5.6. Consider the integrity constraint

$$I \stackrel{\text{def}}{\equiv} \forall x, y.[(\neg P(x) \wedge S(x,y)) \rightarrow P(y)]$$

and the policy $\mathcal{P} = \text{Lcc}[P; S]{:}I$.

All instances of literals in I are $\{\neg P(x), S(x,y), P(y)\}$. The expansion of I w.r.t. $\mathcal{P}$ is then defined as the following set of constraints:

$$\begin{aligned} Exp^{\mathcal{P}}(I) = \{ \; & \forall x, y.[(\neg P(x) \wedge S(x,y)) \rightarrow P(y)], \\ & \forall x, y.[(\neg P(y) \wedge S(x,y)) \rightarrow P(x)], \\ & \forall x, y.[(\neg P(x) \wedge \neg P(y)) \rightarrow \neg S(x,y)] \; \}. \end{aligned}$$

In the case of policy $\mathcal{P}' = \text{Lcc}[S; \emptyset]{:}I$, the expansion of I is defined as

$$Exp^{\mathcal{P}'}(I) = \{\forall x, y.[(\neg P(x) \wedge \neg P(y)) \rightarrow \neg S(x,y)]\}.$$

□

Let $S = \langle S_1, \ldots, S_p \rangle$ be all relations in the Idb and let us fix an Lcc policy $\mathcal{P} = \text{Lcc}[\bar{L}; \bar{K}]{:}I$. In order to compute the definition of minimal w.r.t. $\sqsubseteq$ rough relations, satisfying the constraints I, Edb and Idb we consider the following cases, where $1 \leq i \leq p$:

- if $S_i \notin \bar{L} \cup \bar{K}$, then the positive part of the resulting relation, S_i^+, contains exactly the tuples present in $\text{EDB}^+(S_i) \cup \text{IDB}^+(S_i)$, and the negative part of the resulting relation, S_i^-, contains exactly the tuples present in $\text{EDB}^-(S_i) \cup \text{IDB}^-(S_i)$
- if $S_i \in \bar{L} \cup \bar{K}$, then we consider the set of integrity constraints

$$\begin{aligned} \{\text{FG}(\alpha) :\ & \alpha \in Exp^{\mathcal{P}}(A),\ A \in I \\ & \text{and } S_i \text{ occurs in the consequent of } \alpha\ \}, \end{aligned} \tag{6.10}$$

where FG is the Feferman-Gilmore translation specified in Definition 6.3.1.

We assume that the following integrity constraints, reflecting the contents of EDB and IDB, are implicitly given:

$$\begin{bmatrix} \forall \bar{y}. \left[\left(\bigvee_{\bar{c}: S_i(\bar{c}) \in \text{EDB}^+(S_i) \cup \text{IDB}^+(S_i)} \bar{y} = \bar{c} \right) \rightarrow S_i^+(\bar{y}) \right] \\ \forall \bar{y}. \left[\left(\bigvee_{\bar{c}: S_i(\bar{c}) \in \text{EDB}^-(S_i) \cup \text{IDB}^-(S_i)} \bar{y} = \bar{c} \right) \rightarrow S_i^-(\bar{y}) \right] \end{bmatrix} \tag{6.11}$$

where the empty disjunction is, as usual, interpreted as FALSE.

Now we gather all the constraints in (6.10) and (6.11) with S_i^+ as the consequent into the following single formula:

$$\forall \bar{y}. \left[\left(\bigvee_{1 \le k \le k_i} \exists \bar{z}_{ik}.\phi_{ik}(\bar{R}_k) \right) \rightarrow S_i^+(\bar{y}) \right] \tag{6.12}$$

and all the integrity constraints with S_i^- as the consequent into the following single formula:

$$\forall \bar{y}. \left[\left(\bigvee_{1 \le j \le j_i} \exists \bar{z}_{ij}.\psi_{ij}(\bar{R}_j) \right) \rightarrow S_i^-(\bar{y}) \right]. \tag{6.13}$$

The following definitions of the positive and the negative part of the required minimal rough relations w.r.t. policy $\mathcal{P}$, indicated by the index $\mathcal{P}$, can now be derived:

$$\bar{S}_{\mathcal{P}}^+(\bar{y}) \equiv \text{LFP}\ \bar{S}(\bar{y}). \left[\bigvee_{1 \le k \le k_1} \exists \bar{z}_{1k}.\phi_{1k}(\bar{R}_k), \ldots, \bigvee_{1 \le k \le k_n} \exists \bar{z}_{nk}.\phi_{nk}(\bar{R}_k) \right] \tag{6.14}$$

$$\bar{S}_{\mathcal{P}}^-(\bar{y}) \equiv \text{LFP}\ \bar{S}(\bar{y}). \left[\bigvee_{1 \le j \le j_1} \exists \bar{z}_{1j}.\psi_{1j}(\bar{R}_j), \ldots, \bigvee_{1 \le j \le j_m} \exists \bar{z}_{mj}.\psi_{mj}(\bar{R}_j) \right] \tag{6.15}$$

Observe that the syntactic restrictions placed on the integrity constraints guarantee that the formulas under the fixpoint operators are positive, thus monotone w.r.t. S and, consequently, the fixpoints exist. Observe also, that in the case of non-recursive universal LCC policies the fixpoint operators can be removed and the definitions obtained are classical first-order formulas.[6]

Having computed the suitable parts of the relations in all integrity constraints, one can easily perform a consistency check, indicating whether the integrity constraints can be satisfied by the current contents of the EDB ∪ IDB. Namely, for each relation R one needs to assure that $R^+ \cap R^- = \emptyset$.

The following definition introduces the notion of *rough negation* used to define the semantics of minimized/maximized and varied relations.

Definition 6.5.7. *Let* $\mathcal{P}$ =LCC$[\bar{L}; \bar{K}]$:I *be an* LCC *policy. The* rough negation for the policy $\mathcal{P}$, *denoted by* $\sim_{\mathcal{P}}$, *is defined as follows:*

- $\sim_{\mathcal{P}}$ *satisfies the usual DeMorgan laws for quantifiers, conjunction and disjunction, and:*

$$\sim_{\mathcal{P}} \text{LFP}\, \bar{R}.\alpha(\bar{R}) \stackrel{\text{def}}{\equiv} \text{GFP}\, \bar{R}. \sim_{\mathcal{P}} \alpha(\bar{R})$$

$$\sim_{\mathcal{P}} \text{GFP}\, \bar{R}.\alpha(\bar{R}) \stackrel{\text{def}}{\equiv} \text{LFP}\, \bar{R}. \sim_{\mathcal{P}} \alpha(\bar{R})$$

$$\sim_{\mathcal{P}} \neg S^+ \stackrel{\text{def}}{\equiv} S^+, \quad \sim_{\mathcal{P}} \neg S^- \stackrel{\text{def}}{\equiv} S^-$$

- *if* $S \in \bar{L} \cup \bar{K}$, *then*

$$\sim_{\mathcal{P}} S^+ \stackrel{\text{def}}{\equiv} \neg S^+, \quad \sim_{\mathcal{P}} S^- \stackrel{\text{def}}{\equiv} \neg S^-$$

- *if* $S \notin \bar{L} \cup \bar{K}$, *then*

$$\sim_{\mathcal{P}} S^+ \stackrel{\text{def}}{\equiv} S^-, \qquad \sim_{\mathcal{P}} S^- \stackrel{\text{def}}{\equiv} S^+.$$

□

Lemma 6.5.8. *If the integrity constraints* I *are consistent with* EDB ∪ IDB *then the definitions of minimal and maximal rough relations satisfying constraints in* I *and reflecting the semantics introduced in Section 6.3.4 can be calculated as follows:*[7]

[6] In both cases, however, computing the defined parts of relations can be done in time polynomial in the size of EDB ∪ IDB.

[7] Observe that the LCC policies provide us with direct information about which relations are to be maximized, which are to be minimized and which remain unchanged.

$$S^{+}_{min}(\bar{y}) \equiv \bar{S}^{+}_{\mathcal{P}} \tag{6.16}$$
$$S^{-}_{min}(\bar{y}) \equiv \sim_{\mathcal{P}} S^{+}_{min}(\bar{y}) \equiv\sim_{\mathcal{P}} \bar{S}^{+}_{\mathcal{P}} \tag{6.17}$$
$$S^{-}_{max}(\bar{y}) \equiv \bar{S}^{-}_{\mathcal{P}} \tag{6.18}$$
$$S^{+}_{max}(\bar{y}) \equiv \sim_{\mathcal{P}} S^{-}_{max}(\bar{y}) \equiv\sim_{\mathcal{P}} S^{-}_{max}(\bar{y}), \tag{6.19}$$

where $\bar{S}^{+}_{\mathcal{P}}$ *and* $\bar{S}^{-}_{\mathcal{P}}$ *are defined by formulas (6.14) and (6.15), respectively.* □

In the case of non-recursive universal policies, the fixpoint operators[8] can be removed, as before.

Observe that definitions of varied predicates can now be computed by noticing that these are the minimal w.r.t. $\sqsubseteq$ rough relations satisfying the integrity constraints in the new context of minimized and maximized relations. It then suffices to apply definitions (6.14), (6.15) with minimized and maximized relations replaced by their definitions obtained as (6.16), (6.17), (6.18), and (6.19), as appropriate.

Example 6.5.9. Consider the problem of determining whether a given car on a road is seen from a UAV. We assume that usually large cars are seen. Our database contains relations:

- *Car* containing cars
- *Large* containing large objects
- *See* containing visible objects
- *Ab* standing for abnormal objects, i.e., large but invisible objects.

Define the following integrity constraint I:

$$\forall x.[(Car(x) \wedge Large(x) \wedge \neg See(x)) \rightarrow Ab(x)].$$

We want to minimize abnormality, i.e., to minimize relation Ab, while keeping the relations Car and $Large$ unchanged and See varied. The local closure policy is then

$$L = \text{LCC}[Ab; See].$$

The suitable expansion is defined by the following set of rules:

$$\begin{aligned} Exp^{L}(I) = \{ \ & \forall x.[(Car(x) \wedge Large(x) \wedge \neg See(x)) \rightarrow Ab(x)] \\ & \forall x.[(Car(x) \wedge Large(x) \wedge \neg Ab(x)) \rightarrow See(x)] \ \}. \end{aligned}$$

Now one has to add the rules reflecting the contents of EDB and consider the Feferman-Gilmore translation. This results in the following set of formulas:

[8] Appearing in definitions of $\bar{S}^{+}_{\mathcal{P}}$ and $\bar{S}^{-}_{\mathcal{P}}$.

$$\{\forall x.[(\text{EDB}^+(Ab(x)) \vee (Car^+(x) \wedge Large^+(x) \wedge See^-(x))) \rightarrow Ab^+(x)],$$
$$\forall x.[(\text{EDB}^+(See(x)) \vee (Car^+(x) \wedge Large^+(x) \wedge Ab^-(x))) \rightarrow See^+(x)] \}.$$

According to Lemma 6.5.8 and the discussion following the lemma, we obtain the following characterizations of Ab and See:

$$Ab^+_{min}(x) \equiv \text{EDB}^+(Ab(x)) \vee [Car^+(x) \wedge Large^+(x) \wedge See^-(x)]$$
$$Ab^-_{min}(x) \equiv \neg\text{EDB}^+(Ab(x)) \wedge [Car^-(x) \vee Large^-(x) \vee \neg See^-(x)]$$
$$See^+_{var}(x) \equiv \text{EDB}^+(See(x)) \vee [Car^+(x) \wedge Large^+(x) \wedge Ab^-_{min}(x)]$$
$$See^-_{var}(x) \equiv \text{EDB}^-(See(x)).$$

Note that computing the minimized and varied relations is done by querying the original database in the standard manner. □

Example 6.5.10 (A Surveillance Mission Case Study). Consider a scenario involving a UAV that makes use of contextually closed queries during a surveillance mission. A black car has been reported stolen and the task of the UAV is to locate the car by investigating areas in which the car is suspected to be located. To represent this scenario we make use of the relations:

- $In(x, y)$ (car x is in region y)
- $Color(x, z)$ (the color of car x is z)
- $SuspectIn(y)$ (the stolen car is suspected to be in region y)
- $Investigate(x, y)$ (the UAV should search for car x in region y).

Using these relations we construct a crisp logical theory (6.20) expressing the behavior we wish the UAV to exhibit. All black cars that are in a suspect region should be investigated. If a car is known to have some color other than black it is not necessary to look for it in any region. Finally, when we know that the searched car is not in a region, there is no point going there looking for it.

$$\begin{aligned}
&\forall x, y.\Big[(In(x,y) \wedge SuspectIn(y) \wedge Color(x, \mathsf{black})) \\
&\qquad\qquad\qquad \rightarrow Investigate(x,y)\Big] \wedge \\
&\forall x, y, z.\Big[(Color(x,z) \wedge z \neq \mathsf{black}) \rightarrow \neg Investigate(x,y)\Big] \wedge \\
&\forall x, y.\Big[\neg In(x,y) \rightarrow \neg Investigate(x,y)\Big].
\end{aligned} \tag{6.20}$$

Additionally an intensional rule (6.21) is added to the IDB expressing the fact that if we know the region a car is in, it can not simultaneously be in some other region.

$$\forall x, y_1, y_2.[In(x, y_1) \wedge y_1 \neq y_2] \rightarrow \neg In(x, y_2). \tag{6.21}$$

Continuing the example, we construct a specific scenario by adding facts to the approximate knowledge base. Given three cars, c1, c2 and c3, three regions, r1, r2 and r3, and two colors, black and red, we add the facts expressed in (6.22). A black car c1 is known to be in region r1, the car c2 is red but we do not know in which region it is, and nothing is known about the third car c3. Furthermore, the stolen car is believed to be located somewhere in region r1 or r2.

$$\begin{aligned}&In^{+}(\mathsf{c1},\mathsf{r1})\wedge\\&Color^{+}(\mathsf{c1},\mathsf{black})\wedge Color^{+}(\mathsf{c2},\mathsf{red})\wedge\\&SuspectIn^{+}(\mathsf{r1})\wedge SuspectIn^{+}(\mathsf{r2}).\end{aligned}\tag{6.22}$$

The current knowledge base does not contain any information about which cars and what regions are interesting for the UAV, but this is information that would be invaluable when determining appropriate strategies to search regions for target vehicles. To acquire such information, a contextually closed query can be formulated which takes account of current context. In this case, new information specific to regions of interest can be generated nonmonotonically.

To do this, the closure policy associated with the contextually closed query will minimize the number of suspected regions in order to avoid searching regions that we have no specific reason to believe the stolen car to be in, while varying what cars and regions the UAV should investigate to obtain information about possible actions to take. Consequently we construct the policy of minimizing $SuspectIn$ while varying $Investigate$ and fixing the remaining relations In and $Color$.

In order to ask contextually closed queries we first compute the expansion of (6.20), where the only formula containing an occurrence of a minimized or varied relation that is not already in the consequent is the first one. Thus, in this case, the expansion of (6.20) is obtained by replacing the first conjunct by

$$\begin{aligned}&\forall x,y.[In(x,y)\wedge SuspectIn(y)\wedge Color(x,\mathsf{black})\\&\qquad\qquad\rightarrow Investigate(x,y)]\\&\forall x,y.[In(x,y)\wedge\neg Investigate(x,y)\wedge Color(x,\mathsf{black})\\&\qquad\qquad\rightarrow\neg SuspectIn(y)]\end{aligned}\tag{6.23}$$

We now obtain syntactic definitions for $SuspectIn$ and $Investigate$ according Lemma 6.5.8 and obtain the results shown in (6.24). The positive part of the minimized $SuspectIn$ relation is simply those tuples explicitly stored as positive in the extensional database, while the negative part contains the rest of the tuples, while the definition of the varied relation $Investigate$ is a bit more complex.

$$
\begin{aligned}
SuspectIn^{+}_{min}(x) &\equiv \text{EDB}^{+}(SuspectIn(x))\\
SuspectIn^{-}_{min}(x) &\equiv \text{EDB}^{-}(SuspectIn(x))\\
Investigate^{+}_{var}(x,y) &\equiv \text{EDB}^{+}(Investigate(x,y))\vee\\
&\quad [In^{+}(x,y)\wedge SuspectIn^{+}(y)\wedge\\
&\quad\ Color^{+}(x,\mathsf{black})]\\
Investigate^{-}_{var}(x,y) &\equiv \text{EDB}^{-}(Investigate(x,y))\vee\\
&\quad [In^{-}(x,y)\vee \exists z.[Color^{+}(x,z)\wedge z\neq \mathsf{black}].
\end{aligned}
\tag{6.24}
$$

To evaluate a query containing the minimized or varied relations it suffices to replace those occurrences with their syntactic definitions given in (6.24) and pass the modified query to the intensional database layer. Evaluating the definitions in our examples produces the tuples in (6.25), including new tuples produced by the IDB rule (6.21).

$$
\begin{aligned}
In^{+}(x,y) &: \{\langle \mathsf{c1},\mathsf{r1}\rangle\}\\
In^{-}(x,y) &: \{\langle \mathsf{c1},\mathsf{r2}\rangle,\langle \mathsf{c1},\mathsf{r3}\rangle\}\\
SuspectIn^{+}(y) &: \{\mathsf{r1},\mathsf{r2}\}\\
SuspectIn^{-}(y) &: \{\mathsf{r3}\}\\
Investigate^{+}(x,y) &: \{\langle \mathsf{c1},\mathsf{r1}\rangle\}\\
Investigate^{-}(x,y) &: \{\langle \mathsf{c1},\mathsf{r2}\rangle,\langle \mathsf{c1},\mathsf{r3}\rangle,\langle \mathsf{c2},\mathsf{r1}\rangle,\\
&\quad\ \langle \mathsf{c2},\mathsf{r2}\rangle,\langle \mathsf{c2},\mathsf{r3}\rangle\}.
\end{aligned}
\tag{6.25}
$$

Although the IDB rule excluded the possibility of c1 being anywhere else than in r1, it remains unknown which regions the other cars are in. Minimizing *SuspectIn* removes r3 from the set of suspected regions since there is no reason to believe otherwise, while varying *Investigate* prompts the UAV to search for c1 in region r1 since we know it is a black car located in a region which we suspect the stolen car to be in. In addition, the UAV concludes that it is not necessary to look for c1 anywhere else, using the IDB rule and the part of the theory stating that it should not investigate a region, looking for a car it knows is not there. Car c2 can be in any of the regions but there is no point looking for it as it has the color red, different from black. Finally, it remains unknown, even after applying the closure policy, if searching for the car c3 in any of the regions is necessary.

Now, assume the UAV takes action, flying over region r1 looking for c1, and that it finds the car but it is not the stolen car we are looking for. It updates the knowledge base by removing r1 from the list of suspected regions (from *SuspectIn*) and adding the fact that it did not find c3, expressed by $In^{-}(\mathsf{c3},\mathsf{r1})$. Using the same syntactic definitions of relations, we reevaluate the queries in light of these new facts.

$$\begin{aligned} In^+(x,y) &: \{\langle \mathsf{c1}, \mathsf{r1} \rangle\} \\ In^-(x,y) &: \{\langle \mathsf{c1}, \mathsf{r2} \rangle, \langle \mathsf{c1}, \mathsf{r3} \rangle, \langle \mathsf{c3}, \mathsf{r1} \rangle\} \\ SuspectIn^+(y) &: \{\mathsf{r2}\} \\ SuspectIn^-(y) &: \{\mathsf{r1}, \mathsf{r3}\} \\ Investigate^+(x,y) &: \emptyset \\ Investigate^-(x,y) &: \{\langle \mathsf{c1}, \mathsf{r2} \rangle, \langle \mathsf{c1}, \mathsf{r3} \rangle, \langle \mathsf{c2}, \mathsf{r1} \rangle, \\ &\quad\ \langle \mathsf{c2}, \mathsf{r2} \rangle, \langle \mathsf{c2}, \mathsf{r3} \rangle, \langle \mathsf{c3}, \mathsf{r1} \rangle\}. \end{aligned} \tag{6.26}$$

The *In* tuples in (6.26) has changed to incorporate the fact that c3 has not yet been found, and the r1 tuple in the *SuspectIn* relation has moved to reflect the fact that no stolen car was found there, but the varied *Investigate* relation has changed too. The UAV has already searched region r1 for c1, and it concludes that it is no longer necessary to investigate whether c3 is in r1, but it is still unknown if the UAV should look for c3 in one of the other regions.

Notice that without changing the definitions, the query results have changed to reflect the new knowledge situation. This will stay true until we modify the closure policy or the logical theory describing the mission, in which case the definitions must be recalculated. As long as the policy and theory stay the same, we can cache the calculated definitions, improving efficiency.

In its current state of uncertainty, the UAV might either explain the two remaining possibilities to a mission operator, asking for new information or advice on which action to take, or continue on itself, e.g. by systematically searching for c3, first in region r2 and then in r3. Assuming the latter alternative, and that the stolen car is in fact located in one of the regions, the UAV will find it and successfully complete the mission. □

6.6 Complexity Issues

6.6.1 Reducing the Database Size

The database size can be reduced by removing particular columns or rows from the database tables. Let us first briefly discuss the problem of removing columns.

Rough set methodology offers a means to create reducts (see, e.g., Section 14.2), where reducts are obtained by removing certain columns from the original tables.

From the logical point of view one has to project the whole language into its subset and then to approximate the original theory. The logical tools for such an approach are provided in Chapter 8, in particular in Section 8.5.2.

Another important methodology of reducing the database size is to apply non-monotonic techniques. For instance one can represent typical cases as a default and keep in the database only the information about exceptions. A tractable approach to default reasoning is developed in Chapter 10. Another method could depend on providing theories describing abnormal behavior of objects and then to minimize the abnormality due to circumscriptive policies, as defined in Section 5.3. Observe though that minimization policies are expressible in our approach by using the contextually closed queries, as defined in this chapter.

6.6.2 Reducing the Complexity of Fixpoint Queries

Observe that we often present rough relations by providing their positive and negative parts. There are at least the following two immediate, yet important applications of this representation:

- reducing complexity of calculating rough relations, thus also the complexity of the querying mechanism
- supporting anytime methods by providing always meaningful answers within the assumed time frame, which is important, e.g., in the case of some critical real-time applications.

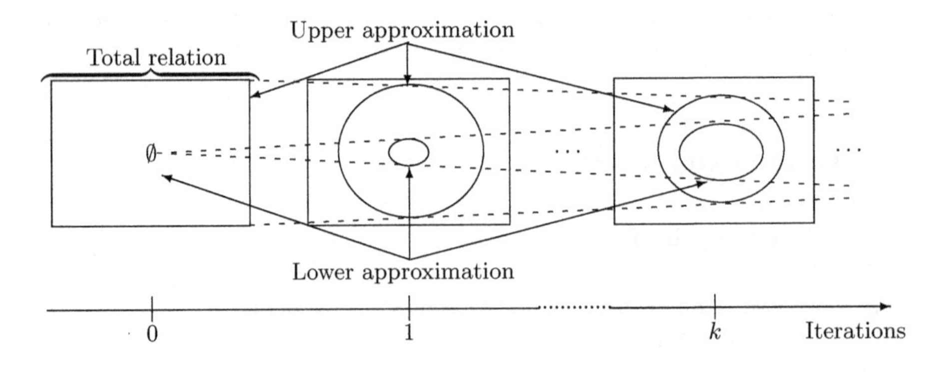

Fig. 6.2. Computing approximations of least and greatest fixpoints.

The main idea that can be applied here is to compute both the positive part, R^+, of a relation R together with its negative part R^-. Both processes are monotone in the sense that the calculated sets R^+ and R^- can only grow in each computation step. Thus the computed boundary region can only become smaller. The process can be stopped when the assumed approximation measure is reached or (in the case of anytime methods) the time limit is reached. One can thus save a considerable amount of computations as, for instance, one often does not have to go through the whole database computing first-order queries.

Observe that the complexity of the approach can also be reduced in applications where a given approximation measure is sufficient and relations are defined by means of fixpoints. Namely, one can perform iterations necessary to compute least and greatest fixpoints until the size of the boundary region becomes acceptable or a time limit is exceeded. This follows from the fact that the least fixpoint, corresponding to the lower approximation of a relation, can be computed starting from the empty relation (and then it can only get fixed or increase). On the other hand, the greatest fixpoint, corresponding to the upper approximation of the relation, can be computed starting from the total relation (and then it can only get fixed or decrease) - see Figure 6.2. Obviously, everything in the lower approximation of a relation is in its positive part and everything which is not in its upper approximation is in its negative part. In many cases a considerable amount of iterations can then be saved.

6.7 Bibliographic Notes

The methodology of contextually closed queries is relatively new. It has been developed in [50], as an alternative approach to dealing with the open world assumption. In particular Sections 6.1-6.5 are heavily based on [50]. Some other important classes of LCC policies, including so-called semi-Horn policies are provided there, too. Example 6.5.10 is from [63].

Other approaches to closing the world locally were considered, e.g., in [57, 71]. It should be noted that the approach we consider here subsumes that of [57, 71].

The syntax of the language of intensional databases is similar to the approach using DATALOG$^{\neg\neg}$ (see, e.g., [1]). However, in order to provide the semantics for intensional rules we use the Feferman-Gilmore translation (see, e.g., [31]).

The notion of information ordering was considered by Fitting and van Benthem (see, e.g., [31]) in the context of three-valued logics.

The fact that any relation computable in PTIME can be expressed by means of the least fixpoint of a formula of the form required in Lemma 6.4.1 provided that the database domain is ordered, can be found, e.g., in [68].

An alternative, logic programming-based approach to rough knowledge databases has been developed by Vitória, Damásio and Małuszyński (see, e.g., [229, 230, 231]).

7

Combining Rough and Crisp Knowledge

7.1 Introduction

This chapter presents a framework for specifying, constructing, and managing a particular class of approximate knowledge structures for use with intelligent artifacts, ranging from simpler devices such as personal digital assistants to more complex ones such as unmanned aerial vehicles. The basic structure for the concepts presented is that of an approximation transducer which takes approximate relations as input, and generates a (possibly more abstract) approximate relation as output. This is done by combining the approximate input relations with a crisp local logical theory representing dependencies between the input and output relations.

Approximation transducers can be combined to produce approximation trees which allow for representation of complex approximate knowledge structures characterized by the properties of elaboration tolerance, groundedness in the application domain, modularity, and context dependency. Approximation trees can be grounded through the use of primitive concepts which can be generated with supervised machine learning techniques. Changes in definitions of primitive concepts or in the local logical theories used by transducers result in changes in the knowledge stored in approximation trees by increasing or decreasing precision in the knowledge in a qualitative manner.

The inference mechanism associated with the use of approximation trees is based on rough knowledge databases, as presented in Chapter 6. By placing certain syntactic restrictions on the local theories used in transducers, the computational processes used in the query/answering and generation mechanism for approximation trees remain in PTIME.

In the philosophical literature, Quine has used the phrase *web of belief* to capture the intricate and complex dependencies and structures which make up human beliefs. In this chapter, we lay the ground work for what might properly be called *webs of approximate knowledge*. One way to view this idea

P. Doherty et al.: *Knowledge Representation Techniques*, Studfuzz **202**, 129–142 (2006)
www.springerlink.com

is as starting with *webs of imprecise knowledge* and gradually incrementing these initial webs with additional approximate and sometimes crisp facts and knowledge. Through this process, a number of concepts, relations and dependencies between them become less imprecise, approximating their crisp counterparts. This is a continual process where the precision in meaning of concepts is continually modified in a change-tolerant manner. Approximate definitions of concepts are the rule rather than the exception even though crisp definitions of concepts are a special case included in the framework.

Specifically, webs of approximate knowledge are constructed from primitive concepts together with approximation transducers in a recursive manner. An approximation transducer provides an approximate definition of one or more output concepts in terms of a set of input concepts and consists of three components:

1. an input consisting of one or more approximate concepts, some of which might be primitive
2. an output consisting of one or more new and possibly more abstract concepts defined partly in terms of the input concepts
3. a local logical theory specifying constraints or dependencies between the input concepts and the output concepts. The theory may also refer to other concepts not expressed in the input.

The local logical theory specifies dependencies or constraints an expert of the application domain would be able to specify. Generally the form of the constraints would be in terms of some necessary and some sufficient conditions for the output concept. The local theory is viewed as a set of crisp logical constraints specified in the language of first-order logic. The local theory serves as a logical template. During the generation of the approximate concept which is output by the transducer, the crisp relations mentioned in the local theory are substituted with the actual approximate definitions of the input. Either lower or upper approximations of the input concepts may be used in the substitution. The resulting output specifies the output concept in terms of newly generated lower and upper approximations. The resulting output relation may then be used as input to other transducers creating approximation trees. The resulting tree represents a web of approximate knowledge capturing intricate and complex dependencies among an agent's conceptual vocabulary.

As an example of a transducer that might be used in the unmanned aerial vehicle domain, we can imagine defining a transducer for the approximate concept of two vehicles being *connected* in terms of *visible connection*, *small distance*, and *equal speed*. The latter three input concepts could be generated from supervised machine learning techniques where the data is acquired from a library of videos previously collected by the UAV on earlier traffic monitoring missions. As part of the local logical theory, an example of a constraint might state that "if two vehicles are *visibly connected*, are at a *small distance* from each other and have *equal speed* then they are *connected*".

Approximation trees are fluid knowledge structures. Changes in the definition of primitive concepts will trickle through the trees via the dependencies and connections, modifying some of the other concept definitions. Changes to the local theories anywhere in the tree will modify those parts of the tree related to the respective output concepts for the local theories. This is a form of elaboration tolerance. These structures are approximate in three respects:

1. the primitive concepts are approximate. They usually consist of upper and lower approximations induced from the sample data
2. the output concepts inherit or are influenced by the approximate aspects of the concepts input to their respective transducers
3. the output concepts also inherit the incompletely specified sufficient and necessary conditions in the local logical theory specified in part with the input concepts.

It is important to point out that the transducers represent a technique for combining both approximate and crisp knowledge. The flow of knowledge through a transducer generally increases the precision of the output concept. The definition can continually be elaborated upon both directly, by modifying the local theory, and indirectly via the modification of concept definitions on which it is recursively dependent or through retraining of the primitive concepts through various machine learning techniques.

7.2 An Introductory Example

In this section, we provide an example for a single approximation transducer describing some simple relationships between objects on a road. Assume we are provided with the following rough relations:

- $V(x, y)$ – there is a visible connection between objects x and y
- $S(x, y)$ – the distance between objects x and y is small
- $E(x, y)$ – objects x and y have equal speed.

We can assume that these relations were acquired using a supervised machine learning technique where sample data was generated from video logs provided by an UAV when flying over a particular road system populated with traffic, or that the relations were defined as part of an approximation tree using other approximation transducers.

Suppose we would like to define a new relation C denoting that its arguments, two objects on the road, are connected. It is assumed that we, as knowledge engineers or domain experts, have some knowledge of this concept. Consider, for example, the following local theory $T(C; V, S, E)$ approximating C:[1]

$$\forall x, y.[V(x, y) \rightarrow C(x, y)] \tag{7.1}$$

$$\forall x, y.[C(x, y) \rightarrow (S(x, y) \wedge E(x, y))]. \tag{7.2}$$

The former provides a sufficient condition for C and the latter a necessary condition. Imprecision in the definition is caused by the following facts:

- the input relations V, S and E are imprecise (rough)
- the theory $T(C; V, S, E)$ does not describe relation C precisely, as there are many possible models for C.

We then accept the least model for C w.r.t. the theory $T(C; V^+, S^+, E^+)$ as the lower approximation of C and the greatest model for C w.r.t. the theory $T(C; V^\oplus, S^\oplus, E^\oplus)$ as the upper approximation of C.

It can now easily be observed (and, in fact, be computed efficiently), that one can generate the following definitions of the lower and upper approximations of C:

$$\forall x, y.[C^+(x, y) \equiv V^+(x, y)] \tag{7.3}$$

$$\forall x, y.[C^\oplus(x, y) \equiv (S^\oplus(x, y) \wedge E^\oplus(x, y))]. \tag{7.4}$$

Relation C can then be used, e.g., while querying the rough knowledge database containing this approximation tree or for defining new approximate concepts, provided that it is coherent with the database contents. In this case, the coherence conditions, which guarantee the consistency of the generated relation with the rest of the database (approximation tree), are expressed by the following formulas:

$$\forall x, y.[V^+(x, y) \rightarrow (S^+(x, y) \wedge E^+(x, y))]$$

$$\forall x, y.[V^\oplus(x, y) \rightarrow (S^\oplus(x, y) \wedge E^\oplus(x, y))].$$

The coherence conditions can also be generated automatically in an efficient manner provided certain syntactic constraints are applied to the local theories in an approximation transducer.

[1] Note that semicolon in $T(C; V, S, E)$ is used to separate target relations from input relations used for approximation.

7.3 Approximation Transducers

As stated in the introduction, an approximation transducer provides a means of generating or defining an approximate relation (the output) in terms of other approximate relations (the input) using various dependencies between the input and the output.[2] The set of dependencies is in fact a logical theory where each dependency is represented as a logical formula in a first-order logical language. Syntactic restrictions can be placed on the logical theory to insure efficient generation of output.

Since we are dealing with approximate relations, both the input and output are defined in terms of upper and lower approximations. It is not necessary to restrict the logical theory to just the relations specified in the input and output for a particular transducer. Other relations may be used since they are assumed to be defined or definitions can be generated simultaneously with the generation of the particular output in question. In other words, it is possible to define an approximation network rather than a tree, but for this presentation, we will stick to the tree-based approach. The network approach is particularly interesting because it allows for limited forms of feedback across abstraction levels in the network.

The main idea is depicted in Figure 7.1. Suppose one would like to define an approximation of a relation R in terms of a number of other approximate relations $R_1, \ldots, R_k$. It is assumed that $R_1, \ldots, R_k$ consist of either primitive relations acquired via a machine learning phase or approximate relations that have been generated recursively via other transducers or combinations of transducers.

The local theory $T(R; R_1, \ldots, R_k)$ is assumed to contain logical formulas relating the input to the output and can be acquired through a knowledge acquisition process with domain experts or even through the use of inductive logic programming techniques. Generally the formulas in the logical theory are provided in the form of rules representing some sufficient and necessary conditions for the output relation in addition to possibly other conditions. The local theory should be viewed as a logical template describing a dependency structure between relations.

The actual transduction process which generates the approximate definition of relation R uses the logical template and contextualizes it with the actual approximate relations provided as input. The transduction process results in a definition of both the upper and lower approximation of R as follows,

- The lower approximation is defined as the least model for R w.r.t. the theory $T^+(R; R_1, \ldots, R_k)$

[2] The technique also works for one or more approximate relations being generated as output, but for clarity of presentation, we describe the techniques using a single output relation.

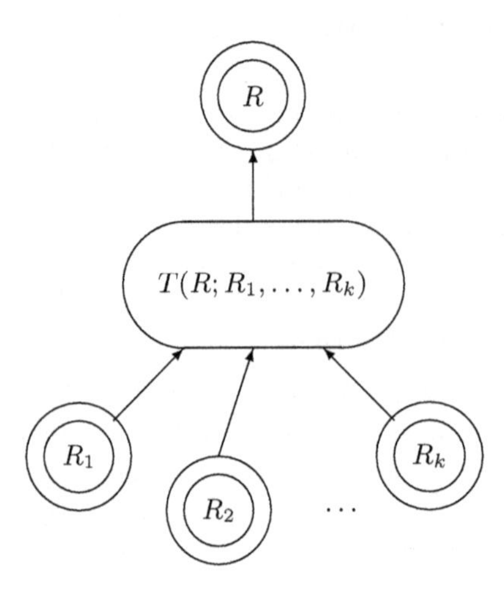

Fig. 7.1. Transformation of rough relations by first-order theories.

- and the upper approximation is defined as the greatest model for R w.r.t. the theory $T^{\oplus}(R; R_1, \ldots, R_k)$,

where $T^{+}(R; R_1, \ldots, R_k)$ and $T^{\oplus}(R; R_1, \ldots, R_k)$ denote theories obtained from T by replacing crisp relations by their corresponding approximations. As a result one obtains an approximation of R defined as a rough relation. Note that appropriate syntactic restrictions are placed on the theory so coherence conditions can be generated which guarantee the existence of the least and the greatest model of the theory and its consistency with the approximation tree in which its transducer is embedded. For details, see Section 7.4.

Implicit in the approach is a notion of abstraction hierarchies where one can recursively define more abstract approximate relations in terms of less abstract approximations by combining different transducers. The result is one or more approximation trees. This intuition has some similarity with the idea of layered machine learning. The technique also provides a great deal of locality and modularity in representation although it does not force this on the user since networks violating locality can be constructed. One can also view an approximation transducer or sub-tree of approximation transducers as simple or complex agents responsible for the management of particular relations and their dependencies. This idea is covered in Chapter 9.

The ability to continually apply machine learning techniques to the primitive relations in the network and to continually modify the logical theories which are constituent parts of transducers provides a great deal of elaboration tolerance and elasticity in the knowledge representation structures.

7.4 Approximation Transducer Semantics and Computation Mechanism

Our specific target is to define a new relation, say R, in terms of some additional relations $R_1, \ldots, R_n$ and a local logical theory $T(R; R_1, \ldots, R_n)$ representing knowledge about R and its relation to $R_1, \ldots, R_n$. The output of the transduction process results in a definition of R^+, the lower approximation of R, as the least model of $T^+(R; R_1, \ldots, R_n)$ and $R^\oplus$, the upper approximation of R, as the greatest model of $T^\oplus(R; R_1, \ldots, R_n)$. The following problems must be addressed:

- is $T(R; R_1, \ldots, R_n)$ consistent with the database?
- does the least and greatest relation R^+ and $R^\oplus$ exist, satisfying

  $T^+(R; R_1, \ldots R_n)$ and $T^\oplus(R; R_1, \ldots R_n)$,

  respectively?
- is the complexity of the mechanisms used to answer the above questions and to calculate suitable approximations R^+ and $R^\oplus$ reasonable from a pragmatic perspective?

In general, consistency is not guaranteed. Moreover, the above problems are generally NPTIME-complete (over finite models). However, a rich class of formulas can be isolated for which the consistency problem and the other problems can be resolved in PTIME. In what follows, we show that a subset of semi-Horn formulas guarantees the following:

- the coherence conditions for $T(R; R_1, \ldots, R_n)$ can be computed and checked in polynomial time
- the least and the greatest relations R^+ and $R^\oplus$, satisfying $T^+(R; R_1, \ldots, R_n)$ and $T^\oplus(R; R_1, \ldots, R_n)$, respectively, always exist provided that the coherence conditions are satisfied
- the time and space complexity of calculating suitable approximations R^+ and $R^\oplus$ is polynomial w.r.t. the size of the database and that of calculating their symbolic definitions is polynomial in the size of the theory $T(R; R_1, \ldots, R_n)$.

In view of these positive results, we will restrict the set of formulas used in local theories in transducers to (finite) conjunctions of semi-Horn rules as defined in Section 2.6. All theories considered in the rest of the chapter are then assumed to consist of semi-Horn rules. We also accept the following notational convention.

Convention 7.4.1. For the sake of simplicity we assume from now on that a theory defines only one intensional rough relation. We shall use notation $T(R; R_1, \ldots, R_n)$ to indicate that R is approximated by T (i.e., we assume in such a case that $R_1, \ldots, R_n$ are extensional or intensional and R is intensional).

We also write $T^+(R; R_1, \ldots, R_n)$ (or T^+, in short) to denote theory T with all symbols R_i occurring positively substituted by $R_i{}^+$ and occurring negatively, by R_i^-. Similarly we write $T^\oplus(R; R_1, \ldots, R_n)$ (or $T^\oplus$, in short) to denote theory T with all symbols R_i, $1 \le i \le n$, occurring positively substituted by $R_i{}^\oplus$ and occurring negatively by $R_i{}^\ominus$.

We often write rules without initial universal quantifiers ($\forall x$ of (2.2) or (2.3)), understanding that the rules are always universally quantified. □

The following lemmas (Lemma 7.4.2 and 7.4.3) provide us with a formal justification of Definition 7.4.4 which follows. Let us first deal with non-recursive rules.[3]

Lemma 7.4.2. Assume that $T(R; R_1, \ldots, R_n)$ consists of rules of the following forms:

$$\forall \bar{x}.[R(\bar{x}) \rightarrow \Phi_i(R_1, \ldots, R_n)] \tag{7.5}$$

$$\forall \bar{x}.[\Psi_j(R_1, \ldots, R_n) \rightarrow R(\bar{x})] \tag{7.6}$$

for $i \in I$, $j \in J$, where I, J are finite, nonempty sets and for all $i \in I$ and $j \in J$, formulas Φ_i and Ψ_j do not contain occurrences of R. Then there exist the least and the greatest R satisfying (7.5) and (7.6). The least such R is defined by the formula:

$$R(\bar{x}) \equiv \bigvee_{j \in J} \Psi_j(R_1, \ldots, R_n) \tag{7.7}$$

and the greatest such R is defined by the formula

$$R(\bar{x}) \equiv \bigwedge_{i \in I} \Phi_i(R_1, \ldots, R_n) \tag{7.8}$$

provided that the following coherence condition[4] is satisfied in the database:

[3] In fact, Lemma 7.4.2 follows easily from Lemma 7.4.3 by observing that fixpoint formulas (7.12), (7.13) and (7.14) reduce in this case to first-order formulas (7.7), (7.8) and (7.9), respectively. However, reductions to classical first-order formulas are worth a separate treatment as these are less complex and easier to deal with.

[4] The coherence conditions reflects the intuition that a lower approximation of a concept is a subset of its upper approximation.

$$\forall \bar{x}. \left[\bigvee_{j \in J} \Psi_j(R_1, \ldots, R_n) \rightarrow \bigwedge_{i \in I} \Phi_i(R_1, \ldots, R_n) \right]. \tag{7.9}$$

□

Then in the case of recursive theories we can prove the following lemma.

Lemma 7.4.3. Assume that $T(R; R_1, \ldots, R_n)$ consists of the following rules:

$$\forall \bar{x}.[R(\bar{x}) \rightarrow \Phi_i(R, R_1, \ldots, R_n)] \tag{7.10}$$

$$\forall \bar{x}.[\Psi_j(R, R_1, \ldots, R_n) \rightarrow R(\bar{x})] \tag{7.11}$$

for $i \in I$, $j \in J$, where I, J are finite, nonempty sets. Then there exist the least and the greatest R satisfying formulas (7.10) and (7.11). The least such R is defined by the formula:

$$R(\bar{x}) \equiv \text{LFP}\, R(\bar{x}).[\bigvee_{j \in J} \Psi_j(R, R_1, \ldots, R_n)] \tag{7.12}$$

and the greatest such R is defined by the formula:

$$R(\bar{x}) \equiv \text{GFP}\, R(\bar{x}).[\bigwedge_{i \in I} \Phi_i(R, R_1, \ldots, R_n)] \tag{7.13}$$

provided that the following coherence condition holds:

$$\forall \bar{x}. \left[\text{LFP}\, R(\bar{x}).[\bigvee_{j \in J} \Psi_j(R, R_1, \ldots, R_n)] \rightarrow \text{GFP}\, R(\bar{x}).[\bigwedge_{i \in I} \Phi_i(R, R_1, \ldots, R_n)] \right]. \tag{7.14}$$

□

The following definition provides us with a semantics of semi-Horn rules used as local theories in approximation transducers.

Definition 7.4.4. *Let B be a rough relational database, $R_1, \ldots, R_n$ be relation symbols, R be an intensional relation symbol, and let $T(R; R_1, \ldots, R_n)$ be a first-order theory expressed by rules of the form (7.10) and/or (7.11) (respectively (7.5) and/or (7.6)).*

By an approximation transducer *with input $R_1, \ldots, R_n$, output R and the* local transducer theory *T we understand a mapping providing lower and upper approximations of R on the basis of input relations and T as follows:*

- *the lower approximation of R is defined as the least relation R satisfying $T(R; R_1, \ldots, R_n)$, i.e., the relation defined by formula* (7.7)$^+$ *or* (7.12)$^+$*, respectively, with $R_1, \ldots, R_n$ substituted as described in Convention 7.4.1*
- *the upper approximation of R is defined as the greatest relation R satisfying $T(R; R_1, \ldots, R_n)$, i.e., the relation defined by formula* (7.8)$^\oplus$ *or* (7.13)$^\oplus$ *with $R_1, \ldots, R_n$ substituted as described in Convention 7.4.1,*

provided that the respective coherence conditions (7.9)$^+$ *or* (7.14)$^+$*, for the lower approximation, and* (7.9)$^\oplus$ *or* (7.14)$^\oplus$*, for the upper approximation, are satisfied in database B.*

By an approximation tree *we mean a tree built using approximation transducers.* □

Observe that we place a number of restrictions on this definition that can be relaxed, such as restricting the use of relation symbols in the local theory of the transducer to be crisp. This excludes use of references to constituent components of other rough relations. In addition, since the output relation of a transducer can be represented explicitly in the rough relational database, approximation trees consisting of combinations of transducers are well-defined.

7.5 The Complexity of the Approach

This framework is presented in the context of relational databases with finite domains with some principled generalizations. In addition, both explicit definitions of approximations to relations and associated coherence conditions are expressed in terms of classical first-order or fixpoint formulas. Consequently, computing the approximations and checking coherence conditions can be done in time polynomial in the size of the database.

In addition, the size of explicit definitions of approximations and coherence conditions is linear in the size of the local theories defining the approximations. Consequently, the proposed framework is acceptable from the point of view of a formal complexity analysis. This serves as a useful starting point for efficient implementation of the techniques. It is clear though, that for very large databases of this type, additional optimization methods would be desirable.

7.6 A Congestion Example

In this section, we provide an example from the UAV–traffic domain which defines the concept of traffic congestion using the proposed framework. We shall use the following relations and constants:

- $In(x, l)$ – denotes whether a vehicle x is in a road segment l
- $Speed(x, z)$ – denotes the approximate speed of x, where $z \in \{\mathsf{low}, \mathsf{medium}, \mathsf{high}, \mathsf{unknown}\}$
- $Distance(x, y, z)$ – denotes the approximate distance between vehicles x and y, where $z \in \{\mathsf{small}, \mathsf{medium}, \mathsf{large}, \mathsf{unknown}\}$
- $Between(z, x, y)$ – denotes whether vehicle z is between vehicles x and y
- $Number(x, y, z)$ – denotes the approximate number of vehicles between vehicles x and y occurring in the region of interest, where $z \in \{\mathsf{small}, \mathsf{medium}, \mathsf{large}, \mathsf{unknown}\}$
- $TrafficCong(1)$ – denotes whether there is traffic congestion in the observed road segment l.

We define traffic congestion by the following formula:

$$\begin{aligned}
&TrafficCong(l) \equiv \\
&\quad \exists x, y.[In(x, l) \wedge In(y, l) \wedge Number(x, y, \mathsf{large}) \wedge \\
&\quad \forall z.(Between(z, x, y) \rightarrow Speed(z, \mathsf{low})) \wedge \\
&\quad \forall z.(Between(z, x, y) \rightarrow \exists t.(Distance(z, t, \mathsf{small})))].
\end{aligned} \tag{7.15}$$

Observe that formula (7.15) contains concepts that are not defined precisely ($Speed, Distance, Number$). However, we assume that the underlying database contains approximations of these concepts. We can then use the approximated concepts and replace formula (7.15) with the following two formulas representing the lower and upper approximation of the target concept:[5]

$$\begin{aligned}
&TrafficCong^{+}(l) \equiv \\
&\quad \exists x, y.[In^{+}(x, l) \wedge In^{+}(y, l) \wedge Number^{+}(x, y, \mathsf{large}) \wedge \\
&\quad \forall z.(Between^{\oplus}(z, x, y) \rightarrow Speed^{+}(z, \mathsf{low})) \wedge \\
&\quad \forall z.(Between^{\oplus}(z, x, y) \rightarrow \exists t.Distance^{+}(z, t, \mathsf{small}))]
\end{aligned} \tag{7.16}$$

$$\begin{aligned}
&TrafficCong^{\oplus}(l) \equiv \\
&\quad \exists x, y.[In^{\oplus}(x, l) \wedge In^{\oplus}(y, l) \wedge Number^{\oplus}(x, y, \mathsf{large}) \wedge \\
&\quad \forall z.(Between^{+}(z, x, y) \rightarrow Speed^{\oplus}(z, \mathsf{low})) \wedge \\
&\quad \forall z.(Between^{+}(z, x, y) \rightarrow \exists t.Distance^{\oplus}(z, t, \mathsf{small}))].
\end{aligned} \tag{7.17}$$

These formulas can be automatically generated using the techniques described previously.

It can now be observed that formula (7.15) defines a cluster of situations that can be considered as traffic congestions. Namely, small deviations of data do not have a substantial impact on the target concept. This is a consequence

[5] Observe that $(p \rightarrow q)^{+} \equiv (p^{\oplus} \rightarrow q^{+})$ and $(p \rightarrow q)^{\oplus} \equiv (p^{+} \rightarrow q^{\oplus})$.

of the fact that (7.15) refers to values that are also approximated such as low, small and large. Thus small deviations of vehicle speed or distance between vehicles usually do not change the qualitative classification of these notions.

Let us denote deviations of data by *dev* with suitable indices. Now, assuming that the deviations satisfy the following properties:

$$\begin{aligned}
&\langle x', l' \rangle \in dev_{In}(x, l) \equiv [In^{+}(x, l) \rightarrow In^{+}(x', l')] \qquad (7.18)\\
&x' \in dev_{Speed}(x) \equiv [Speed^{+}(x, \mathsf{low}) \rightarrow Speed^{+}(x', \mathsf{low})]\\
&\langle x', y' \rangle \in dev_{Number}(x, y) \equiv\\
&\quad [Number^{+}(x, y, \mathsf{large}) \rightarrow Number^{+}(x', y', \mathsf{large})]\\
&\langle x', y' \rangle \in dev_{Distance}(x, y) \equiv\\
&\quad [Distance^{+}(x, y, \mathsf{small}) \rightarrow Distance^{+}(x', y', \mathsf{small})]\\
&\langle z', x', y' \rangle \in dev_{Between}(z, x, y) \equiv\\
&\quad [Between^{+}(z, x, y) \rightarrow Between^{+}(z', x', y')],
\end{aligned}$$

one can conclude that:

$$[\mathit{TrafficCong}^{+}(l) \land l' \in dev_{\mathit{TrafficCong}}(l)] \rightarrow \mathit{TrafficCong}^{+}(l'),$$

where $dev_{\mathit{TrafficCong}}(l)$ denotes the set of all situations obtained by deviations of l satisfying conditions expressed by (7.18).

The above reasoning schema is then robust w.r.t. small deviations of input concepts. In fact, any approximation transducer defined using purely logical means enjoys this property since small deviations of data, by not changing basic properties, do not change the target concept.

Note that relation *dev* in the above formulas should also be generated on the basis of particular data.

7.7 On the Approximation Quality of First-Order Theories

So far, we have focused on the generation of approximations to relations using local logical theories in approximation transducers and then building approximation trees from these basic building blocks. This immediately raises the interesting issue of viewing the approximate global theory itself as a conceptual unit. We can then ask what the approximation quality of a theory is and whether we can define qualitative or quantitative measures of the theory's approximation quality. If this is possible, then individual theories can be compared and assessed for their value under different reasoning contexts. One application of such an assessment tool would be to choose approximate theories for an application domain at the proper level of abstraction or detail, moving across the different levels of abstraction relative to the needs of the application. In this section, we provide a tentative proposal to compare the approximation quality of first-order theories.

7.7.1 Comparing Approximation Power of Semi-Horn Theories

Definition 7.7.1. *We say that a* theory $T_2(R)$ better approximates relation R than a theory $T_1(R)$ relative to a database B *and denote this by* $T_1(R) \leq_B T_2(R)$ *provided that, in database* B*, we have* $R_1^+ \subseteq R_2^+$ *and* $R_2^\oplus \subseteq R_1^\oplus$*, where for* $i = 1, 2$, R_i^+ *and* $R_i^\oplus$ *denote the lower and upper approximation of* R *defined by theory* T_i. □

Observe that the notion of a better approximation has a correspondence to information ordering. From the rough set perspective, a theory which better approximates a relation over another theory has the result of decreasing the boundary region,

Example 7.7.2. Let $CL(x, y)$ denote that objects x, y are close to each other, $SL(x, y)$ denote that x, y are on the same lane, $CH(x, y)$ denote that objects x, y can hit each other, and let $HR(x, y)$ denote that the relative speed of x and y is high. We assume that the lower and upper approximations of these relations can be extracted from data or are already defined in a rough database, B. Consider the following two theories approximating the concept $D(x, y)$ which denotes a dangerous situation caused by objects x and y:

- $T_1(D; CL, SL, CH)$ has two rules:

$$\begin{aligned} &\forall x, y.[(CL(x, y) \wedge SL(x, y)) \rightarrow D(x, y)] \\ &\forall x, y.[D(x, y) \rightarrow CH(x, y)] \end{aligned} \tag{7.19}$$

- $T_2(D; CL, SL, HR)$ has two rules:

$$\begin{aligned} &\forall x, y.[CL(x, y) \rightarrow D(x, y)] \\ &\forall x, y.[D(x, y) \rightarrow (HR(x, y) \wedge SL(x, y))]. \end{aligned} \tag{7.20}$$

Using Lemma 7.4.2, we can compute the following definitions of approximations of D:

- relative to theory $T_1(D; CL, SL, CH)$:

$$\begin{aligned} &\forall x, y.[{D^{(1)}}^+(x, y) \equiv (CL^+(x, y) \wedge SL^+(x, y))] \\ &\forall x, y.[{D^{(1)}}^\oplus(x, y) \equiv CH^\oplus(x, y)] \end{aligned} \tag{7.21}$$

- relative to theory $T_2(D; CL, SL, HR)$:

$$\forall x, y.[D^{(2)^+}(x, y) \equiv CL^+(x, y)]$$
$$\forall x, y.[D^{(2)^\oplus} \equiv (HR^\oplus(x, y) \wedge SL^\oplus(x, y))]. \tag{7.22}$$

Obviously $D^{(1)^+} \subseteq D^{(2)^+}$. If we additionally assume that in our domain of discourse (and consequently, in database B) $HR \cap SL \subseteq CH$ applies, we can also obtain the additional relation that $D^{(2)^\oplus} \subseteq D^{(1)^\oplus}$. Thus $T_1 \leq_B T_2$, which means that an agent possessing the knowledge implicit in T_2 is better equipped to approximate concept D than an agent possessing knowledge implicit in T_1. □

7.8 Bibliographic Notes

This chapter is mainly based on [52].

There has been very little work in traditional knowledge representation with the dynamics and management of knowledge structures. Some related work would include the development of belief revision and truth maintenance systems in addition to the notion of *elaboration tolerant* knowledge representation and the use of *contexts* as first-class objects introduced by McCarthy [121]. In these cases, the view pertaining to properties and relations is still quite traditional with little emphasis on the approximate and contextual character of knowledge. The assumed granularity of the primitive components of these knowledge structures, in which these theories and techniques are grounded, is still that of classical properties and relations in a formal logical context.

The concept of web of belief was introduced by Quine in [171].

The idea of layered machine learning is described, e.g., in [209].

It is sometimes convenient to use definitions of upper approximations for computing negative knowledge instead of positive knowledge. Such an approach has proven to be profitable in various applications, as reported, e.g., in [143, 218]

The methodology, where agents synthesizing more complex notions on a higher level, using data preprocessed by agents of lower layers, is strongly advocated in [127].

Proof of Lemma 7.4.2 follows easily, e.g., from Theorem 5.3 of [55] and proof of Lemma 7.4.3, e.g., from Theorem 5.2 of [55].

8

Weakest Sufficient and Strongest Necessary Conditions

8.1 Introduction

In the case of large data sets and knowledge databases one of the major concerns is the ability to react to events or queries in a reasonable and acceptable time. In particular, any real-time reasoning process has to be highly efficient. On the other hand, there is a trade-off between the accuracy of data/knowledge representation and effectiveness of querying knowledge databases and reasoning. In consequence, there is also a trade-off between the accuracy of data/knowledge representation and the response time of autonomous agents reacting on occurring events.

When applying the rough sets or other machine learning or data mining and knowledge discovery techniques one can substantially reduce the amount of data (see, e.g., Section 6.6 and Chapter 14 for some examples of possible approaches). In the current chapter we discuss the problem of approximating knowledge expressed by logical formulas and provide tools that allow one to understand what the approximations of data mean from the point of view of logic. The tools we apply are based on the notions of sufficient and necessary conditions that serve us as the required approximations.

Consider a formula A expressed in some logical language. Assume that one is interested in approximating A in a less expressive language, say L, which allows for more efficient reasoning. A sufficient condition of A, expressed in L, provides a lower approximation of A and a necessary condition of A, expressed in L, provides an upper approximation of A. Thus the weakest sufficient condition provides "the best" lower approximation and the strongest necessary condition provides "the best" upper approximation of A, expressed in the less expressive language.

Let us emphasize that sufficient and necessary conditions are vital for providing solutions to important problems appearing, e.g., in the areas of applications outlined below.

P. Doherty et al.: *Knowledge Representation Techniques*, Studfuzz **202**, 143–158 (2006)
www.springerlink.com

Building Communication Interfaces between Agents

In the case of distributed architectures it is often necessary to exchange information between agents. In large-scale applications it is unavoidable that different agents use different vocabularies or even different ontologies. For example, an agent specialized to supply information about weather usually, for the purpose of internal reasoning, uses some notions that are not known to other agents, e.g., to agents specialized in reasoning about geographical information. Similarly, many concepts known by the geographical information agents are not known by the weather agent. However, in order to communicate, the agents should have an interface built over a common language. We shall call such an interface a *communication interface between agents.*

Assume thus that agent M knows concepts (relations) $\bar{R}, \bar{S}$ and agent N knows concepts (relations) $\bar{S}, \bar{T}$, i.e., the common vocabulary of M and N consists of relations in $\bar{S}$. Suppose N asks query $A(S,T)$ to agent M, where $S \in \bar{S}$ and $T \in \bar{T}$. Of course, agent M does not know the concept T, but still has to provide a meaningful answer. One can consider at least the following policies for building the communication interface between M and N:

- M might approximate query $A(S,T)$ by "projecting out" the concept T it does not know and answer the query. In this case the resulting answer approximates the query in the sense that due to a more expressive language $A(S,T)$ can be more specific than $A'(S)$ which is obtained by removing T
- M might ask N to approximate the query $A(S,T)$ by requiring that N "projects out" T and supplies the approximated query fully understood by M
- M might ask N to approximate concept T and provide the approximation of T (which is a form of explanation provided by N). Based on the approximations of all concepts not understood by M and explained by N, M answers the query.

Of course, the dialog might be much more advanced here. It is worth observing, however, that due to different knowledge possessed by both agents, in each case the answer might be more or less accurate.

Note also that building interfaces between agents using different languages is of a great concern in the area of granular computing as will be discussed in Chapter 12.

Modularization and Information Hiding

In the process of designing software systems or components, *modularization* and *information hiding* are vital tools. Information should often be encapsulated and parts of it should remain hidden due to reasons of security or effectiveness of representing and manipulating knowledge whose contents and even structure dynamically changes in time.

In the case of deliberative components the information hiding is not only a matter of using a stronger or weaker vocabulary. When highly secure software systems interact with less secure systems it is important to detect what information is revealed to the less secure systems. In such a case all approximations occurring in reasoning done with a use of the interface should already be deducible by means of less secure systems. This suggests that at least more secure agents should not approximate the received queries by themselves, but rather demand that less secure agents provide the necessary approximations.

Knowledge Compilation and Theory Approximation

In many practical applications one would like to specify knowledge using highly expressive languages without bothering about the complexity of the formalism and of the reasoning it involves, and on the other hand, to have tools to efficiently manipulate the knowledge and keep the reasoning timely and effortless. For instance, circumscription discussed in Section 5.3, provides one with a highly expressive and natural formalism to represent knowledge. At the same time, reasoning with unrestricted circumscription is intractable. In pragmatic knowledge representation systems one either restricts the expressiveness of the representation language or uses incomplete or approximate reasoning machinery, or accepts nonstandard semantics for standard logical connectives like negation.

The term *knowledge compilation* refers then to the process of approximating more expressive and complex knowledge representation mechanisms by less expressive, but simpler and more efficient mechanisms. From the logical point of view, such a knowledge compilation can be approached in many ways. For instance, one can design specialized algorithms translating the higher level notions into lower level ones. Such algorithms are usually incomplete in the sense that they are not always successful in providing a translation even if a particular higher level concept can be actually represented by means of the lower level concepts. Another approach is to approximate theories and deal with approximated concepts rather than with precise theories and notions. In this chapter we follow the latter approach. One can, however, observe that in other parts of this book some second-order knowledge representation formalisms are reduced to substantially less complex predicate or fixpoint calculus by applying (in general incomplete) second-order quantifier elimination methods, thus the former approach is also used. Also note that weakest sufficient and strongest necessary conditions which are the basis of the solutions presented here, are always expressible in second-order logic, and not always in first-order or fixpoint calculus. We provide complete methods for classes of theories of a restricted syntax, but one can also use second-order quantifier elimination techniques to obtain results for more general cases, but without any guarantee to obtain the required results.

Abduction

Abduction is a form of reasoning "inverting" the more usual deductive reasoning. In deduction one is interested in obtaining facts that logically follow from a given background knowledge. Suppose now that a given conclusion does not follow from given knowledge (theory). One can then ask what are the weakest assumptions that, together with the existing knowledge, would make the conclusion true. Abduction is just the process of finding such assumptions.

Why is abduction vital in many pragmatic AI systems? Consider, for example, an autonomous agent that classifies objects, say vehicles, on the basis of camera images gathered in real time. Suppose the agent is not able to classify a given object. It is then reasonable for the agent to hypothesize that the object is, say, a small car, and ask what information is to be gathered in order to be convinced that the object is indeed a small car. On the basis of abductive reasoning the agent might find out, for example, that the camera image needs to be more accurate and then might generate and execute a plan to actually get more accurate images. Also generating a plan can be supported by asking questions pertaining to what immediate step would make the desired goal true, then what would make the step true and so on. Similar ideas can be applied in generating successor state axioms in robot domains, i.e., generating conditions on the initial state that make fulfilling the final goal feasible.

Another application of abduction depends on finding explanations for phenomena that occur in the observed reality. Suppose an agent observes a fact that does not follow from the current knowledge. It might then ask for explanations, i.e., diagnosis justifying the observed fact. Such a diagnosis can substantially contribute to the available knowledge and allow the agent to act properly.

Reasoning with Reduced Data Sets

One of the most important techniques developed in the context of approximating information systems by rough relations depends on generating reducts. This technique is heavily used in such topics, as machine learning, classifier construction, data mining and knowledge discovery (see Part III of this book). Reducts allow one to approximate large data sets defined with the use of many attributes, by using a relatively small number of the most relevant attributes. This, however, has a substantial impact on the knowledge representation layer, where one deals with theories expressed using those original attributes that are no longer present in the row data. In order to make the reasoning efficient and meaningful, one thus has to project out from the knowledge representation layer all the attributes that do not occur in the obtained reducts. In consequence, one needs to approximate theories serving knowledge representation purposes, too. Again, strongest necessary and weakest sufficient conditions are a powerful tool that can be used to achieve this important goal, as they correspond to lower and upper approximations considered in rough sets.

8.2 Weakest Sufficient and Strongest Necessary Conditions

In the following, we will be dealing with the predicate calculus with equality. Recall that we limit ourselves to finite theories. Since each such theory is logically equivalent to the conjunction of its axioms, in the sequel, we shall never distinguish between a theory T and the sentence which is the conjunction of all axioms of T.

The following are definitions for necessary and sufficient conditions of a formula A relativized to a subset $\bar{P}$ of relation symbols under a theory T.

Definition 8.2.1. *By a* necessary condition of a formula A on the set of relation symbols $\bar{P}$ under theory T *we shall understand any formula B containing only symbols in $\bar{P}$ such that $T \models A \to B$. It is the* strongest necessary condition, *denoted by* $\textsc{Snc}(A;T;\bar{P})$ *if, additionally, for any necessary condition C of A on $\bar{P}$ under T, we have $T \models B \to C$.* □

Definition 8.2.2. *By a* sufficient condition of a formula A on the set of relation symbols $\bar{P}$ under theory T *we shall understand any formula B containing only symbols in $\bar{P}$ such that $T \models B \to A$. It is the* weakest sufficient condition, *denoted by* $\textsc{Wsc}(A;T;\bar{P})$ *if, additionally, for any sufficient condition C of A on $\bar{P}$ under T, we have $T \models C \to B$.* □

The set $\bar{P}$ in Definitions 8.2.1 and 8.2.2 is referred to as the *target language.* If $\bar{P}'$ is a set of relation symbols then $\textsc{Snc}(A;T;-\bar{P}')$ and $\textsc{Wsc}(A;T;-\bar{P}')$ indicate that the target language $\bar{P}$ consists of all relation symbols of the considered language, except for those in $\bar{P}'$.

Figure 8.1 shows the relationships between necessary conditions and sufficient conditions. One can easily observe that:

- $\textsc{Wsc}(A;T;\bar{P})$ is the lower approximation of concept A expressed in terms of a language containing symbols $\bar{P}$; other sufficient conditions expressed in this language are included in $\textsc{Wsc}(A;T;\bar{P})$
- $\textsc{Snc}(A;T;\bar{P})$ is the upper approximation of concept A expressed in terms of a language containing symbols $\bar{P}$; other necessary conditions expressed in this language include $\textsc{Snc}(A;T;\bar{P})$.

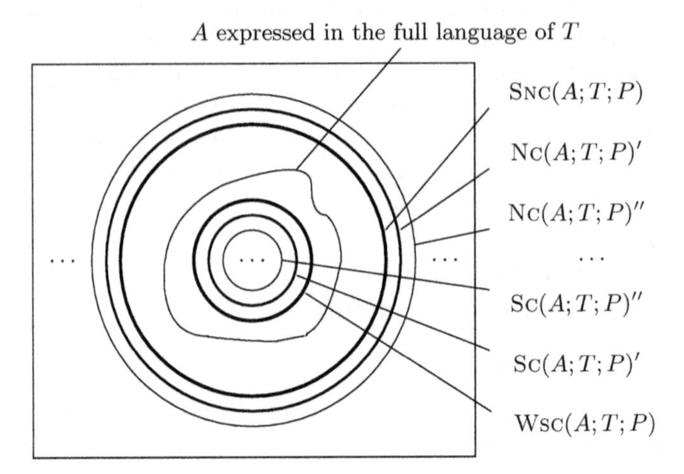

Fig. 8.1. The relationships between necessary and sufficient conditions. Nc(A;T;P) and Sc(A;T;P) stand for necessary and sufficient conditions, respectively.

To provide some additional intuition as to how these definitions can be used, consider the theory

$$T = \{\forall x.[HasWheels(x) \rightarrow CanMove(x)], \\ \forall x.[Car(x) \rightarrow HasWheels(x)]\},$$

and the formula $A \equiv \forall x.CanMove(x)$. Clearly $T \not\models A$. Quite often, it is useful to hypothesize a preferred explanation B for A under a theory T where $T \wedge B \models A$ and B is minimal in the sense of not being overly specific, where the explanation is constrained to a particular subset $\bar{P}$ of symbols in the vocabulary. Clearly, the weakest sufficient condition B for the formula A on $\bar{P}$ under T provides the basis for a minimal preferred explanation of A where $T \models B \rightarrow A$. In the case of $\bar{P} = \{HasWheels\}$, the weakest sufficient condition is $B = \forall x.HasWheels(x)$, as is the case for $\bar{P} = \{HasWheels, Car\}$. Generating abductive hypotheses is just one application of weakest sufficient conditions. As discussed in Section 8.1, there are many other applications which require the generation of weakest sufficient or strongest necessary conditions, several of which are described in Section 8.5.

8.3 The Propositional Case

In this section, we define strongest necessary and weakest sufficient conditions for propositional formulas under propositional theories as formulas with quantification over propositional variables, show how the quantifier elimination techniques of Section 2.9 can be applied and provide complexity results for the technique.

We use the notation $\exists p.A$ and $\forall p.A$, where p is a propositional variable and A is a formula, with the following meaning:

$$\begin{aligned} \exists p.A &\stackrel{\text{def}}{\equiv} A[p := \text{TRUE}] \vee A[p := \text{FALSE}] \\ \forall p.A &\stackrel{\text{def}}{\equiv} A[p := \text{TRUE}] \wedge A[p := \text{FALSE}]. \end{aligned} \tag{8.1}$$

We start with the following lemma.

Lemma 8.3.1. *For any formula A, any set of propositions $\bar{P}$ and theory T,*

1. $\text{SNC}(A; T; \bar{P})$ *is defined by* $\exists \bar{q}.[T \wedge A]$
2. $\text{WSC}(A; T; \bar{P})$ *is defined by* $\forall \bar{q}.[T \to A]$,

where $\bar{q}$ consists of all propositions appearing in T or A, but not in $\bar{P}$. □

The quantifiers over propositions can be automatically eliminated using the DLS algorithm (for references, see Section 2.11). For instance, all eliminations in Example 8.3.4 can be made using this algorithm.

Theorem 2.9.2 reduces in the propositional case to Proposition 8.3.2. It is worth emphasizing here that propositional fixpoint formulas are equivalent to propositional formulas.[1]

Proposition 8.3.2. Assume that the propositional formula A is positive w.r.t. proposition p.

- if propositional formula B is negative w.r.t. p then

$$\exists p.\ [A(p) \to p] \wedge [B(p)] \ \equiv\ B[p := \text{LFP}\, p.A(p)] \tag{8.2}$$

- if B is positive w.r.t. p then

$$\exists p.[p \to A(p)] \wedge [B(p)] \ \equiv\ B[p := \text{GFP}\, p.A(p)]. \tag{8.3}$$

□

[1] In the first iteration towards the fixpoint, one replaces p in A with false. In the next disjunct, p in A is replaced by this result. The fixpoint, a propositional formula, is always reached in a few iterations. Of course, the same result can be obtained by applying equivalence (8.1), but in general equivalences (8.2) and (8.3) provide us with a more efficient method.

Observe that in the case when an input formula is a conjunction of propositional semi-Horn formulas of the form in the lefthand side of (8.2) or a conjunction of formulas of the form in the lefthand side of (8.3), the length of the resulting formula is, in the worst case, $O(n^2)$, where n is the size of the input formula. Otherwise the result might be of exponential length.

Given these results, if $T \wedge A$ or $T \rightarrow A$ in Lemma 8.3.1 can be transformed into an equivalent conjunction of semi-Horn formulas then the propositional equivalent of the $\text{SNC}(A; T; \bar{P})$ and $\text{WSC}(A; T; \bar{P})$ can be generated efficiently. If not, the propositional equivalent can still be generated but not necessarily in a tractable manner.

Remark 8.3.3. Weakest sufficient and strongest necessary conditions for propositional formulas are related to prime implicants and implicates, respectively. Generating prime implicants and implicates for propositional formulas is intractable and, in the general case, the same applies for weakest sufficient and strongest necessary conditions. □

Example 8.3.4. Consider the following examples.

1. $T_1 = \{q \rightarrow (p_1 \wedge p_2)\}$. According to Lemma 8.3.1,
   - $\text{SNC}(q; T_1; \{p_1, p_2\})$ is defined by the formula

     $\exists q.[(q \rightarrow (p_1 \wedge p_2)) \wedge q]$,

     which, according to Proposition 8.3.2, is logically equivalent to

     $\text{GFP}\, q.(p_1 \wedge p_2)$,

     i.e., to $(p_1 \wedge p_2)$,
   - condition $\text{SNC}(q; T_1; \{p_1\})$ is defined by the formula

     $\exists q.\exists p_2.[(q \rightarrow (p_1 \wedge p_2)) \wedge q]$,

     which, according to Proposition 8.3.2, is logically equivalent to p_1 (observe that p_2 is equivalent to the semi-Horn formula $\text{TRUE} \rightarrow p_2$).
2. $T_2 = \{q \rightarrow (p_1 \vee p_2)\}$. We have that
   - $\text{SNC}(q; T_2; \{p_1, p_2\})$ is defined by the formula $\exists q.[(q \rightarrow (p_1 \vee p_2)) \wedge q]$, which, according to Proposition 8.3.2, is logically equivalent to $(p_1 \vee p_2)$.
   - $\text{SNC}(q; T_2; \{p_1\})$ is defined by the formula

     $\exists q.\exists p_2.[(q \rightarrow (p_1 \vee p_2)) \wedge q]$,

     which is logically equivalent to TRUE.
3. $T_3 = \{(p \wedge q) \rightarrow s\}$. The formula $\text{SNC}(p \wedge q; T_3; \{s\})$ is equivalent to

$$\exists p.\exists q.[((p \wedge q) \rightarrow s) \wedge (p \wedge q)],$$

which, according to Proposition 8.3.2, is logically equivalent to s. □

In summary, propositional equivalents of strongest necessary or weakest sufficient conditions can be generated for any propositional formula and theory. In the case that the conjunction of both is in semi-Horn form, the computation is guaranteed to be tractable.

8.4 The First-Order Case

In this section, we generalize the results in section 8.3 to the first-order case using primarily the same techniques, but with quantification over relational symbols.

Lemma 8.4.1. *For any formula A, any set of relation symbols $\bar{P}$ and a closed theory T:*[2]

1. $\text{SNC}(A;T;\bar{P})$ *is defined by* $\exists \bar{X}.[T \wedge A]$
2. $\text{WSC}(A;T;\bar{P})$ *is defined by* $\forall \bar{X}.[T \rightarrow A]$,

where $\bar{X}$ consists of all relation symbols appearing in T or A, but not in $\bar{P}$. □

Observe that a second-order quantifier over the relational variables $\bar{X}$ can be eliminated from any semi-Horn formula w.r.t. $\bar{X}$ (as shown in Section 4.6). In such cases the resulting formula is a fixpoint formula. If the formula is non-recursive, then the resulting formula is a first-order formula. The input formula can also be a conjunction of semi-Horn formulas of the form (2.6) or a conjunction of semi-Horn formulas of the form (2.7). On the other hand, one should be aware that in other cases the reduction is not guaranteed. Thus the elimination of second-order quantifiers is guaranteed for any formula of the form $\exists \bar{X}.[T \wedge A]$, where $T \wedge A$ is a conjunction of semi-Horn formulas w.r.t. all relational variables in $\bar{X}$.[3] Observe also, that in the case when an input formula is a conjunction of semi-Horn formulas of the form (2.6) or a conjunction of formulas of the form (2.7), the length of the resulting formula is, in the worst case, $O(n^2)$, where n is the size of the input formula.

Example 8.4.2. Consider the following examples

[2] In fact, it suffices to assume that the set of free variables of T is disjoint with the set of free variables of A.

[3] For universal quantification, $\forall \bar{X}.A$, one simply negates the formula $(\exists \bar{X}.\neg A)$, and assuming $\neg A$ can be put into semi-Horn form, one eliminates the existential quantifiers and negates the result.

1. $T_4 = \{\forall x.[Ab(x) \to (Bird(x) \wedge \neg Flies(x))]\}$.

   Consider $\textsc{Snc}(Ab(z); T_4; \{Bird, Flies\})$. According to Lemma 8.4.1, it is equivalent to

$$\exists Ab.[\forall x.(Ab(x) \to (Bird(x) \wedge \neg Flies(x))) \wedge Ab(z)]. \tag{8.4}$$

   By Lemma 2.9.2, formula (8.4) is equivalent to $(Bird(z) \wedge \neg Flies(z))$.

2. $T_5 = \{\forall x.[Parent(x) \to \exists z.(Father(x, z) \vee Mother(x, z))]\}$.

   Consider $\textsc{Snc}(Parent(y); T_5; \{Mother\})$. According to Lemma 8.4.1, it is equivalent to

$$\begin{aligned}&\exists Parent.\exists Father.[\forall x.(Parent(x) \to \\ &\quad \exists z.(Father(x, z) \vee Mother(x, z)) \wedge Parent(y)].\end{aligned} \tag{8.5}$$

   Formula (8.5) is not in the form required in Lemma 2.9.2, but the DLS algorithm eliminates the second-order quantifiers and results in the equivalent formula TRUE, which is the required strongest necessary condition. Consider now $\textsc{Snc}(Parent(y) \wedge \forall u, v.(\neg Father(u, v)); T_5; \{Mother\})$. It is equivalent to

$$\begin{aligned}&\exists Parent.\exists Father.[\forall x.(Parent(x) \to \\ &\quad \exists z.(Father(x, z) \vee Mother(x, z)) \wedge \\ &\qquad Parent(y) \wedge \forall u, v.(\neg Father(u, v))],\end{aligned} \tag{8.6}$$

   i.e., after eliminating second-order quantifiers, to $\exists z.Mother(y, z)$. □

In summary, for the non-recursive semi-Horn fragment of first-order logic, the strongest necessary or weakest sufficient condition for a formula A and theory T are guaranteed to be reducible to compact first-order formulas. For the recursive case, the strongest necessary and weakest sufficient conditions are guaranteed to be reducible to fixpoint formulas. When one might want to use strongest necessary or weakest sufficient conditions to query a knowledge database, this case is still tractable. The techniques may still be used for the full first-order case, but neither reduction nor complexity results are guaranteed, although the algorithm will always terminate.

8.5 Applications

In this section, we demonstrate the use of the techniques by applying them to a number of potentially useful application areas.

8.5.1 Communicating Agents

Agents communicating, e.g., via the Internet have to use the same language to understand each other or use mediators to translate between languages of different expressive power.

Assume an agent A wants to ask a query Q to agent B. Suppose the query can be asked using terms $\bar{R}, \bar{S}$ such that the terms from $\bar{S}$ are not in agent B's vocabulary. Let $T(\bar{R}, \bar{S})$ be a theory describing some relationships between $\bar{R}$ and $\bar{S}$. It is then natural for agent A to first compute the approximations given by the weakest sufficient condition $\text{Wsc}(Q; T(\bar{R}, \bar{S}); \bar{R})$ and the strongest necessary condition $\text{Snc}(Q; T(\bar{R}, \bar{S}); \bar{R})$ with the target language restricted to $\bar{R}$ and then to replace the original query with the computed approximations. The new queries might not be as precise as the previous one, but they are the best that can be asked under the given assumptions. The following example illustrates this idea.

Example 8.5.1. Assume an agent Ag wants to select from a database all persons x such that $High(x) \vee Silny(x)$ holds. Assume further, that both agents know the terms $High$ and $Sound$. Suppose that the database agent does not know the term $Silny$.[4] Suppose, further that Ag lives in a world in which the condition $\forall y.[Silny(y) \to Sound(y)]$ holds. It is then natural for Ag to use

$$\text{Wsc}(High(x) \vee Silny(x); \forall y.[Silny(y) \to Sound(y)]; \{High, Sound\})$$
$$\text{Snc}(High(x) \vee Silny(x); \forall y.[Silny(y) \to Sound(y)]; \{High, Sound\})$$

as an approximation to the original query, one that will be understood by the database agent that will process the query. According to Lemma 8.4.1 these conditions are equivalent to

$$\forall Silny.\, \{\forall y.[Silny(y) \to Sound(y)] \to (High(x) \vee Silny(x))\} \tag{8.7}$$
$$\exists Silny.\, \{\forall y.[Silny(y) \to Sound(y)] \wedge (High(x) \vee Silny(x))\}\,. \tag{8.8}$$

By applications of Theorem 2.9.2, formula (8.7) is equivalent to $High(x)$, and (8.8) is equivalent to $High(x) \vee Sound(x)$. Thus, in the given target language and background theory, the set of tuples surely satisfying the original query are those satisfying $High(x)$ and those that might satisfy the original query are those satisfying $High(x) \vee Sound(x)$. □

8.5.2 Theory Approximation

The concept of approximating more complex theories by simpler theories has been studied mainly in the context of approximating arbitrary propositional

[4] In Polish "Silny" means "Strong," but it is assumed that the database agent does not know the Polish language.

theories by propositional Horn clauses. Note that strongest necessary and weakest sufficient conditions relativized to a subset of relation symbols provide us with approximations of theories expressed in a richer language by theories expressed in a less expressive language. This leads to a generalization of existing results which allows us to approximate any finite propositional or first-order theory which is semi-Horn w.r.t. the eliminated propositions or relational symbols.

Example 8.5.2. Consider the following well-known theory, denoted by T:

$$(CompSci \wedge Phil \wedge Psych) \rightarrow CogSci \tag{8.9}$$

$$ReadsMcCarthy \rightarrow (CompSci \vee CogSci) \tag{8.10}$$

$$ReadsDennett \rightarrow (Phil \vee CogSci) \tag{8.11}$$

$$ReadsKosslyn \rightarrow (Psych \vee CogSci). \tag{8.12}$$

Reasoning with this theory was shown to be quite complicated due to the large number of cases. On the other hand, one would like to check, for instance, whether a computer scientist who reads Dennett and Kosslyn is also a cognitive scientist. Reasoning by cases shows that this is true. One can, however, substantially reduce the theory and make the reasoning more efficient. In the first step one notices that *Phil* and *Psych* are not used in the query, thus they might appear redundant in the reasoning process. On the other hand, these notions appear in disjunctions in clauses (8.11) and (8.12). In this context we might consider

$$\textsc{Snc}(CompSci \wedge ReadsDennett \wedge ReadsKosslyn; T; -\{Phil, Psych\}) \tag{8.13}$$

where, as usual, $-\{Phil, Psych\}$ denotes all symbols in the language, other than *Phil* and *Psych*. Performing simple calculations one obtains the following formula equivalent to (8.13):

$$\begin{aligned}(8.10) \wedge\ & [CompSci \wedge ReadsDennett \wedge ReadsKosslyn] \\ & \wedge [(CompSci \wedge (ReadsDennett \wedge \neg CogSci) \\ & \wedge (ReadsKosslyn \wedge \neg CogSci)) \rightarrow CogSci]\end{aligned} \tag{8.14}$$

which easily reduces to

$$\begin{aligned}(8.10) \wedge CompSci \wedge ReadsDennett \wedge \\ ReadsKosslyn \wedge (\neg CogSci \rightarrow CogSci).\end{aligned} \tag{8.15}$$

Thus the strongest necessary condition for the original formula is

$$CompSci \wedge ReadsDennett \wedge ReadsKosslyn$$

which implies $CogSci$ and, consequently, the formula also implies $CogSci$.

Assume that one wants to compute the weakest sufficient condition of being a computer scientist in terms of

$$\{ReadsDennett, ReadsKosslyn, ReadsMcCarthy, CogSci\}.$$

We then consider

$$\text{WSC}(CompSci; T; -\{Phil, Psych, CompSci\}). \tag{8.16}$$

After eliminating quantifiers over $Phil, Psych, CompSci$ from the second-order formulation of the weakest sufficient condition, one obtains the following formula equivalent to (8.16):

$$ReadsMcCarthy \wedge \neg CogSci.$$

Thus the weakest condition that, together with theory T, guarantees that a person is a computer scientist is that the person reads McCarthy and is not a cognitive scientist. □

8.5.3 Abduction

The weakest sufficient condition corresponds to the weakest abduction.

Example 8.5.3. Consider the theory

$$T = \{\forall x.[HasWheels(x) \rightarrow CanMove(x)],\\ \forall x.[Car(x) \rightarrow HasWheels(x)]\}.$$

Assume one wants to check whether an object can move. There are three interesting cases:

1. the target language is $\{HasWheels\}$; and we consider

   $$\text{WSC}(CanMove(x); T; \{HasWheels\}),$$

   which is equivalent to $\forall CanMove, Car.[T \rightarrow CanMove(x)]$
2. the target language is $\{Car\}$ and we consider

   $$\text{WSC}(CanMove(x); T; \{Car\}),$$

   which is equivalent to $\forall HasWheels, Car.[T \rightarrow CanMove(x)]$

3. the target language is $\{HasWheels, Car\}$ and we consider

   $\mathrm{WSC}(CanMove(x); T; \{HasWheels, Car\})$,

   which is equivalent to $\forall CanMove.[T \rightarrow CanMove(x)]$.

After eliminating second-order quantifiers we obtain the following results:

1. $\mathrm{WSC}(CanMove(x); T; \{HasWheels\}) \equiv HasWheels(x)$
2. $\mathrm{WSC}(CanMove(x); T; \{Car\}) \equiv Car(x)$
3. $\mathrm{WSC}(CanMove(x); T; \{HasWheels, Car\}) \equiv$
   $\forall x.[Car(x) \rightarrow HasWheels(x)] \rightarrow HasWheels(x)$.

The first two conditions are rather obvious. The third one might seem a bit strange, but observe that $\forall x.[Car(x) \rightarrow HasWheels(x)]$ is an axiom of theory T. Thus, in the third case, after simplification we have

$$\mathrm{WSC}(CanMove(x); T; \{HasWheels, Car\}) \equiv HasWheels(x).$$

□

8.5.4 Generating Successor State Axioms

Successor state axioms are of great importance when using the situation calculus to reason about action and change in robotics domains. Automatic generation of successor state axioms is a useful technique and can be done using the weakest sufficient conditions.

Example 8.5.4. Consider the problem of generating successor state axioms in a robot domain. Observe that a first-order formulation of the problem is natural and compact. We thus apply first-order logic rather than the propositional calculus. We introduce the following relations:

- $move(o, i, j)$ - the robot is performing the action of moving the object o from location i to location j
- $at(o, i)$ - initially, the object o is in the location i
- $at1(o, j)$ - after the action $move(o, i, j)$, the object is in location j
- $atR(i)$ - initially, the robot is at location i
- $atR1(j)$ - after the action $move(o, i, j)$, the robot is at location j
- $h(o)$ - initially, the robot is holding the object o
- $h1(o)$ - after the action, the robot is holding the object o.

Assume that the background theory contains the following axioms, abbreviated by T:

$$\forall o.(at(o,1)) \wedge \forall o.(\neg at(o,2))$$
$$\forall o.[h(o) \equiv h1(o)]$$
$$\forall o,i,j.[(atR(i) \wedge at(o,i) \wedge h(o) \wedge move(o,i,j)) \rightarrow (atR1(j) \wedge at1(o,j))].$$

The goal is to find the weakest sufficient condition on the initial situation ensuring that the formula $at1(package,2)$ holds. Thus we consider

$$\text{WSC}(at1(package,2);T;\{h,at,atR,move\}). \tag{8.17}$$

The approach we propose is based on the observation that

$$\text{WSC}(at1(package,2);T;\{h,at,atR\}) \equiv$$
$$\forall h1 \forall at1 \forall atR1.(T \rightarrow at1(package,2)).$$

After some simple calculations which can be performed automatically using the DLS algorithm we thus obtain that (8.17) is equivalent to

$$[\forall o.at(o,1) \wedge \forall o.\neg at(o,2)] \rightarrow [h(package) \wedge$$
$$\exists i.(atR(i) \wedge at(package,i) \wedge move(package,i,2))]$$

which, in the presence of axioms of theory T, reduces to

$$[h(package) \wedge \exists i.(atR(i) \wedge at(package,i) \wedge move(package,i,2))] \tag{8.18}$$

and, since $at(package,i)$ holds in the theory T only for i equal to 1, formula (8.18) reduces to

$$h(package) \wedge atR(1) \wedge move(package,1,2).$$

Thus, the weakest condition on the initial state, making sure that after the execution of an action the package is in location 2, expresses the requirement that the robot is in location 1, holds the package and that it executes the action of moving the package from location 1 to location 2. □

8.6 Bibliographic Notes

This chapter is mainly based on [56].

In [108], Lin proposed the notion of weakest sufficient and strongest necessary conditions for propositional theories. It has been extended for first-order logic in [56]. Strongest necessary and weakest sufficient conditions have many potential uses and applications ranging from generation of abductive hypotheses to approximation of theories. In fact, special cases of strongest necessary and weakest sufficient conditions, namely strongest postconditions and weakest preconditions, have had widespread usage as a basis for programming language semantics [44].

The concept of approximating more complex theories by simpler theories has been studied in [32, 97], mainly in the context of approximating arbitrary propositional theories by propositional Horn clauses. The concept of approximate theories is also discussed in [121].

Section 8.5.1 is based on [56]. Further development of the method is provided in [62]. Examples given in 8.3.4 and 8.5.4 were considered in [108] and, in the presented form, in [56]. Example 8.5.2 was originally considered in [97].

Abduction has gained a great deal of interest in many fields of philosophy (see [86]) and AI (see [74, 94, 95]).

Observe that the techniques for computing strongest necessary and weakest sufficient conditions can be applied in computing interpolants thus also to solve a variety of important pragmatic problems, discussed in [21].

9

CAKE: Computer Aided Knowledge Engineering

9.1 Introduction

Knowledge engineering often involves the development of modeling tools and inference mechanisms (both standard and non-standard) which are targeted for use in practical applications, where expressiveness in representation must be traded off for efficiency in use. Some representative examples of such applications would be the structuring and querying of knowledge on the semantic web, or the representation and querying of epistemic states used with softbots, robots or smart devices. In these application areas, declarative representations of knowledge enhance the functionality of such systems and also provide a basis for insuring the pragmatic properties of modularity and incremental composition. On the other hand, the mechanisms developed should be tractable, but at the same time, expressive enough to represent such aspects as default reasoning, or approximate or incomplete representations of the environments in which the entities in question are embedded or used, be they virtual or actual.

Equally important are the tools used to do the modeling. Although difficult to evaluate formally, such modeling tools should provide straightforward methods which ensure the modularity and incremental composition of the knowledge structures being designed in addition to guaranteeing formal semantics and transparency of usage.

In many applications one requires an efficient representation and query mechanism for the knowledge structures and epistemic states used by robots or softbots, in particular for applications where planning in the context of incomplete states and approximate knowledge is a necessity. We have focused on a generalization of deductive databases and query languages where the generalization involves the use of rough knowledge databases and where queries can be non-monotonically contextualized to locally close only parts of the database since a closed-world assumption is not feasible. This approach provides

P. Doherty et al.: *Knowledge Representation Techniques*, Studfuzz **202**, 159–179 (2006)
www.springerlink.com

us with a reasonably efficient query mechanism and a reasonably expressive query language for querying approximate knowledge structures. These techniques have been discussed in Chapters 6 and 7. In such knowledge structures, both positive and negative knowledge must be stored explicitly to ensure the open-world assumption.

In the approach we pursue here, we view a (generalized) database as a loosely coupled confederation of granules, where each granule is responsible for managing all or part of a relation or property. In fact, several granules may contribute locally to the definition of a relation. In addition, each relation is viewed as a partial or approximate object represented in terms of positive and negative information. Granules may be composed and abstractions of these compositions (called knowledge modules) can be constructed where the module is viewed externally as the manager of a specific relation, hiding the complexity of generating its extension. Knowledge modules may be defined recursively in terms of other modules or as combinations of modules and explicit types of granules.

Querying such confederations of dynamic knowledge structures can be done in a number of ways using a number of querying techniques. For instance, certain granules may manage and compute default rules, while others may adjudicate between several default granules when there is a conflict. Other granules may manage a local context which locally closes or minimizes, maximizes or fixes several different relations.

These mechanisms are intended to be used in environments where knowledge or information is distributed, often times locally inconsistent, and where granules can compose and decompose dynamically in order to represent knowledge structures and query them in a flexible and tractable manner. In order to construct such knowledge structures and granule confederations in a principled and straightforward manner, we propose a diagrammatic technique for building representations and doing inference which insures formal correctness. The diagrammatic technique and its semantics will be the focus of this chapter.

We call the method CAKE, an acronym which stands for *Computer Aided Knowledge Engineering.* CAKE provides us with a means for constructing and visualizing the complex dependencies between granules. It can be naturally viewed as an extension of well-known entity-relationship diagrams designed for representing relations in relational databases. It also provides tools to represent a complex querying mechanism for generalized deductive databases, which is expressive enough to model numerous knowledge representation paradigms, including defaults and many circumscription policies (see Chapter 10).

CAKE enjoys two important properties. Firstly, it has a simple well-defined semantics. Secondly, it is tractable: any reasoning process that can be represented using CAKE is computable in polynomial time. This makes our formalism attractive from the standpoint of practical applications.

CAKE allows one to:

- visualize the dependencies between granules
- group granules into knowledge modules
- represent voting mechanisms
- automatically generate queries to underlying databases.

The central concept of the CAKE method is that of a *knowledge diagram* (or *diagram*, for short). Knowledge diagrams correspond to granules and knowledge modules. In the CAKE method we deal with CAKE *granules* and *voting granules* grouped in *knowledge modules.* A CAKE granule stores information about a relation. Each granule is *responsible* for delivering a single relation, though a relation may be distributed among many CAKE granules. A CAKE granule can store its own facts as well as rules defining the relation or imposing some constraints. A granule which only stores data is called a *database granule.* The rules define a computation mechanism which allow one to compute the relation. Such a mechanism is called a *granule's method.* A knowledge module gathers some CAKE granules and possibly other modules. Observe that various CAKE granules might deliver contradictory information concerning a given relation. Such a conflict should somehow be resolved. One could accept a voting mechanism based on a principle, according to which whenever a fact is claimed to hold and, at the same time, not to hold within a CAKE granule (knowledge module), then the fact is assumed unknown by the granule (knowledge module, respectively). However, in such a case one tends not to distinguish between unknown and contradictory information. We thus do not remove inconsistencies, but rather develop mechanisms that allow one for dealing with inconsistencies. One of the basic tools here depends on the following simple encoding of possible situations:

$R(\bar{a})$ is TRUE if and only if only $R^+(\bar{a})$ holds
$R(\bar{a})$ is FALSE if and only if only $R^-(\bar{a})$ holds
$R(\bar{a})$ is UNKNOWN if and only if neither $R^+(\bar{a})$ nor $R^-(\bar{a})$ holds
$R(\bar{a})$ is INCONSISTENT if and only if both $R^+(\bar{a})$ and $R^-(\bar{a})$ hold.

The underlying querying mechanism we consider allows us to compute all the above facts in time polynomial in the size of the database (see Chapter 6).

There are many reasonable solutions to deal with inconsistencies. For instance one might find a source of information more reliable then the other sources and give it some priority. In order to represent such solutions voting granules are introduced. Voting granules provide user-defined methods for solving conflicts. The following example illustrates these ideas.

Example 9.1.1. Consider a database containing a relation $C(x, y)$ denoting that a place x on a map is connected via a sequence of roads with a place

y, directly or indirectly. Suppose, however, that our database does not have complete information about all indirect connections. One can provide a CAKE granule responsible for delivering information concerning the indirect connections, using the following rule:

$$[\exists z.C(x,z) \wedge C(z,y)] \rightarrow C(x,y). \tag{9.1}$$

The situation becomes more complicated for cases where the database is distributed and refers to many, not necessarily disjoint maps, some of them, for example, older than the others. One could then define CAKE granules responsible for delivering information from distributed sources. Assume there are two databases and two CAKE granules A_1 and A_2 responsible for delivering the relation C from the respective data sources (see Figure 9.5, page 166). It might now happen that one database, served by A_1, contains the fact $C(J,M)$ and the other, served by A_2, contains $C(M,K)$. Observe that the information about the indirect connection between J and K has to be computed. It could then be useful to define a new CAKE granule, say AG, accessing information form distributed sources and using the rule (9.1) to combine the obtained information. However the combined information might appear inconsistent. For instance, suppose that CAKE granule A_2 has the information $\neg C(J,K)$. On the other hand, AG computes that $C(J,K)$ holds. Thus AG, when asked whether $C(J,K)$, answers TRUE and the same happens when it is asked whether $\neg C(J,K)$ since this an answer provided by A_2.

In order to allow one to deal with such contradictions, we introduce *adjudicating granules*. In general, adjudicating granules serve to combine information from various sources, in particular to adjudicate contradictions and to prioritize the information sources. □

9.2 The Language

9.2.1 Syntax Rules

We extend the usual first-order vocabulary by introducing the following two sets of names:

- AGNAME - a finite *set of* CAKE *granule names,*
- KNNAME - a finite *set of knowledge module names*
- LCCNAME - a finite *set of local closure policy names.*

CAKE granules and knowledge modules are also called *components*. Any component can be *responsible* for delivering some relations. In order to avoid ambiguity, we assume that a relation R delivered by a component named

C is denoted by $C.R$. Thus we shall deal with the language extending the classical first-order language defined in Section 2.6, by assuming that the syntactic category $\langle$Atomic Formula$\rangle$ is replaced by the syntactic category $\langle$Cake Atomic Formula$\rangle$, representing Cake *atomic formulas* and defined by means of the following rule:

$\langle\textsc{Cake Atomic Formula}\rangle ::=$
$\quad\langle\textsc{Rel}\rangle^{+}([\langle\textsc{Terms}\rangle]\{,\langle\textsc{Terms}\rangle\}) \;||$
$\quad\langle\textsc{Rel}\rangle^{\oplus}([\langle\textsc{Terms}\rangle]\{,\langle\textsc{Terms}\rangle\}) \;||$
$\quad\langle\textsc{Rel}\rangle^{-}([\langle\textsc{Terms}\rangle]\{,\langle\textsc{Terms}\rangle\}) \;||$
$\quad\langle\textsc{Rel}\rangle^{\ominus}([\langle\textsc{Terms}\rangle]\{,\langle\textsc{Terms}\rangle\}) \;||$
$\quad\langle\textsc{Rel}\rangle^{\pm}([\langle\textsc{Terms}\rangle]\{,\langle\textsc{Terms}\rangle\}) \;||$
$\quad[\langle\textsc{KnName}\rangle.]\langle\textsc{Cake Atomic Formula}\rangle \;||$
$\quad[\langle\textsc{AgName}\rangle.]\langle\textsc{Cake Atomic Formula}\rangle$

where $\langle$Terms$\rangle$ is restricted to constants and variables only. Recall that R^{+} and $R^{\oplus}$ denote lower and upper approximations of positive facts, i.e., facts of the form $R(\bar{a})$. Similarly, R^{-} and $R^{\ominus}$ denote lower and upper approximations of negative facts, i.e., facts of the form $\neg R(\bar{a})$, and $R^{\pm}$ denotes unknown facts about R.

First-order formulas built in this manner are used to define rules in Cake diagrams.

Any expression of the form $N.R$ is called a *reference* to R while N is called the *prefix of the reference.*

We also introduce Cake *labels* defined by syntactic category $\langle$Cake Label$\rangle$ according to the following rule:

$\langle\textsc{Cake Label}\rangle ::=$
$\quad\langle\textsc{AgName}\rangle : \langle\textsc{Rel}\rangle([\langle V_{\mathsf{I}}\rangle]\{,\langle V_{\mathsf{I}}\rangle\})\{,\langle\textsc{Rel}\rangle([\langle V_{\mathsf{I}}\rangle]\{,\langle V_{\mathsf{I}}\rangle\})\} \;||$
$\quad\langle\textsc{KnName}\rangle : \langle\textsc{Rel}\rangle([\langle V_{\mathsf{I}}\rangle]\{,\langle V_{\mathsf{I}}\rangle\})\{,\langle\textsc{Rel}\rangle([\langle V_{\mathsf{I}}\rangle]\{,\langle V_{\mathsf{I}}\rangle\})\}$

Cake labels are used to declare names of components and relations the components are responsible for.

Let $L \in \langle$LccName$\rangle$. Given a universal Lcc policy L (see Definition 6.5.3 in Section 6.5) we allow formulas of the form $L\{A\}$, where A is an arbitrary formula. We treat the Lcc policies as macro definitions. This is possible, since in the case of universal Lcc policies one can compute the definitions of the relations changed by the Lcc policy. The macro application $L\{A\}$ replaces respective relation symbols in A by the corresponding definitions.

In order to simplify the notation, whenever it does not lead to ambiguities, we allow granule references without prefixes. Then one refers directly to the relation delivered by the granule without using the dotted notation. This happens, e.g., when in a given context there is only one granule responsible for delivering a relation. We also sometimes omit variables of relations in CAKE diagrams, when these are known from the context.

9.2.2 Context Conditions

We assume the following context conditions:

- references of the form $M.R$ are allowed if R is a relation appearing within the component M
- for $M \in$ AGNAME $\cup$ KNNAME, if $M.R$ is a valid expression then M references a unique component in a given component.

9.3 Diagrams

The diagram in Figure 9.1 represents a *diagram of a* CAKE *granule responsible for delivering the rough relation* $R(\bar{x})$, where N is the granule's name. The part of the diagram below the dashed line is called the *positive part of the diagram*, the part between the dashed line and the solid line is called the *negative part of the diagram* and the part above the solid line is called the *context of the diagram*. The part containing the component's name and the relation's name is called the *label of the diagram*. Observe that the *boundary part of the diagram* is given implicitly and contains all facts that are neither in the positive part of the diagram nor in the negative part of the diagram.

For clarity, we mark the positive and negative part of a diagram using + and − which is placed on the lefthand side of the diagram.

Let D be a diagram labelled by $N : R(\bar{x})$. The rôle of the diagram's parts is the following:

- the label introduces the granule's name N and declares the relation, R, which the granule delivers together with the names of its arguments $\bar{x}$
- the context defines the granule's LCC policies LCC$_1, \ldots$ LCC$_m$ together with their names $l_1, \ldots, l_m \in \langle$LCCNAME$\rangle$
- the positive part of the diagram represents facts that are assumed to be TRUE
- the negative part of the diagram represents facts that are assumed to be FALSE.

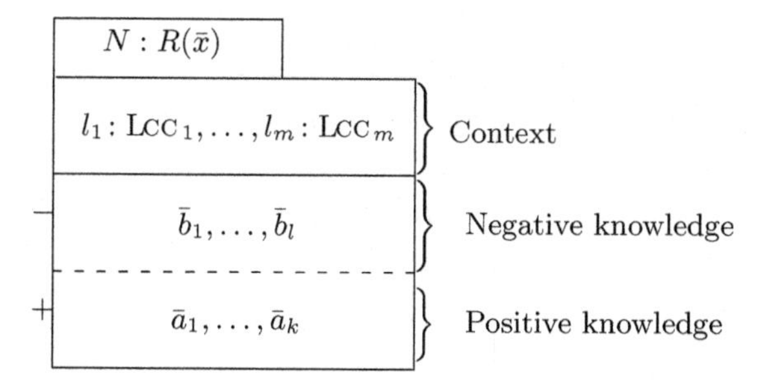

Fig. 9.1. Diagram representing a CAKE granule.

Consider a diagram labelled by $N : R(\bar{x})$. We assume the notational convention according to which the conjunction of formulas of respective parts of the diagram are denoted by $N^{\circledcirc}(R(\bar{x}))$ or $N^{\circledcirc}(R)$ if the arguments $\bar{x}$ of R are known from the context, or $N^{\circledcirc}$ if the relation R is also known, where the superscript $\circledcirc$ indicates the part of the diagram as follows

$$\circledcirc = \begin{cases} + & \text{in the case of the positive part of the diagram} \\ \pm & \text{in the case of the (implicit) boundary part of the diagram} \\ \oplus & \text{in the case of the positive-boundary part of the diagram} \\ \ominus & \text{in the case of the negative-boundary part of the diagram} \\ - & \text{in the case of the negative part of the diagram.} \end{cases}$$

The elements $a_1, \ldots, a_k$ are in the positive part of the diagram and the elements $b_1, \ldots, b_l$ in the negative part of the diagram. This describes that the following conjunction holds:

$$R^+(\bar{a}_1) \wedge \ldots \wedge R^+(\bar{a}_k) \wedge R^-(\bar{b}_1) \wedge \ldots \wedge R^-(\bar{b}_l).$$

Any object in the underlying universe that is outside of the positive and negative parts of a diagram is assumed to be in the boundary part of the diagram.

The relation R can also be defined by means of any first-order formulas representing R^+, R^- rather than by explicitly writing tuples $\bar{a}_i, \bar{b}_j$.

As in the case of first-order queries (see Chapter 4, Section 4.2), a first-order formula A defines a respective part of the diagram representing R in the database B to be the least set of tuples of the form $\langle a_1, \ldots, a_n \rangle$, where

$$a_1, \ldots, a_n \in \text{DOM} \text{ and } B \models A(\langle x_1, \ldots, x_n \rangle := \langle a_1, \ldots, a_n \rangle).$$

The granules' rules are given by providing the body part of each rule without the head. The heads of such rules are known from the context. Consider the

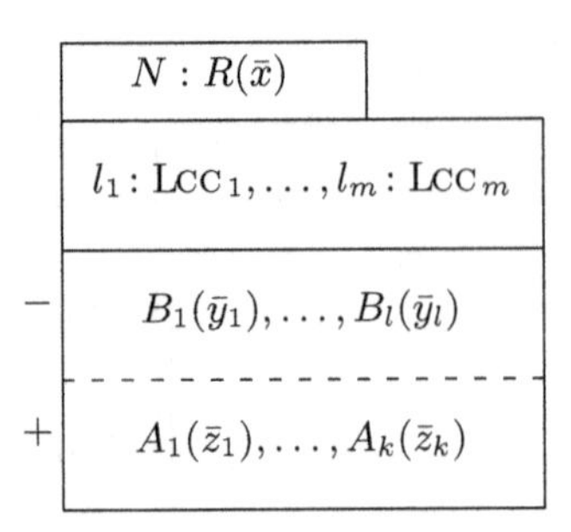

Fig. 9.2. Diagram representing granule's rules.

CAKE granule diagram shown in Figure 9.2. The following rules are defined by the diagram:

$$[\exists \bar{u}.(N.A_1(\bar{z}_1) \vee \ldots \vee N.A_k(\bar{z}_k))] \rightarrow N.R^+(\bar{x}) \qquad (9.2)$$
$$[\exists \bar{v}.(N.B_1(\bar{y}_1) \vee \ldots \vee N.B_l(\bar{y}_l))] \rightarrow N.R^-(\bar{x}),$$

where $\bar{u} = [(\bar{z}_1 \cup \ldots \cup \bar{z}_k) - \bar{x}]$ and $\bar{v} = [(\bar{y}_1 \cup \ldots \cup \bar{y}_l) - \bar{x}]$.

The formulas $N.A_1(\bar{z}_1), \ldots, N.A_k(\bar{z}_k), N.B_1(\bar{y}_1), \ldots, N.B_l(\bar{y}_l)$ are called *methods of N*.[1]

In cases where a granule requires some input relations in order to compute the relation it is responsible for, we require that the connections between inputs and the diagram representing the relation are defined by *arrows*, as shown in Figure 9.3.

An arrow from a component N to a component M is called an *input arrow of component* M and an *output arrow of component* N.

It is assumed that no arrow can cross a component's border, i.e., all input arrows lead to the whole component and all output arrows come from the whole component, not from its subcomponents.

Example 9.3.1. Consider a CAKE granule A responsible for delivering relation $R(\bar{x})$. Then the CAKE granule B shown in Figure 9.4 simulates the closed world assumption applied to $A.R(\bar{x})$. This follows from the fact, that granule B classifies any object which is in the negative or boundary part of granule A as negative information about R. Thus any information about relation R unknown by A becomes false from the point of view of B. □

[1] Recall that these formulas can be arbitrary first-order formulas built over the syntactic category ⟨CAKE ATOMIC FORMULA⟩.

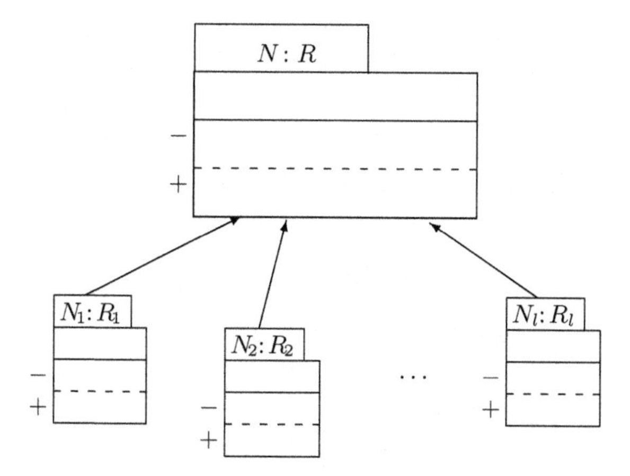

Fig. 9.3. Diagram representing input relations for a CAKE granule.

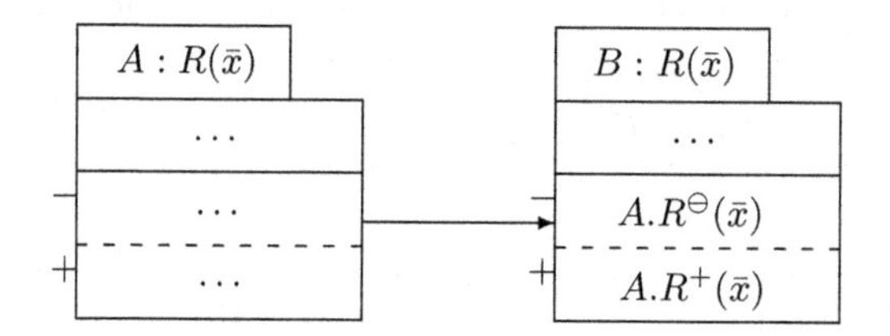

Fig. 9.4. Diagram representing the closed world assumption of Example 9.3.1.

Example 9.3.2 (Example 9.1.1 continued). Consider the case discussed in Example 9.1.1. We now have three CAKE granules (see Figure 9.5). Granules A_1 and A_2 deliver information to granule AG, which is indicated by arrows and rules in the diagram of AG.

Observe that the following rule is attached to AG (this rule is obtained in the same manner as formulas listed in (9.2); see also Figure 9.2):

$$[A_1.C^+(x,y) \vee A_2.C^+(x,y) \vee \exists z.(AG.C^+(x,z) \wedge AG.C^+(z,y))] \rightarrow AG.C^+(x,y).$$

Granule AG computes positive knowledge about the direct and indirect connections between places. □

Knowledge modules are collections of CAKE granules and other knowledge modules. *Knowledge module diagrams* are represented by dashed boxes as shown in Figure 9.6. Each knowledge module has a name ($Name$) followed by the list of relations the module is responsible for. Knowledge modules can be

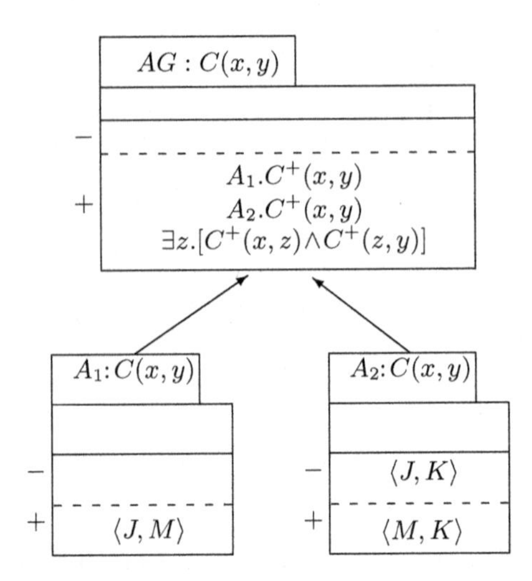

Fig. 9.5. Diagram representing CAKE granules considered in Example 9.3.2.

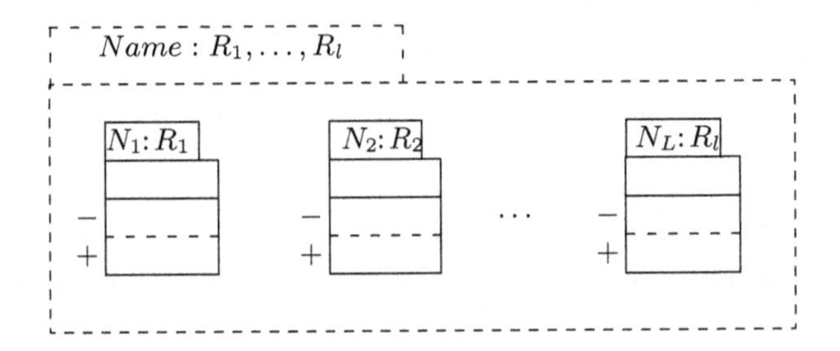

Fig. 9.6. Diagram representing a knowledge module.

nested. If a knowledge module contains components responsible for different relations it is treated simply as a collection of those components. Components directly included in a component C are called *subcomponents* of C.

As in the case of CAKE granules, inputs to knowledge modules are indicated by arrows.

We define the *positive part*, the *boundary part* and the *negative part of a knowledge module* to consist of positive facts, unknown facts and negative facts of relations delivered by the knowledge module. If M is a knowledge module and R is a relation delivered by M, then $M.R^+, M.R^\pm$ and $M.R^-$ denote the respective parts of R. If a module delivers a single relation, terms for the positive, boundary and negative parts of the knowledge module are also used to indicate the suitable parts of the relation.

Observe that in a module there may be many CAKE granules, which used together deliver inconsistent information. A mechanism is required, which we call a *voting mechanism* for computing the final answer for any relation in a module. Voting is represented by a special, distinguished CAKE granule called the *adjudicating granule.* In each knowledge module there may be at most one such granule for any relation served by the module. *Adjudicating granule diagrams* are represented as in Figure 9.7. The answer determined by the granule is the answer that the module returns.

In the absence of an adjudicating granule for a relation served by a module, the module is assumed to act according to the following principle.

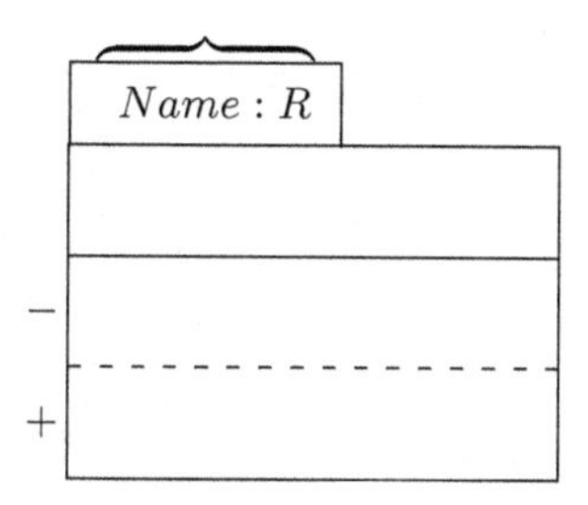

Fig. 9.7. Diagram representing an adjudicating granule.

Definition 9.3.3. *Assume that a module is asked a query about relation R. In the absence of an adjudicating granule for R, the following* standard voting mechanism *is assumed:*

- *if at least one granule or knowledge module for R contained in the module answers* TRUE *to the query and none of the granules and knowledge modules for R answers* FALSE*, the final answer to the query is* TRUE
- *if at least one granule or knowledge module for R contained in the module answers* FALSE *to the query and none of the granules and knowledge modules for R answers* TRUE*, the final answer to the query is* FALSE
- *otherwise, the answer to the query is* UNKNOWN. □

Example 9.3.4 (Example 9.1.1 continued). Consider the diagram, shown in Figure 9.8, corresponding to the situation described in Examples 9.1.1 and 9.3.2.

Observe that granules A_1 and A_2 are now grouped into a knowledge module N. Collecting granules A_1 and A_2 in one knowledge module resolves the potential for contradictory information possessed by the granules, as in this case the default voting mechanism of N would assure that the granule AG receives

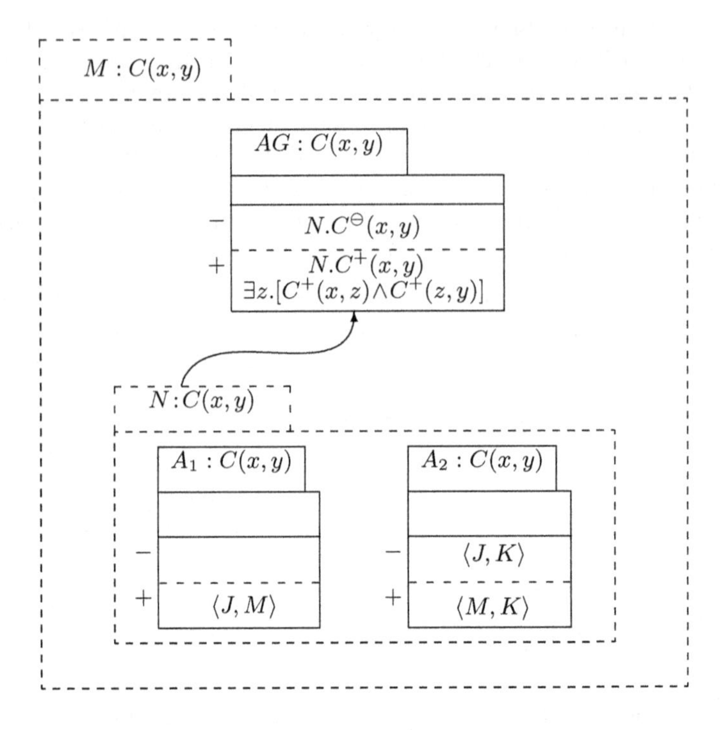

Fig. 9.8. Diagram corresponding to Example 9.3.4.

consistent information. On the other hand, the information computed by AG may be inconsistent. This, in fact happens, as the tuple $\langle J, K\rangle$ is both in the positive and negative part of the diagram of AG. Thus the granule returns inconsistent information. However, in the module M there is no explicit adjudicating granule. Thus the standard voting mechanism is accepted and all the inconsistent tuples are in the boundary part of M.

The case in which the database granules have a priority over the granule AG is illustrated in Figure 9.9, where an adjudicating granule AV is introduced. In this case $\langle J, K\rangle$ is in the positive part of AG and in the negative part of N. The adjudicating granule AV will answer that $C(J, K)$ is FALSE, due to the rule in the negative part of AV and the fact that none of the rules in the positive part of AV can be applied. □

9.4 Visibility and Binding Rules

The following definition distinguishes between declarations of variables and relations and references to variables and relations.

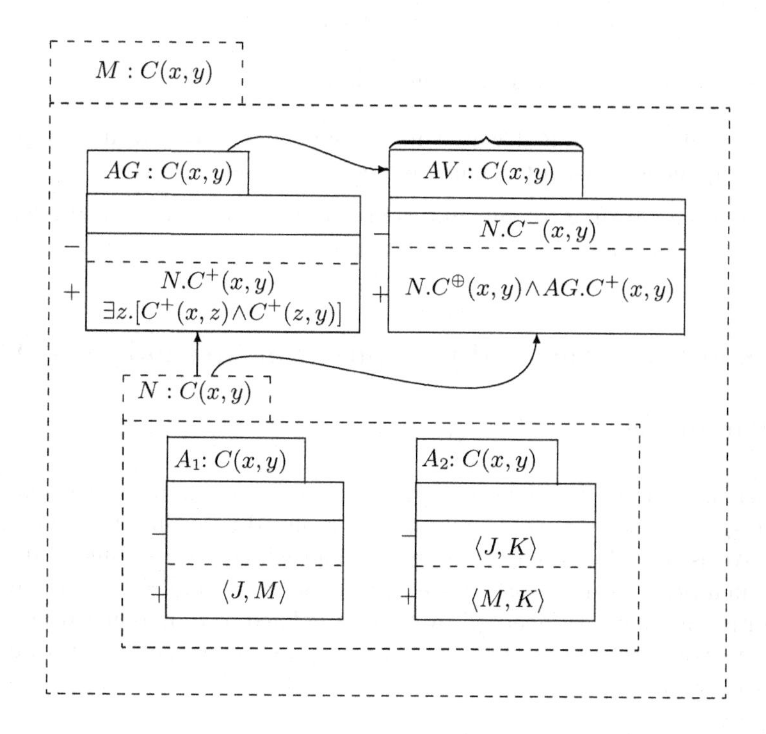

Fig. 9.9. Diagram illustrating the prioritized voting process of Example 9.3.4.

Definition 9.4.1. *An occurrence of a relation (individual variable) is called a* declaration of the relation (variable) *if it appears in a label of a diagram, otherwise it is called a* reference to the relation (variable). □

The following visibility rules are assumed:

- a relation R (Lcc policy L) is *visible in a component* C if one of the following conditions holds:
  - R (respectively L) is declared in C
  - C contains a component D and relation R is declared (respectively L appears in the context part) in D
  - there is an arrow from a component D to C and the relation R is declared (respectively L appears in the context part) in D.

  In the two latter cases the relation R (Lcc policy L) can be accessed from D via the remote access, denoted by $D.R$ (respectively by $D.L$)
- a variable x is *visible in a component* C if x is declared in C or in any component that includes C

- a component name is visible only outside of the component.

We also assume the following binding rules:

- if a relation name R declared in a component C occurs in any part of a component C, then the name refers to the relation $C.R$
- no diagram leading to ambiguities in binding or with undefined references is considered valid.

9.5 The Semantics and Computation Method for CAKE

9.5.1 Introduction

The semantics of CAKE diagrams is given by means of an inductive definition. The respective parts of the definition are grouped according to the diagram type. We assume that diagrams represent approximation transducers and that the semantics of rules is as defined in Sections 6.3 and 7.4. Formally, the semantics of diagrams, denoted by $\|D\|$, is defined as a mapping from rough structures of a given vocabulary SIG into rough structures of possibly another vocabulary SIG$'$:

$$\|\cdot\| : \text{STRUC}[\text{SIG}] \longrightarrow \text{STRUC}[\text{SIG}'].$$

In the sequel we show how to compute first-order or fixpoint definitions of all the regions in relations. This supplies us with a formal semantics of the diagrams.

9.5.2 The Semantics of LCC Policies

Consider the following universal LCC policy:

$$L : \text{LCC}[L_1, \ldots, L_p; K_1, \ldots, K_r]\text{:IC}.$$

In Section 6.5 direct definitions of minimal (respectively maximal) and varied relations are provided. The meaning of formula $L\{A\}$ is then the following:

- for each relation of $L_1, \ldots, L_p$ occurring in formula A create a CAKE granule and place in its positive and negative parts righthand sides of definitions given in Lemma 6.5.8, respectively
- for each relation of $K_1, \ldots, K_r$ occurring in formula A create a CAKE granule and place in its positive and negative parts righthand sides of rules given by definitions (6.14), (6.15) with minimized and maximized relations replaced by their definitions, as appropriate

- add all necessary arrows between granules (indicating the input relations to the new granules)
- replace any occurrence of a relation symbol of $L_1, \ldots, L_p$ in formula A by a reference to a suitable CAKE granule defined in items above.

9.5.3 Attaching Rules to Diagrams

Through this section we always assume that the rules attached to the diagrams are universally quantified over free variables.

Interpretation of Occurrences of Relation Names in Diagrams

Assume we have a formula in negation normal form.[2]. Then any positive occurrence of any relation symbol, say R, refers to R^+ and any negative occurrence $\neg R$ of R refers to R^-. We also accept the convention according to which all references to boundary regions of relations are eliminated from formulas. Table 9.1 describes the convention.

Table 9.1. Rules for eliminating references to boundary regions.

Occurrence in a diagram	Actual Meaning
$R^{\oplus}$	$\neg R^-$
$\neg R^{\oplus}$	R^-
$R^{\ominus}$	$\neg R^+$
$\neg R^{\ominus}$	R^+
$R^{\pm}$	$\neg R^+ \wedge \neg R^-$
$\neg R^{\pm}$	$R^+ \vee R^-$

Attaching Rules to CAKE Granules and Adjudicating Granules

Assume we are given a granule or an adjudicating granule diagram labelled by $N : R(\bar{x})$ and containing parts as in Figure 9.2. Then we attach to the diagram rules (9.2), modified by LCC's, as discussed in Section 9.2.1.

Now the definition of the boundary region of the relation defined by N is the following

$N.R^{\pm}(\bar{x}) \equiv (\neg N.R^+(\bar{x}) \wedge \neg N.R^-(\bar{x}))$.

[2] Recall from Chapter 2 that any formula is easily transformed to this form.

Attaching Rules to Knowledge Modules

Assume we are given a knowledge module diagram labelled by

$M : R_1(\bar{x}_1), \ldots, R_k(\bar{x}_k)$.

Let, for $1 \leq i \leq k$, M_i be the set of all subcomponents of M responsible for delivering the relation R_i and assume M_i does not contain an adjucating granule for R_i. Then we attach to the diagram the following set of rules, for any $1 \leq i \leq k$:

$$\left[\bigvee_{N \in M_i} N.R_i{}^-(\bar{x}_i) \wedge \neg \bigvee_{N \in M_i} N.R_i{}^+(\bar{x}_i)\right] \rightarrow M.R_i{}^-(\bar{x}) \tag{9.3}$$
$$\left[\bigvee_{N \in M_i} N.R_i{}^+(\bar{x}_i) \wedge \neg \bigvee_{N \in M_i} N.R_i{}^-(\bar{x}_i)\right] \rightarrow M.R_i{}^+(\bar{x}).$$

In the case where M contains an adjudicating granule A responsible for delivering the relation R_i, then we attach to the diagram the following rules:

$$A.R_i^+(\bar{x}_i) \rightarrow M.R_i^+(\bar{x}_i) \tag{9.4}$$
$$A.R_i^-(\bar{x}_i) \rightarrow M.R_i^-(\bar{x}_i).$$

The definition of the boundary region of the relation defined by M is obtained as in the case of CAKE granule diagrams, i.e., it is given by the following equivalence:

$$M.R^{\pm}(\bar{x}) \equiv (\neg M.R^+(\bar{x}) \wedge \neg M.R^-(\bar{x})).$$

9.5.4 Obtaining the Explicit Definitions of Relations

We now provide a tractable fixpoint semantics for CAKE. All knowledge diagrams in this section are assumed to be stratified according to the following definition.

Definition 9.5.1. *A knowledge diagram is* stratified *if the set of rules attached to the diagram (see Section 9.5.3) is stratified.* □

The case of non-stratified diagrams is to be dealt with by applying the well-founded semantics (see Section 9.5.5). As indicated in Section 4.5.2, stratified semantics and well-founded semantics agree on stratified DATALOG$^\neg$ programs. However, we deal with stratified diagrams separately since they enjoy nicer computational properties and can be directly implemented using versions of PROLOG that allow for stratified negation.

The Case of a Single Stratum

Consider first the simplest case when a set of rules consists of a single stratum. Let $S, \ldots, T$ be all relations appearing in the heads of rules attached to CAKE granules, adjucating granules and knowledge modules, and let B be the conjunction of the bodies of the rules.[3] Then the following simultaneous fixpoint formula defines relations $S, \ldots, T$:

$$\text{LFP}\, S, \ldots, T.\, B. \tag{9.5}$$

The boundary regions of the relations are then obtained using the suitable definitions.

Observe that the formula (9.5) represents a vector of relations. We refer to the particular relations $S, \ldots, T$ as the *S-coordinate*,..., *T-coordinate* of (9.5).

The General Case

Let the set of rules consist of strata $P_1, \ldots, P_n$. Consider a stratum P^i, where $1 \leq i \leq n$. Let $S^i, \ldots, T^i$ be all relations appearing in the heads of rules attached to CAKE granules, adjucating granules and knowledge modules, and let B^i be the conjunction of the bodies of the rules appearing in stratum i. In such a case one applies a method given in Section 4.5 on stratified DATALOG$^\neg$. The corresponding definitions of relations are then computed inductively in order given by strata, according to rules (9.6) and (9.7) provided below.

- First one computes relations $S^1, \ldots, T^1$ as the simultaneous fixpoint

  $$\text{LFP}\, S^1, \ldots, T^1.B^1. \tag{9.6}$$

- Having computed definitions of all relations $S^k, \ldots, T^k$, for $1 \leq k < i \leq n$, one computes $S^i, \ldots, T^i$ as

  $$\text{LFP}\, S^i, \ldots, T^i.B^i, \tag{9.7}$$

  where it is assumed that names of relations appearing in B^i and computed in strata $1, \ldots, i-1$ are replaced by the obtained fixpoint definitions of the relations.

Remark 9.5.2. Let $\mathcal{R}$ be a stratified set of rules, B be the conjunction of bodies of rules, and assume that S,..., T are all relations appearing in the heads of the rules. Given a particular stratification, the explicit definitions of these relations, obtainable by the above computations, will be denoted by

[3] Observe that in this case stratification ensures that bodies of rules attached to all diagrams are positive w.r.t. all relations appearing in the heads of the rules.

$$\text{LFP}\, S, \ldots, T.B. \tag{9.8}$$

In the sequel, an expression of the form (9.8) will be referred to as a *simultaneous fixpoint expression* and particular relations $S, \ldots, T$ as the S-coordinate,..., T-coordinate of (9.8). □

9.5.5 Computing the Relations

Observe that the definitions of relations obtained in Section 9.5.4 are expressed by means of fixpoint formulas. Using Theorem 4.3.1 one can now provide a tractable method for computing the relations. Moreover, if the database is linearly ordered, then any PTIME query can be modelled by knowledge diagrams, since recursion within the diagrams is allowed.

It should be emphasized, however, that in practice one should use known optimization techniques developed for DATALOG, DATALOG$^\neg$ and fixpoint queries.

In the case of non-stratified CAKE diagrams we do not obtain explicit definitions of relations. In order to compute the relations, we apply the well-founded semantics (see Section 4.5.2).[4] In such a case, CAKE rules are to be expressed by means of DATALOG$^\neg$ rules, which, in the presence of ordering on the domain, can easily be done, since first-order quantifiers can be expressed as DATALOG$^\neg$ rules, as shown below.

Assume an ordering on a database domain DOM is given by its least element, denoted by 0, together with a successor relation $S(x, y)$, meaning that y is an immediate successor of x. Assume a given formula is in the PNF form, i.e., all quantifiers appear in its prefix and is closed, i.e, contains no free variables.[5]

In order to remove quantifiers we proceed from the innermost to outermost quantifiers:

1. consider a subformula of the form $\exists x.A(x, \bar{y})$, where $\bar{y}$ are all free variables of A. We introduce a fresh relation symbol, say $R_A(x, \bar{y})$, with the intuitive meaning that $A(z, \bar{y})$ holds for domain element z accessible from x through zero or more applications of the immediate successor relation S. We introduce the following rules:

   $$R_A(x, \bar{y}) \leftarrow A(x, \bar{y})$$
   $$R_A(x, \bar{y}) \leftarrow S(x, z), R_A(z, \bar{y}),$$

   where z is a fresh variable symbol. Now $[\exists x.A(x, \bar{y})] \equiv R_A(0, \bar{y})$.

---

[4] Observe that subsequences $\{I_{2i}\}_{i \geq 0}$ and $\{I_{2i+1}\}_{i \geq 0}$, defined in Section 4.5.2, correspond respectively to lower approximations of positive and negative parts of the computed relations.

[5] CAKE rules are implicitly universally quantified over all free variables, thus each CAKE rule is, in fact, a closed formula.

2. consider a subformula $\forall x.A(x, \bar{y})$, where $\bar{y}$ are all free variables of A. If $\forall x$ appears in the quantifier prefix of the whole formula, and is preceded by universal quantifiers only, then we remove the quantifier together with all preceding quantifiers.

   In the opposite case, due to the equivalence $\forall x.A(x, \bar{y}) \equiv \neg\exists x.(\neg A(x, \bar{y}))$, it is now sufficient to introduce rules for $\exists x.(\neg A(x, \bar{y}))$, defining say $R_{\neg A}$, and refer to $\neg R_{\neg A}(0, \bar{y})$ instead of $\forall x.A(x, \bar{y})$.

When all quantifiers are removed, we apply the usual propositional reasoning to obtain DATALOG$^{\neg}$ rules.

Observe that the rules obtained in the above procedure can be unsafe. However, we accept unsafe rules, as discussed in Remark 4.4.1.

The following example illustrates this procedure.

Example 9.5.3. Consider the formula

$$\forall x.\exists y.[(R(x, y) \vee \neg Q(y, x)) \wedge T(x, y)]. \tag{9.9}$$

The innermost occurrence of a quantifier is $\exists y$. We introduce a fresh relation symbol, say $V(y, x)$, and the following rules:

$$V(y, x) \leftarrow [(R(x, y) \vee \neg Q(y, x)) \wedge T(x, y)] \tag{9.10}$$
$$V(y, x) \leftarrow S(y, z), V(z, x). \tag{9.11}$$

By propositional reasoning one can verify that rule (9.10) is equivalent to the following two DATALOG$^{\neg}$ rules:

$$V(y, x) \leftarrow R(x, y), T(x, y) \tag{9.12}$$
$$V(y, x) \leftarrow \neg Q(y, x), T(x, y). \tag{9.13}$$

Formula (9.9) reduces now to $\forall x.V(0, x)$. The quantifier $\forall x$ is now replaced by introducing a fresh relation symbol, say $W(x)$, and the following rules:

$$W(x) \leftarrow \neg V(0, x) \tag{9.14}$$
$$W(x) \leftarrow S(x, z), W(z). \tag{9.15}$$

Formula (9.9) is equivalent to $\neg W(0)$ and, in order to compute it, we need DATALOG$^{\neg}$ rules (9.11), (9.12), (9.13), (9.14), (9.15). □

9.5.6 Remarks on Approximation Forests

One of the cases where a more efficient treatment of fixpoints is possible is that of the approximation forests defined below.

Definition 9.5.4. *By an* approximation tree *we mean any diagram forming a tree. By an* approximation forest *we mean any diagram consisting of a set of approximation trees.* □

In such a case there are no cycles between the components and one can consider each component separately, when defining the appropriate rules. The computation process can in this case be handled starting from the leaves of the trees until reaching their roots. Such a solution leads to more compact "local" definitions of the relations and "local" computations that can easily be parallelized.

Of course, there is a close relationship between approximation forests and approximation transducers and trees introduced in Chapter 7, as shown in the following example.

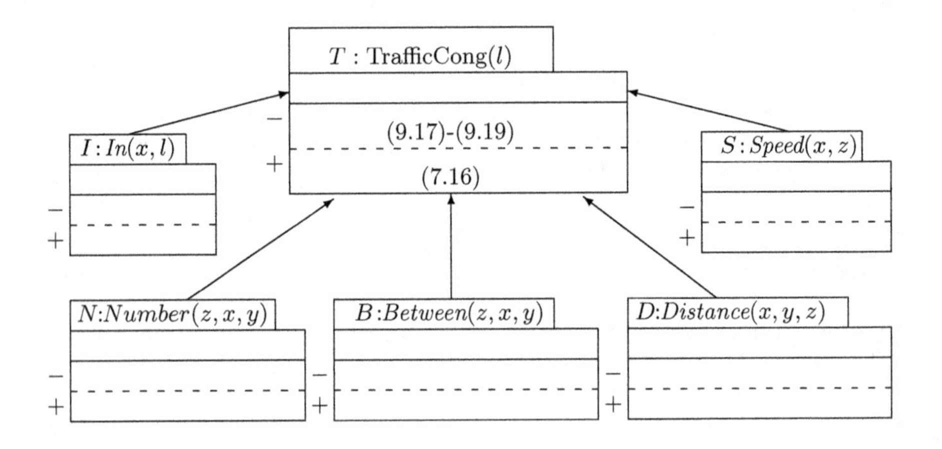

Fig. 9.10. Diagram representing a knowledge structure considered in the congestion example.

Example 9.5.5. In Section 7.6 we considered a congestion example, and obtained, among others, the characterization of $TrafficCong^{+}(l)$ by formula (7.16), and the characterization of $TrafficCong^{\oplus}(l)$ by formula (7.17). CAKE diagrams consist of positive and negative parts and the boundary part is represented implicitly. However, using the equivalence $R^{-} \equiv \neg R^{\oplus}$, one can easily obtain the characterization on the negative part of a relation on the basis of its boundary positive part.

Accordingly, on the basis of formula (7.17), we obtain

$$TrafficCong^{-}(l) \equiv \tag{9.16}$$

$$\neg\exists x, y.[In^{\oplus}(x, l) \wedge In^{\oplus}(y, l) \wedge Number^{\oplus}(x, y, \mathsf{large}) \wedge \quad (9.17)$$
$$\forall z.(Between^{+}(z, x, y) \rightarrow Speed^{\oplus}(z, \mathsf{low})) \wedge \quad (9.18)$$
$$\forall z.(Between^{+}(z, x, y) \rightarrow \exists t.Distance^{\oplus}(z, t, \mathsf{small}))]. \quad (9.19)$$

Consequently, one can obtain a CAKE diagram for the congestion example, as presented in Figure 9.10. □

9.6 Bibliographic Notes

There is a long line of research concerning the use of various diagrams in the process of software design and development. Such tools, known under a common name CASE,[6] range from relational database design tools with various forms of entity-relationship diagrams among them (see, e.g., [174, 214]), through structural design (see, e.g., [42]) to object-oriented modelling and design (see, e.g., UML[7] [24, 23]).

The standard voting principle reflects solutions known from the field of *paraconsistent logics*, i.e., logics that allow handling of contradictory facts without inferring the contradiction. Many voting mechanisms can be developed using solutions proposed by the paraconsistent logics community. For literature on paraconsistent logics see, e.g., [8, 25, 33, 169]. Also Belnap's approach [17] can be modelled in CAKE. Voting has also been studied in the context of building classifiers (see, e.g., [87, 128]).

[6] Computer Aided Software Engineering.
[7] Unified Modelling Language.

10

Formalization of Default Logic Using CAKE

10.1 Introduction

In this chapter, we formalize a subset of default logic using the CAKE method.[1] The goal of this chapter is to do a case study showing how the CAKE method can be used to model a particular type of reasoning commonly used in knowledge representation and important in many applications. This will be done by representing two basic versions of default logic: *rough default logic* and *rough default logic with strong prerequisites*. The main difference between the two versions that will be modeled lies in different treatment of the prerequisite of a default while determining the default's applicability. In the former, a default can be applied if its prerequisite is believed (not contradicting known information). In the latter, we may require that the prerequisite of a default (or a part of it) has to be known, rather than believed, to make the default applicable. The possibility of using both versions substantially increases the expressive power of the resulting logic. We also show that both rough default logic and rough default logic with strong prerequisites can be naturally extended to their prioritized versions by slightly changing the voting policy mechanism used.

It is important to note that the underlying semantics for the considered versions of default logic differ from traditional approaches such as Reiter's default logic.

The following assumes familiarity with the material contained in Chapters 5 and 9.

A default is said to be *disjunction-free* if its prerequisite and justification are conjunctions of literals and its consequent is a single literal. A default theory is said to be *disjunction-free* if all its defaults are disjunction-free and all its axioms are ground literals.

[1] Recall that default logic has been presented in Section 5.2, Chapter 5.

P. Doherty et al.: *Knowledge Representation Techniques*, Studfuzz **202**, 181–212 (2006)
www.springerlink.com

In the sequel, we restrict ourselves to finite disjunction-free default theories with consistent sets of axioms.[2] We assume that the axioms of a theory are ground literals extracted from an underlying extensional database(s).[3]

We will start by modeling and discussing a number of standard default theories found in the literature and provide a means of comparing our underlying semantics with the more traditional approaches found in the literature. We will then provide the technical framework and formal semantics based on the use of CAKE.

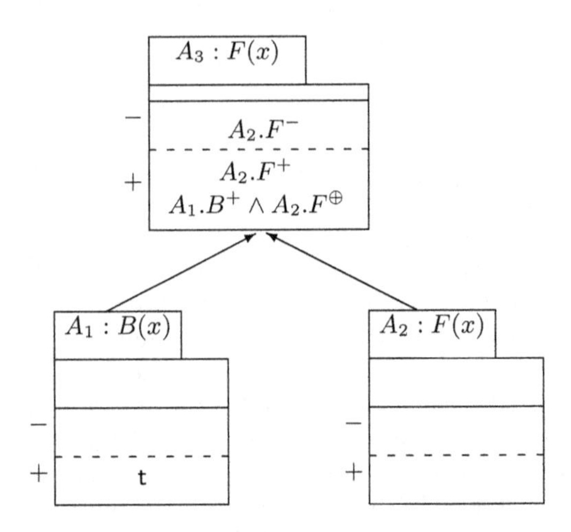

Fig. 10.1. Diagram corresponding to the theory of Example 10.1.1.

Example 10.1.1. Consider the theory $T = \langle\{B(\mathsf{t})\}, \{B(x) : F(x)/F(x)\}\rangle$. This is the standard "Bird" theory with B, F, t denoting *Bird*, *Flies* and Tweety, respectively. Figure 10.1 shows a CAKE diagram corresponding to the theory T.[4]

The bottom granules, A_1 and A_2, are CAKE granules representing data from an assumed extensional database. These granules are responsible for the relations B and F, respectively. Note that t is in the positive part of the diagram

[2] Note that the task of determining whether a set of axioms of a given disjunction-free default theory is consistent is trivial.

[3] That is, ground literals of the form $R(\overline{a})$ (respectively $\neg R(\overline{a})$), where $\overline{a}$ is in the positive (respectively negative) part of the diagram of the extensional database granule responsible for the relation R.

[4] Recall that for the sake of simplicity we often omit variables of relations in CAKE diagrams, when these are known from the context.

of A_1 and in the boundary region of the diagram of A_2. The granule A_3 represents the default rule of the theory. It is responsible for the relation F which occurs in the default consequent. There are three methods associated with this granule. The first two, namely $A_2.F^-$ and $A_2.F^+$, allows the granule A_3 to use knowledge of the granule A_2 while computing the relation F. Using these methods the CAKE granule A_3 can infer that, for any object x, x lies in the negative (respectively positive) part of his diagram, provided that x lies in the negative (respectively positive) part of the diagram of A_2. The third method, i.e., $A_1.B^+ \wedge A_2.F^\oplus$, represents the default of the theory. Note that the method is placed in the positive part of the diagram. It is to be viewed as the following rule:

> "For any object x, if x is in the positive part of the diagram of A_1 and x is in the positive/boundary part of the diagram of A_2, infer that x is in the positive part of the diagram of A_3."

The following set of rules is associated with the above diagram:[5]

$$A_2.F^-(x) \rightarrow A_3.F^-(x)$$
$$\{A_2.F^+(x) \vee [A_1.B^+(x) \wedge A_2.F^\oplus(x)]\} \rightarrow A_3.F^+(x)$$
$$A_1.B^+(\mathrm{t}).$$

Eliminating the reference to the boundary region in $A_2.F^\oplus(x)$, i.e., replacing $A_2.F^\oplus(x)$ by $\neg A_2.F^-(x)$ (see Chapter 9, Section 9.4), results in the following modified rule set:

$$A_2.F^-(x) \rightarrow A_3.F^-(x) \tag{10.1}$$
$$\{A_2.F^+(x) \vee [A_1.B^+(x) \wedge \neg A_2.F^-(x)]\} \rightarrow A_3.F^+(x) \tag{10.2}$$
$$A_1.B^+(\mathrm{t}). \tag{10.3}$$

Clearly, the partition $\{(10.3)\}, \{(10.1), (10.2)\}$ provides a stratification of the above set of rules.

The relations $A_3.F^-$, $A_3.F^+$ and $A_1.B^+$ occurring in the heads of the rules are specified by a simultaneous fixpoint expression of the form

$$\mathrm{LFP}\, A_3.F^-, A_3.F^+, A_1.B^+.\mathcal{B}, \tag{10.4}$$

where $\mathcal{B}$ is the conjunction of the rules' bodies.[6]

Below, we compute the relations characterized by (10.4), using a generalization of Algorithm 4.3.2 suggested in the end of Section 4.3. The successive lines consist of sets of literals obtained after performing all successive iterations.

[5] Recall that all free variables occurring in the rules are implicitly universally quantified.

[6] See Chapter 9, Section 9.5.4.

$\{\}$
$\{A_1.B^+(\mathsf{t})\}$
$\{A_1.B^+(\mathsf{t}), A_3.F^+(\mathsf{t})\}.$

Suppose the query we are interested in is $F(\mathsf{t})$. It is immediately seen that t satisfies the $A_3.F^+$-coordinate of (10.4). Consequently, the answer to the query is TRUE. □

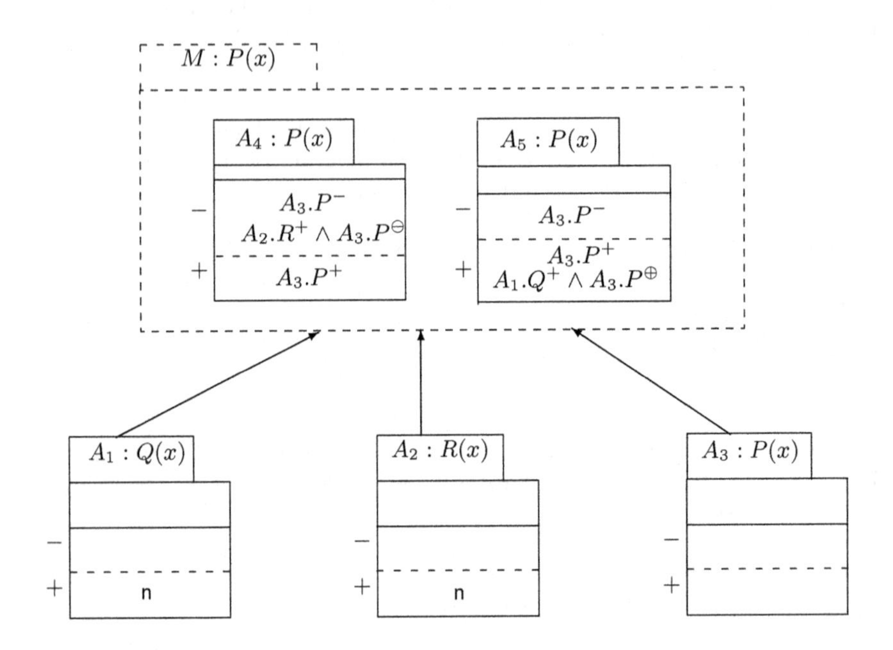

Fig. 10.2. Diagram corresponding to the theory of Example 10.1.2.

Example 10.1.2. Consider the theory

$$T = \left\langle \{Q(\mathsf{n}), R(\mathsf{n})\}, \left\{ \frac{R(x) : \neg P(x)}{\neg P(x)}, \frac{Q(x) : P(x)}{P(x)} \right\} \right\rangle.$$

This is the standard "Nixon" diamond theory with R, P, Q, n standing for *Republican*, *Pacifist*, *Quaker* and Nixon, respectively (see Example 5.2.5). Figure 10.2 shows a CAKE diagram corresponding to the theory T.

The granules A_1, A_2 and A_3 represent data from an assumed extensional database, whereas A_4 and A_5 represent the defaults of the theory.[7] Since

[7] Observe that the method $A_2.R^+ \wedge A_3.P^{\ominus}$ used by the granule A_4 is placed in the negative part of its diagram. Accordingly, it has the following reading: "for

both A_4 and A_5 make inferences concerning the relation P, they have been grouped into a single default module.[8] It is the module, not an individual granule, that is responsible for default inferences about P.

The following set of rules is associated with the above diagram:[9]

$$A_1.Q^+(\mathsf{n})$$
$$A_2.R^+(\mathsf{n})$$
$$\{A_3.P^-(x) \vee [A_2.R^+(x) \wedge \neg A_3.P^+(x)]\} \rightarrow A_4.P^-(x)$$
$$A_3.P^+(x) \rightarrow A_4.P^+(x)$$
$$A_3.P^- \rightarrow A_5.P^-(x)$$
$$\{A_3.P^+(x) \vee [A_1.Q^+(x) \wedge \neg A_3.P^-(x)]\} \rightarrow A_5.P^+(x)$$
$$[A_4.P^-(x) \vee A_5.P^-(x)] \wedge \neg[A_4.P^+(x) \vee A_5.P^+(x)] \rightarrow M.P^-(x)$$
$$[A_4.P^+(x) \vee A_5.P^+(x)] \wedge \neg[A_4.P^-(x) \vee A_5.P^-(x)] \rightarrow M.P^+(x).$$

The last two rules represent the standard voting mechanism used by the module M.

The above set of rules is clearly stratified by the partition P_1, P_2, P_3, where P_1 consists of the first two rules, P_3 consists of the last two rules and P_2 contains the remaining rules. The relations $A_1.Q^+$, $A_2.R^+$, $A_4.P^-$, $A_4.P^+$, $A_5.P^-$, $A_5.P^+$, $M.P^-$ and $M.P^+$, occurring in the heads of the above rules, are defined by the fixpoint expression given by

$$\text{LFP } A_1.Q^+, A_2.R^+, A_4.P^-, A_4.P^+, A_5.P^-, A_5.P^+, M.P^-, M.P^+.\mathcal{B} \tag{10.5}$$

where $\mathcal{B}$ denotes the conjunction of the rules' bodies.

Below, we compute the relations characterized by the expression (10.5).

$$\{\}$$
$$\{A_1.Q^+(\mathsf{n}), A_2.R^+(\mathsf{n})\}$$
$$\{A_1.Q^+(\mathsf{n}), A_2.R^+(\mathsf{n}), A_4.P^-(\mathsf{n}), A_5.P^+(\mathsf{n})\}.$$

Suppose the query of interest is $P(\mathsf{n})$. It is immediately seen that the granule A_4 answers FALSE and the granule A_5 answers TRUE to the query. However, since it is the default module that is responsible for default inferences

any object x, if x is in the positive part of the diagram of A_2 and x is in the negative/boundary part of the diagram of A_3, infer that x is in the negative part of the diagram of A_4."

[8] In Chapter 4, a collection of CAKE granules was referred to as a knowledge module. Here we use the term "a default module," since all granules included in such a module represent defaults.

[9] The relations $A_3.P^{\ominus}$ and $A_3.P^{\oplus}$, occurring in the diagram, have been replaced by $\neg A_3.P^+$ and $\neg A_3.P^-$, respectively.

about P, and since n satisfies neither the $M.P^+$-coordinate of (10.5) nor the $M.P^-$-coordinate of (10.5), we conclude that the answer to the query $P(\mathsf{n})$ is UNKNOWN. □

The examples considered so far are typified by the fact that the CAKE granules that represent default rules obtain all relevant input data from an assumed extensional database. The next example shows that the situation can be more complex.

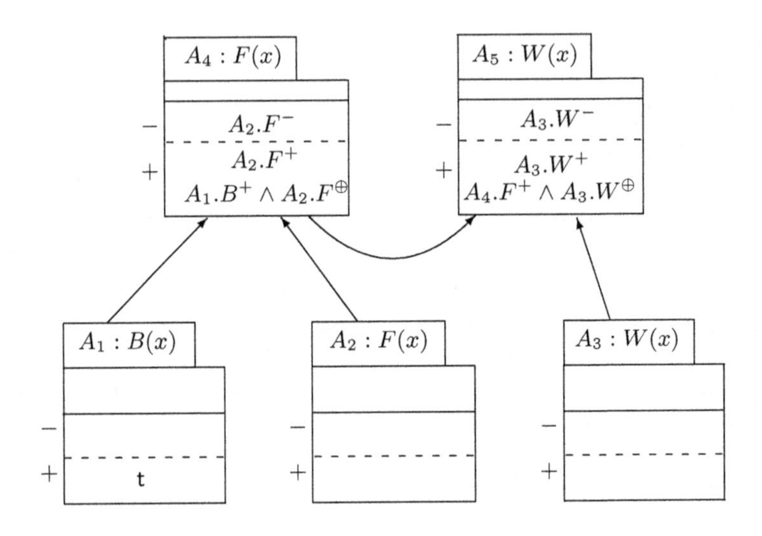

Fig. 10.3. Diagram corresponding to the theory of Example 10.1.3.

Example 10.1.3. Consider the theory T given by

$$T = \left\langle \{B(\mathsf{t})\}, \left\{ \frac{B(x) : F(x)}{F(x)}, \frac{F(x) : W(x)}{W(x)} \right\} \right\rangle,$$

where B, F, W and t denote *Bird*, *Flies*, *HasWings* and Tweety, respectively.

The corresponding CAKE diagram for T is provided in Figure 10.3.

Note that there is an arrow between A_4 and A_5. This is because A_4 is responsible for the relation F that is used by A_5 while computing the relation W. In this case, not all the base data input to the default granules comes from an extensional database as in the previous examples.

The following set of rules are associated with the diagram.

$$A_1.B^+(\mathtt{t})$$
$$A_2.F^-(x) \rightarrow A_4.F^-(x)$$
$$\{A_2.F^+(x) \vee [A_1.B^+(x) \wedge \neg A_2.F^-(x)]\} \rightarrow A_4.F^+(x)$$
$$A_3.W^-(x) \rightarrow A_5.W^-(x)$$
$$\{A_3.W^+(x) \vee [A_4.F^+(x) \wedge \neg A_3.W^-(x)]\} \rightarrow A_5.W^+(x).$$

The partition P_1, P_2, where P_1 consists of the first rule and P_2 contains the remaining rules, provides a stratification of the above set.

The relations $A_1.B^+$, $A_4.F^-$, $A_4.F^+$, $A_5.W^-$ and $A_5.W^+$, occurring in the heads of the rules, are defined by the fixpoint expression given by

$$\text{LFP}\, A_1.B^+, A_4.F^-, A_4.F^+, A_5.W^- A_5.W^+.\mathcal{B} \tag{10.6}$$

where $\mathcal{B}$ denotes the conjunction of the rules' bodies.

The computation of the relations characterized by (10.6) is given below.

$$\{\}$$
$$\{A_1.B^+(\mathtt{t})\}$$
$$\{A_1.B^+(\mathtt{t}), A_4.F^+(t)\}$$
$$\{A_1.B^+(\mathtt{t}), A_4.F^+(\mathtt{t}), A_5.W^+(\mathtt{t})\}.$$

Suppose the query we are interested in is $W(\mathtt{t})$. Since $\mathtt{t}$ satisfies the $A_5.W^+$-coordinate of fixpoint (10.6), we conclude that the answer to the query $W(\mathtt{t})$ is TRUE. □

The next example shows that this version of default logic does not admit ungrounded conclusions.

Example 10.1.4. Consider the theory

$$T = \left\langle \emptyset, \left\{ \frac{B(x) : F(x)}{F(x)}, \frac{F(x) : B(x)}{B(x)} \right\} \right\rangle,$$

where B and F stand for *Bird* and *Flies*, respectively. A CAKE diagram for the theory T is provided in Figure 10.4.

The following set of rules is associated with the diagram:

$$A_2.F^-(x) \rightarrow A_3.F^-(x)$$
$$\{A_2.F^+(x) \vee [A_4.B^+(x) \wedge \neg A_2.F^-(x)]\} \rightarrow A_3.F^+(x)$$
$$A_1.B^-(x) \rightarrow A_4.B^-(x)$$
$$\{A_1.B^+(x) \vee [A_3.F^+(x) \wedge \neg A_1.B^-(x)]\} \rightarrow A_4.B^+(x).$$

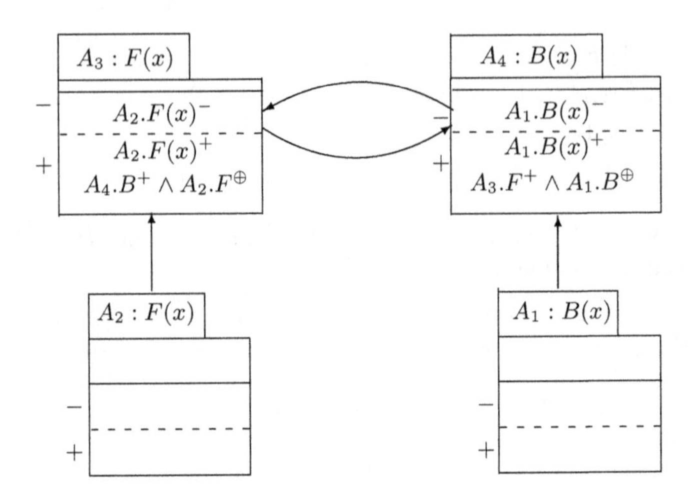

Fig. 10.4. Diagram corresponding to the theory of Example 10.1.4.

The set of rules shown above is stratified using a single partition P_1, where P_1 consists of all the rules.

The relations occurring in the heads of the rules are specified by the fixpoint expression of the form

$$\text{LFP}\, A_3.F^-, A_3F^+, A_4.B^-, A_4.B^+.\mathcal{B}, \tag{10.7}$$

where $\mathcal{B}$ denotes the conjunction of the rules' bodies.

Suppose that the query we are interested in is $B(\mathsf{t})$. It is immediately seen that all the relations characterized by the expression (10.7) have empty extensions.[10] Accordingly, the answer to the query is UNKNOWN. □

Example 10.1.5. Consider the theory

$$T = \left\langle \{L(\mathsf{j}, \mathsf{b})\}, \left\{ \frac{L(x,y) : H(y)}{H(y)} \right\} \right\rangle,$$

where L, H, j and b denote *Likes*, *Happy*, John and Bill, respectively. The CAKE diagram corresponding to the theory T is shown in Figure 10.5.

The rules associated with the diagram are the following.

[10] Recall that the extension of an n-ary relation is the set of all n-tuples satisfying this relation.

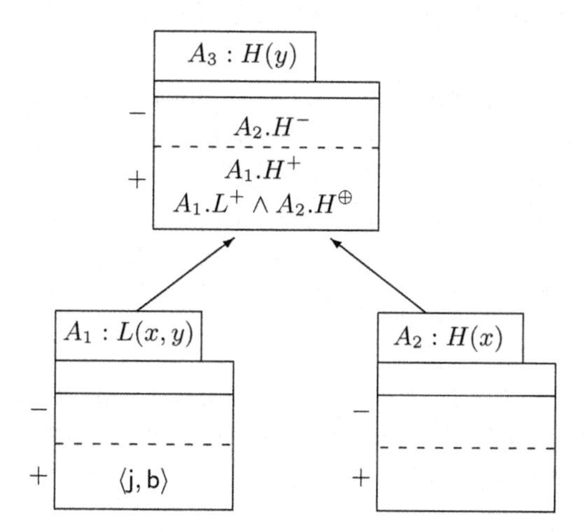

Fig. 10.5. Diagram corresponding to the theory of Example 10.1.5.

$$A_1.L^+(\mathsf{j},\mathsf{b})$$
$$A_2.H^-(x) \rightarrow A_3.H^-(x)$$
$$\exists y.\, \{A_2.H^+(x) \vee [A_1.L^+(x,y) \wedge \neg A_2.H^-(x)]\} \rightarrow A_3.H^+(x).$$

The partition P_1, P_2, where P_1 consists of the first rule and P_2 contains the remaining rules provides a stratification of the above set.

The relations occurring in the heads of the rules are defined by the fixpoint expression given by

$$\text{LFP}\, A_1.L^+, A_3.H^-, A_3.H^+.\mathcal{B}, \tag{10.8}$$

where, as usual, $\mathcal{B}$ is the conjunction of all the rules' bodies.

The computation for the relations mentioned in formula (10.8) are provided below.

$$\{\}$$
$$\{A_1.L^+(\mathsf{j},\mathsf{b})\}$$
$$\{A_1.L^+(\mathsf{j},\mathsf{b}), A_3.H^+(\mathsf{b})\}.$$

Suppose that the query under consideration is $H(\mathsf{b})$. Since b satisfies the $A_3.H^+$-coordinate of (10.8), the answer to the query $H(\mathsf{b})$ is TRUE. □

10.2 Computing Normal Default Theories

We refer to the version of default logic exemplified in the previous section as *rough default logic.* In this section, we begin the presentation of the technical details by first considering normal default theories.[11]

We begin with some terminology. Two defaults are said to be *similar* if their consequents contain the same relation symbol. The following rather detailed definition describes the construction of CAKE diagrams for arbitrary disjunction-free normal default theories.

Definition 10.2.1. *Let $T = \langle W, D \rangle$ be a normal default theory over a language $\mathcal{L}$. A* normal default CAKE diagram *corresponding to a normal default theory T is constructed as follows.*

1. *For each relation symbol R occurring in $\mathcal{L}$, we construct a* CAKE *granule, called a* database granule, *representing R. If $R(\overline{t})$ (respectively $\neg R(\overline{t})$) is in W, the tuple $\overline{t}$ occurs in the positive (respectively negative) part of the diagram of the granule. All other tuples of constants are assumed to be in the boundary region of the diagram.*[12] *Each database granule is labeled by name: $R(-)$, where name is a unique name of the granule and R is the relation symbol it is responsible for.*
2. *For each default d from D, we construct a* CAKE *granule, called a* default granule, *representing d. The granule is labeled by name: R, where name is a unique name of the granule and R is the relation symbol occurring in the consequent of d. Methods used by default granules will be specified later.*
3. *All default granules, say $A_1, \ldots, A_k$, where $k > 1$, representing similar defaults are grouped into a single knowledge module, referred to as a* default module. *The module is labeled by name: R, where name is a unique name of the module and R is the relation symbol the granules $A_1, \ldots, A_k$ are responsible for. In what follows, any default granule which is not embedded in a default module will be referred to as an* independent default granule. *Otherwise, the granule will be referred to as a* dependent default granule.
4. *Let A be an independent default granule representing a default d and suppose that the consequent of d contains a relation symbol R. Assume further that $R_1, \ldots, R_k$ are all the relation symbols occurring in the prerequisite of d. For each $1 \le i \le k$, we create an arrow from a database granule representing R_i to A and an arrow from a module (or independent de-*

[11] Recall that all the theories considered in this chapter are disjunction-free default theories with consistent sets of axioms.

[12] Recall that the boundary part of a diagram is not explicitly drawn.

fault granule if it exists) representing R_i. We also create an arrow from a database granule representing R to A.

5. *Let M be a module representing a relation R. Let $A_1, \ldots, A_n$ be default granules of M and suppose that A_i corresponds to a default d_i. Assume further that $R_1, \ldots, R_k$ are all the relation symbols occurring in the prerequisites of $d_1, \ldots, d_n$. For each $1 \leq i \leq k$, we create an arrow from a database granule representing R_i to M and, an arrow from a module (or independent default granule if it exists) representing R_i. We also create an arrow from a database granule representing R to M.*
6. *For each default granule (dependent or not) we now assign the methods it will employ. Assume that the granule represents a default of the form $l_1 \wedge \cdots \wedge l_n : l/l$, where $n \geq 0$, and suppose that $R_1, \ldots, R_n, R$ are the relation symbols occurring in $l_1, \ldots, l_n, l$, respectively. First of all, there are two methods, namely $A_k.R^+$ and $A_k.R^-$, where A_k is the database granule responsible for the relation R. These methods, which will be referred to as database methods, are placed in the positive and negative parts of the diagram, respectively. In addition, there is one method corresponding to the default represented by the granule. This method, referred to as a* default method, *is placed in the positive part of the granule's diagram if l is a positive literal and in the negative part, otherwise. The default method is constructed as follows.*
   a) *Suppose first that none of $R_1, \ldots, R_n$ is represented either by a default module or by an independent default granule. Then the default method is given by*
   $$A_1.X_1 \wedge \ldots \wedge A_n.X_n \wedge A.X, \tag{10.9}$$
   where $A_1, \ldots, A_n, A$ are database granules responsible for $R_1, \ldots, R_n, R$, respectively, each X_i is ${R_i}^+$ (if l_i is positive) or ${R_i}^-$ (if l_i is negative) and X is $R^{\oplus}$ (if l is positive) or $R^{\ominus}$ (if l is negative).
   b) *Suppose now that at least one of $R_1, \ldots, R_n$ is represented by a default module or by an independent default granule. Let $R_i, \ldots R_j$ be all relations from $R_1, \ldots, R_n$ that are represented by default modules (or independent default granules) and suppose that $M_i, \ldots, M_j$ are the corresponding modules (or independent default granules). Then the default method is (10.9) with $A_i, \ldots, A_j$ replaced by $M_i, \ldots, M_j$, respectively.* □

The set of rules corresponding to a diagram of a default theory is specified using CAKE's standard methodology see Chapter 9, Section 9.5.3). We also use the standard voting policy to resolve conflicts, i.e., adjudicating granules are not included in default modules.

The following example shows that the rules attached to diagrams of normal default theories are not always stratified.

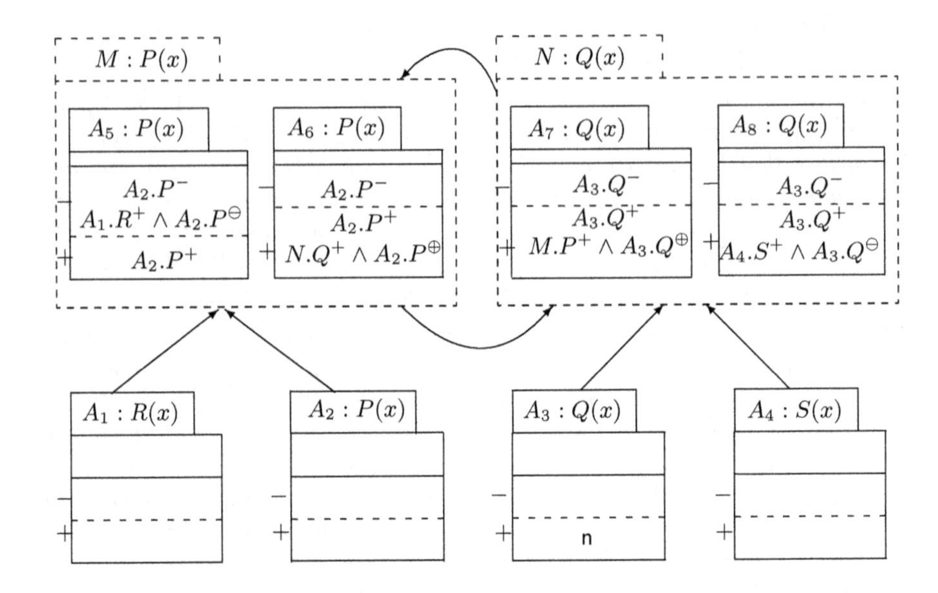

Fig. 10.6. Diagram corresponding to the theory of Example 10.2.2.

Example 10.2.2. Consider the theory $T = \langle W, D \rangle$, where $W = \emptyset$ and

$$D = \left\{ \frac{Q(x) : P(x)}{P(x)}, \frac{R(x) : \neg P(x)}{\neg P(x)}, \frac{P(x) : Q(x)}{Q(x)}, \frac{S(x) : \neg Q(x)}{\neg Q(x)} \right\}.$$

The diagram of this theory is shown in Figure 10.6. The set of rules attached to the diagram is:

$$A_3.Q^+(\mathsf{n}) \tag{10.10}$$
$$[A_2.P^-(x) \vee (A_1.R^+(x) \wedge \neg A_2.P^+(x))] \rightarrow A_5.P^-(x) \tag{10.11}$$
$$A_2.P^+(x) \rightarrow A_5.P^+(x) \tag{10.12}$$
$$A_2.P^-(x) \rightarrow A_6.P^-(x) \tag{10.13}$$
$$[A_2.P^+(x) \vee (N.Q^+(x) \wedge \neg A_2.P^-(x))] \rightarrow A_6.P^+(x) \tag{10.14}$$
$$A_3.Q^-(x) \rightarrow A_7.Q^-(x) \tag{10.15}$$
$$[A_3.Q^+(x) \vee (M.P^+(x) \wedge \neg A_3.Q^-(x))] \rightarrow A_7.Q^+(x) \tag{10.16}$$
$$A_3.Q^-(x) \rightarrow A_8.Q^-(x) \tag{10.17}$$
$$[A_3.Q^+(x) \vee (A_4.S^+(x) \wedge \neg A_3.Q^-(x))] \rightarrow A_8.Q^+(x) \tag{10.18}$$
$$[A_5.P^-(x) \vee A_6.P^-(x)] \wedge \neg[A_5.P^+(x) \vee A_6.P^+(x)] \rightarrow M.P^-(x) \tag{10.19}$$
$$[A_5.P^+(x) \vee A_6.P^+(x)] \wedge \neg[A_5.P^-(x) \vee A_6.P^-(x)] \rightarrow M.P^+(x) \tag{10.20}$$
$$[A_7.Q^-(x) \vee A_8.Q^-(x)] \wedge \neg[A_7.Q^+(x) \vee A_8.Q^+(x)] \rightarrow N.Q^-(x) \tag{10.21}$$
$$[A_7.Q^+(x) \vee A_8.Q^+(x)] \wedge \neg[A_7.Q^-(x) \vee A_8.Q^-(x)] \rightarrow N.Q^+(x). \tag{10.22}$$

It can easily be observed that the above rules are not stratified. □

In what follows, we mainly deal with stratified default theories as defined below. However, non-stratified default theories will be briefly discussed in Section 10.6.

Definition 10.2.3. *By a* stratified default theory *(normal or not) we understand a theory whose* CAKE *diagram leads to a stratified set of rules.* □

The following definition of provability will be used shortly.

Definition 10.2.4. *Let T be a stratified normal default theory and suppose that $\mathbf{R}$ is the set of rules corresponding to the diagram of T. Let $\mathcal{B}$ be the conjunction of the rules' bodies from $\mathbf{R}$ and assume that $R_1, \ldots, R_k$ are all relation symbols occurring in the heads of the rules. Consider the fixpoint expression given by*

$$\mathrm{LFP}\, R_1, \ldots, R_k.\mathcal{B}. \tag{10.23}$$

A ground literal of the form $P(\overline{c})$ (respectively $\neg P(\overline{c})$) is said to be provable *in T if and only if $\overline{c}$ satisfies the R_i-coordinate of (10.23), R_i is of the form $A.P^+$ (respectively $A.P^-$) and A is the name associated with a database granule, an independent default granule or a default module.* □

The central concept of default logic is that of an extension of a default theory. In rough default logic this notion can be defined as follows.

Definition 10.2.5. *Let T be a stratified normal default theory. An* extension *of T is the set of all ground literals which are provable in T.* □

In view of earlier examples, one may suspect that a ground literal $l(\overline{t})$ belongs to an extension of a stratified normal default theory T if and only if $l(\overline{t})$ is a member of every extension of T in Reiter's default logic. However, as the next example shows, this is not always the case.

Example 10.2.6. Consider the theory $T = \langle W, D \rangle$, where

$$W = \{Q(\mathsf{n}), R(\mathsf{n})\}$$

$$D = \left\{ \frac{Q(x) : P(x)}{P(x)}, \frac{R(x) : \neg P(x)}{\neg P(x)}, \frac{P(x) : V(x)}{V(x)}, \frac{\neg P(x) : V(x)}{V(x)} \right\}.$$

This is an extended version of the standard "Nixon" diamond theory where R, P, Q, V, n denote *Republican*, *Pacifist*, *Quaker*, *Voter* and Nixon, respectively. The CAKE diagram corresponding to the theory T is shown in Figure 10.7.

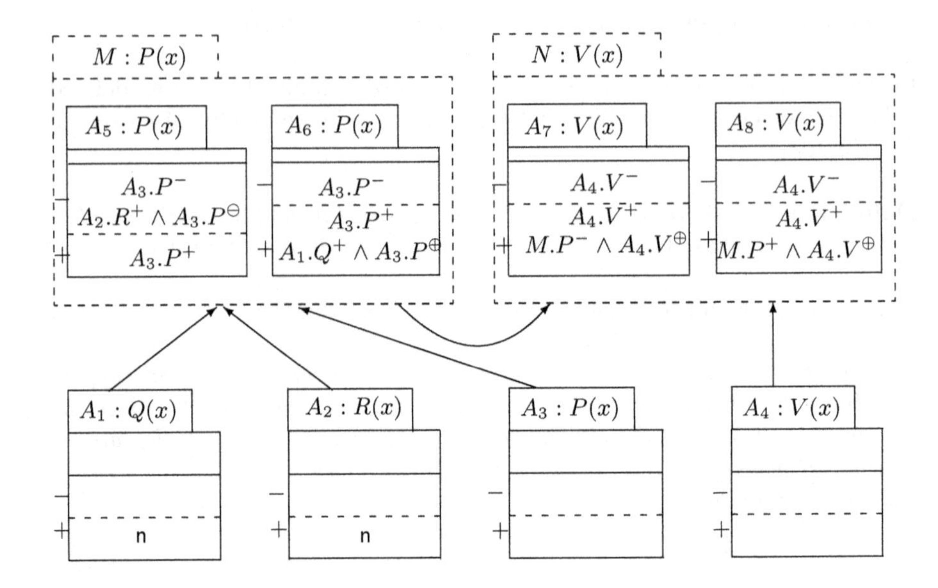

Fig. 10.7. Diagram corresponding to the theory of Example 10.2.6.

In Reiter's default logic T has two extensions, E_1 and E_2, given by

$$E_1 = Cn(\{Q(\mathsf{n}), R(\mathsf{n}), V(\mathsf{n}), P(\mathsf{n})\})$$
$$E_2 = Cn(\{Q(\mathsf{n}), R(\mathsf{n}), V(\mathsf{n}), \neg P(\mathsf{n})\}).$$

Note that the literal $V(\mathsf{n})$ is a member of both E_1 and E_2. However, in rough default logic the literal $V(\mathsf{n})$ does not belong to the extension. To see this, consider the set of rules associated with the diagram of the theory T:

$$A_1.Q^+(\mathsf{n})$$
$$A_2.R^+(\mathsf{n})$$
$$\{A_3.P^-(x) \vee [A_2.R^+(x) \wedge \neg A_3.P^+(x)]\} \rightarrow A_5.P^-(x)$$
$$A_3.P^+(x) \rightarrow A_5.P^+(x)$$
$$A_3.P^-(x) \rightarrow A_6.P^-(x)$$
$$\{A_3.P^+(x) \vee [A_1.Q^+(x) \wedge \neg A_3.P^-(x)]\} \rightarrow A_6.P^+(x)$$
$$A_4.V^-(x) \rightarrow A_7.V^-(x)$$
$$\{A_4.V^+(x) \vee [M.P^-(x) \wedge \neg A_4.V^-(x)]\} \rightarrow A_7.V^+(x)$$
$$A_4.V^-(x) \rightarrow A_8.V^-(x)$$
$$\{A_4.V^+(x) \vee [M.P^+(x) \wedge \neg A_4.V^-(x)]\} \rightarrow A_8.V^+(x)$$
$$[A_5.P^-(x) \vee A_6.P^-(x)] \wedge \neg[A_5.P^+(x) \vee A_6.P^+(x)] \rightarrow M.P^-(x)$$
$$[A_5.P^+(x) \vee A_6.P^+(x)] \wedge \neg[A_5.P^-(x) \vee A_6.P^-(x)] \rightarrow M.P^+(x)$$

$$[A_7.V^-(x) \vee A_8.V^-(x)] \wedge \neg[A_7.V^+(x) \vee A_8.V^+(x)] \rightarrow N.V^-(x)$$
$$[A_7.V^+(x) \vee A_8.V^+(x)] \wedge \neg[A_7.V^-(x) \vee A_8.V^-(x)] \rightarrow N.V^+(x).$$

The last four rules represent the standard voting mechanism used by the modules M and N.

The above set of rules is clearly stratified.

The relations occurring in the heads of the above rules are defined by the following fixpoint expression:

$$\text{LFP}\, \overline{X}.\mathcal{B}, \tag{10.24}$$

where $\overline{X}$ is the tuple $A_1.Q^+, A_2R^+, A_5.P^-, A_5.P^+, A_6.P^-, A_6.P^+, A_7.V^-, A_7.V^+, A_8.V^-, A_8.V^+, M.P^-, M.P^+, N.V^-, N.V^+$ and $\mathcal{B}$ is the conjunction of all the rules' bodies.

The computation of the relations defined by the formula (10.24) is given below.

$$\{\}$$
$$\{A_1.Q^+(\mathsf{n}), A_2.R^+(\mathsf{n})\}$$
$$\{A_1.Q^+(\mathsf{n}), A_2.R^+(\mathsf{n}), A_5.P^-(\mathsf{n}), A_6.P^+(\mathsf{n})\}.$$

Now, it is immediately seen that n does not satisfy the $N.V^+$-coordinate of (10.24). Accordingly, $V(\mathsf{n})$ is not provable in the theory T and hence $V(\mathsf{n})$ does not belong to the extension of T. This result might seem a bit counterintuitive in the framework of classical two-valued logic (no matter whether $P(x)$ or $\neg P(x)$ holds, $V(x)$ should be concluded). However, here we deal with three-valued logic and $P(x)$ might additionally be UNKNOWN. In this context the result makes perfect sense. □

The above examples should provide some intuitions as to how one models and uses normal default theories in CAKE. The crucial observation is that defaults whose consequents refer to the same relation symbol are grouped in a module. In contrast to Reiter's logic, where these defaults can interact individually, in this formalism the interaction is achieved via modules.

The next two results follow straightforwardly.

Theorem 10.2.7. *In rough default logic every stratified normal default theory has one extension.* □

Theorem 10.2.8. *In rough default logic every stratified normal default theory with a consistent set of axioms has a consistent extension.* □

Observe that the property of semi-monotonicity generally does not hold in rough default logic. To see this, consider the normal theory

$$T = \left\langle \{Q(\mathsf{n}), R(\mathsf{n})\}, \left\{ \frac{Q(x) : P(x)}{P(x)} \right\} \right\rangle.$$

The extension of this theory is $E = \{Q(\mathsf{n}), R(\mathsf{n}), P(\mathsf{n})\}$.

Now, if we add a new default $\frac{R(x) : \neg P(x)}{\neg P(x)}$, the extension of the expanded theory is $E' = \{Q(\mathsf{n}), R(\mathsf{n})\}$.

10.3 Default Logic with Strong Prerequisites

Both standard default logic and the approach introduced in section 10.2 lack tools to distinguish between facts *known* to be true, i.e., derivable from axioms, and those *believed* to be true, i.e., derivable by applying defaults. In consequence, both these formalisms are unable to properly deal with many practically occurring settings. The next example will help to illustrate this.

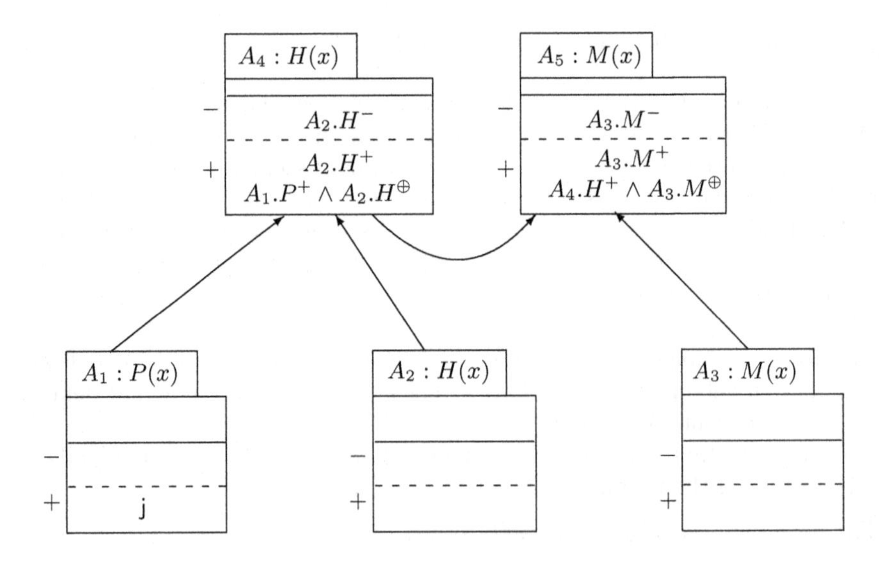

Fig. 10.8. Diagram corresponding to the theory of Example 10.3.1.

Example 10.3.1. Suppose we are given the following facts:

John is a person.
Normally, a person is honest.
Normally, a person known to be honest can be safely lent money.

Given these facts, we are prepared to believe that John is honest, but we do not want to infer that he can be safely lent money. The reason, of course, is that John's honesty is a default conclusion rather, than an iron-clad fact.

This commonsense theory cannot be properly formalized according to the intuitions stated above in the framework of default logic introduced in section 10.2. One can use the following default theory T given by

$$W = \{P(\mathsf{j})\} \qquad D = \left\{ \frac{P(x) : H(x)}{H(x)}, \frac{H(x) : M(x)}{M(x)} \right\},$$

where P, H and M and j stand for "Person," "Honest," "Can be safely lent money" and "John," respectively. In Reiter's default logic $M(\mathsf{j})$ is, in fact, provable. No distinction is made between $H(\mathsf{j})$ as a fact and $H(\mathsf{j})$ inferred by default.

The CAKE diagram for theory T is provided in Figure 10.8.

It is easily checked that, contrary to our intuition, the literal $M(\mathsf{j})$ is a member of the extension of T.

Fortunately, the CAKE method provides us with a simple way to block the unwanted conclusion which also makes intuitive sense. All that has to be done is to remove the arrow from A_4 to A_5, create an additional arrow from A_2 to A_5 and replace the method $A_4.H^+ \wedge A_3.M^\oplus$, assigned to A_5, by $A_2.H^+ \wedge A_3.M^\oplus$ (see Figure 10.9).

In this representation, the second default is inapplicable for John because his honesty cannot be inferred directly from the database. □

We now formalize the above idea in a systematic manner by introducing a new variant of rough default logic. This variant, called *rough default logic with strong prerequisites* (SP *rough default logic*, for short), explicitly distinguishes between facts known to be true and those believed to be true. Facts known to be true are simply those atomic facts that are represented explicitly in an assumed extensional database.

We extend the language of rough default logic by adding a new operator **K** which can be applied to literals. A formula of the form **K**l is to be read as "l is known to be true."[13]

As stated previously, we limit ourselves to disjunction-free default theories. However, in SP rough default logic any conjunct occurring in the prerequisite of a default can be preceded by the operator **K**. More specifically, we consider defaults of the form

$$\frac{B_1 \wedge \cdots \wedge B_n : l}{l}$$

[13] It should be emphasized that the **K** operator we use here has the flavor of syntactic sugar rather than of knowledge operators commonly employed in modal logics.

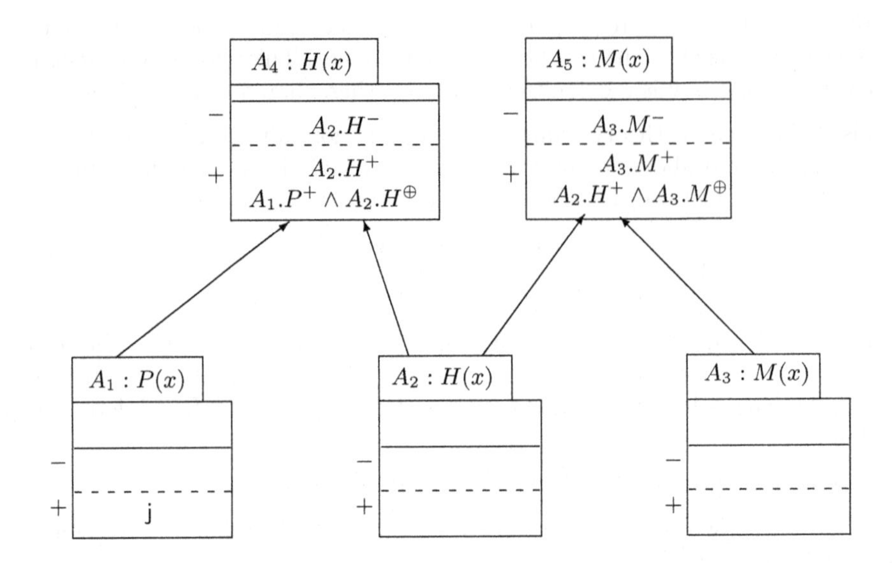

Fig. 10.9. Modified diagram corresponding to the theory of Example 10.3.1.

where each B_i is either a literal or a literal preceded by the operator $\mathbf{K}$ and l is a literal.

The formal difference between rough and SP rough default logic lies in the definition of a diagram. In the latter, any conjunct of the form $\mathbf{K}l$ occurring in the prerequisite of a default must be supported by the axioms of the theory under consideration, rather than by the consequent of an applied default.

Definition 10.3.2. *Let T be a disjunction-free normal default theory of* SP *rough default logic. A diagram corresponding to T is constructed as before (see Definition 10.2.1) with points 4, 5 and 6 replaced by:*

4. *Let A be an independent default granule representing a default d and suppose that the consequent of d contains a relation symbol R. Assume further that $R_1, \ldots, R_k$ are all the relation symbols occurring in the prerequisite of d and let $R'_1, \ldots R'_l$ be those elements from $R_1, \ldots, R_k$ that are not preceded by the operator* $\mathbf{K}$. *For each $1 \leq i \leq k$, we place an arrow from a database granule representing R_i to A and, for each $1 \leq i \leq l$, we place an arrow from a module (or independent default granule if it exists), representing R'_i to A. We also place an arrow from a database granule representing R to A.*
5. *Let M be a module representing a relation R. Let $A_1, \ldots, A_n$ be default granules of M and suppose that A_i corresponds to a default d_i. Assume further that $R_1, \ldots, R_k$ are all the relation symbols occurring in the pre-*

requisites of $d_1, \ldots, d_n$ *and* $R'_1, \ldots, R'_l$ *are those elements from* $R_1, \ldots,$ R_k *that are not preceded by the operator* **K**. *For each* $1 \leq i \leq k$, *we place an arrow from a database granule representing* R_i *to* M *and, for each* $1 \leq i \leq l$, *we place an arrow from a module (or independent granule if it exists) representing* R'_i *to* M. *In addition, we place an arrow from a database granule representing* R *to* M.

6. *For each default granule we assign methods it will employ. Assume that the granule represents a default of the form* $B_1 \wedge \cdots \wedge B_n : l/l$, *where* $n \geq 0$, *and suppose that* $R_1, \ldots, R_n, R$ *are relation symbols occurring in* $l_1, \ldots, l_n, l$, *respectively. Assume further that* $B'_1, \ldots, B'_k$ *are those conjuncts of* $B_1 \wedge \cdots \wedge B_n$ *which are not preceded by* **K** *and let* $R'_1, \ldots, R'_k$ *be the relation symbols occurring in* $B'_1, \ldots, B'_k$. *First of all, there are two database methods, namely* $A'.R^+$ *and* $A'.R^-$, *where* A' *is the database granule responsible for the relation* R. *These methods are placed in the positive and the negative part of the diagram, respectively. In addition, there is one default method corresponding to the default represented by the granule. This method is placed in the positive part of the granule's diagram if* l *is positive and in the negative part otherwise. The default method is specified as follows.*

   a) *Suppose first that none of* $R'_1, \ldots, R'_k$ *is represented either by a default module or by an independent default granule. Then the default method used by the granule is given by*

   $$A_1.X_1 \wedge \cdots \wedge A_n.X_n \wedge A'.X, \tag{10.25}$$

   where $A_1, \ldots, A_n, A'$ *are database granules responsible for* $R_1, \ldots,$ R_n, R, *respectively, each* X_i *is* $R_i{}^+$ *(if* l_i *is positive) or* $R_i{}^-$ *(if* l_i *is negative) and* X *is* $R^{\oplus}$ *(if* l *is positive) or* $R^{\ominus}$ *(if* l *is negative).*

   b) *Suppose now that at least one of* $R'_1, \ldots, R'_k$ *is represented by a default module or by an independent default granule. Let* $R'_i, \ldots R'_j$ *be all relation symbols from* $R'_1, \ldots, R'_k$ *that are represented by default modules (independent default granules) and suppose that* $M_i, \ldots, M_j$ *are the corresponding modules (independent default granules). Then the default method used by the granule is (10.25) with* $A_i, \ldots, A_j$ *replaced by* $M_i, \ldots, M_j$, *respectively.* □

The set of rules corresponding to a diagram of a normal default theory of SP rough default logic is specified as usual. We also use the standard voting policy.

Theorem 10.3.3. *The set of rules* R *corresponding to a diagram* $\mathcal{D}$ *of a normal default theory of* SP *default logic is stratified.* □

The notion of provability in a normal default theory and the notion of an extension of a normal default theory are specified as before (see Definitions 10.2.4 and 10.2.5).

Example 10.3.4. Suppose we are given the following:

> John is a person.
> Normally, a person is honest.
> Normally, if a person is believed to be honest, she/he can be safely lent a small amount of money.
> Normally, if a person is known to be honest, she/he can be safely lent a large amount of money.

In SP rough default logic these common-sense facts can be naturally represented by the following theory T:

$$W = \{P(\mathsf{j})\}$$

$$D = \left\{ \frac{P(x) : H(x)}{H(x)}, \frac{H(x) : SM(x)}{SM(x)}, \frac{\mathbf{K}H(x) : LM(x)}{LM(x)} \right\}.$$

where j, P, H, SM and LM stand for "John," "Person," "Honest," "Can be safely lent a small amount of money" and "Can be safely lent a large amount of money," respectively.

The CAKE diagram corresponding to the theory T is provided in Figure 10.10.

The set of rules corresponding to the diagram of the considered theory is the following.

$$\begin{array}{l}
A_1.P^+(\mathsf{j}) \\
A_2.H^-(x) \rightarrow A_5.H^-(x) \\
\{A_2.H^+(x) \vee [A_1.P^+(x) \wedge \neg A_2.H^-(x)]\} \rightarrow A_5.H^+(x) \\
A_3.SM^-(x) \rightarrow A_6.SM^-(x) \\
\{A_3.SM^+(x) \vee [A_5.H^+(x) \wedge \neg A_3.SM^-(x)]\} \rightarrow A_6.SM^+(x) \\
A_4.LM^-(x) \rightarrow A_7.LM^-(x) \\
\{A_4.LM^+(x) \vee [A_2.H^+(x) \wedge \neg A_4.LM^-(x)]\} \rightarrow A_7.LM^+(x).
\end{array}$$

The partition P_1, P_2, where P_1 consists of the first rule and P_2 contains the remaining rules, provides a stratification of the above set.

The relations occurring in the heads of the above rules are defined by the fixpoint expression given by

$$\mathrm{LFP}\,\overline{X}.\mathcal{B}, \tag{10.26}$$

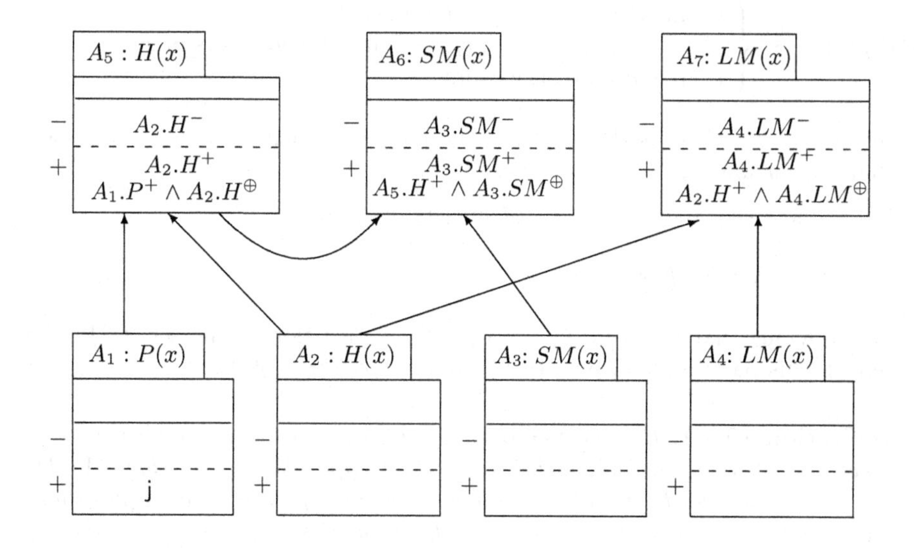

Fig. 10.10. Diagram corresponding to the theory of Example 10.3.4.

where $\overline{X}$ is the tuple $A_1.P^+$, $A_5.H^-$, $A_5.H^+$, $A_6.SM^-$, $A_6.SM^+$, $A_7.LM^-$, $A_7.LM^+$ and $\mathcal{B}$ is the conjunction of all the rules' bodies.

The computation of the relations specified by the expression (10.26) are provided below.

$\{\}$
$\{A_1.P^+(\text{j})\}$
$\{A_1.P^+(\text{j}), A_5.H^+(\text{j})\}$
$\{A_1.P^+(\text{j}), A_5.P^+(\text{j}), A_6.SM^+(\text{j})\}$.

The queries we are interested in are $SM(\text{j})$ and $LM(\text{j})$. It is immediately observed that, according to intuition, $SM(\text{j})$ is in the extension of the theory T, whereas $LM(\text{j})$ is not. □

10.4 Prioritized Default Logic

Both rough default logic and SP rough default logic can be easily extended to their prioritized versions (see Chapter 5, Section 5.2.6).

We start with some terminology, where $<$ is a strict partial order on a finite set X.

For all $x, y \in X$, y is said to be a *successor* of x if and only if $x < y$. With each $x \in X$ we associate a natural number, denoted by $depth(x)$, as follows:

$$depth(x) \stackrel{\text{def}}{=} \begin{cases} 1+\max\{depth(x_1),\dots,depth(x_k)\}, \\ \qquad\qquad \text{where } x_1,\dots,x_k \in X \text{ are all successors of } x \\ 1 \quad \text{if } x \text{ has no successors.} \end{cases}$$

Intuitively, $depth(x)$ is the length of the longest sequence $x_0, x_1, \dots, x_k$ such that $x_0 = x$ and $x_i < x_{i+1}$, for $i = 0, 1, \dots, k-1$.

We define a sequence $X_1, \dots X_n$ of subsets of X by

$X_i = \{x \in X : depth(x) = i\}$.

It is obvious that the sets $X_1, \dots X_n$ form a partition of the set X. In the sequel, this partition will be referred to as $<$*-partition* of X.

Definition 10.4.1. *By a* normal prioritized theory of rough default logic *(respectively* SP rough default logic*) we mean a triple $\langle W, D, <\rangle$, where $\langle W, D\rangle$ is a normal default theory of rough default logic (respectively* SP *rough default logic) and $<$ is a strict partial order on D. Intuitively, $d_1 < d_2$ means that d_2 has higher priority than d_1.* □

The only difference between rough default logic (respectively SP rough default logic) and the prioritized version of rough default logic (respectively the prioritized version of SP rough default logic) lies in the use of a different voting policy.

Definition 10.4.2. *Let M be a default module associated with a relation symbol R and suppose that $d_1, \dots, d_n$ are all the defaults of the considered theory containing R in their consequents. Assume further that $D_1, \dots, D_k$ is a $<$-partition of the set $\{d_1, \dots, d_n\}$.*

The modified voting process starts by applying the standard voting policy to all the granules representing the defaults from D_1. If the answer is TRUE *or* FALSE, *this is the final answer of the modified voting process. If the answer is* UNKNOWN *because at least one granule representing the defaults from D_1 answered* TRUE *and at least one of those granules answered* FALSE, *the final answer of the modified voting process is also* UNKNOWN. *Otherwise, i.e., if all the granules representing defaults from D_1 answered* UNKNOWN, *the standard voting policy is applied to the granules representing the defaults from D_2. This may continue for $D_3, \dots, D_k$ unless at least one granule answers* TRUE *or* FALSE. □

Remark 10.4.3. Note that the voting policy can be modified in different ways if one is not satisfied with the policy described above. □

In order to specify diagrams for prioritized default logic theories, we will require the following additional notation.

Suppose that $A_1, \ldots, A_n$ are default granules responsible for a relation R. We write $V^+(A_1, \ldots, A_n; R)$, $V^-(A_1, \ldots, A_n; R)$ and $V^\pm(A_1, \ldots, A_n; R)$ as the abbreviations for the formulas:[14]

$$[A_1.R^+ \vee \cdots \vee A_n.R^+] \wedge \neg[A_1.R^- \vee \cdots \vee A_n.R^-]$$
$$[A_1.R^- \vee \cdots \vee A_n.R^-] \wedge \neg[A_1.R^+ \vee \cdots \vee A_n.R^+]$$
$$A_1.R^\pm \wedge \cdots \wedge A_n.R^\pm.$$

Definition 10.4.4. *Let $\langle W, D < \rangle$ be a prioritized normal default theory of rough default logic (respectively* Sp *rough default logic). The diagram corresponding to T is specified as before, see Definition 10.2.1 (respectively Definition 10.3.2) with the following additions.*

7. *Assume that granules $A_1, \ldots, A_n$, representing defaults $d_1, \ldots, d_n$, respectively, are grouped in a default module M. Let R be the relation symbol occurring in the consequents of $d_1, \ldots, d_n$ and let $D_1, \ldots, D_k$ be a $<$-partition of the set $\{d_1, \ldots, d_n\}$. If $k > 1$, then the adjudicating granule, labeled by name:R, is introduced into the module M.[15] As usual, "name" is a unique name of the introduced granule. For each $1 \leq i \leq n$, we put an arrow to the adjudicating granule from A_i.*
8. *Methods used by the new adjudicating granule are specified as follows. Assume that, for each $1 \leq j \leq k$, $d_{j1}, d_{j2}, \ldots d_{ji_j}$ are all the defaults from the set D_j. The following methods are put in the negative part of the diagram of the granule:*

$$V^-(A_{11}, \ldots, A_{1i_1}; R)$$
$$V^\pm(A_{11}, \ldots, A_{1i_1}; R) \wedge V^-(A_{21}, \ldots, A_{2i_2}; R)$$
$$\ldots$$
$$V^\pm(A_{11}, \ldots, A_{1i_1}; R) \wedge \cdots \wedge V^\pm(A_{(k-1)1}, \ldots, A_{(k-1)i_{k-1}}; R) \wedge V^-(A_{k1}, \ldots, A_{ki_k}; R).$$

The following methods are put in the positive part of the diagram of the granule:

$$V^+(A_{11}, \ldots, A_{1i_1}; R)$$
$$V^\pm(A_{11}, \ldots, A_{1i_1}; R) \wedge V^+(A_{21}, \ldots, A_{2i_2}; R)$$
$$\ldots$$

[14] Observe that $V^+(A_1, \ldots, A_n; R)$ (respectively $V^-(A_1, \ldots, A_n; R)$) holds if the standard voting policy about R applied to the granules $A_1, \ldots, A_n$ returns the answer True (respectively False). The formula $V^\pm(A_1, \ldots, A_n; R)$ holds if and only if the answer of all the granules is Unknown.

[15] Note that $k > 1$ implies that there are priorities among the defaults represented in the module M.

$$V^{\pm}(A_{11},\ldots,A_{1i_1};R)\wedge\cdots\wedge V^{\pm}(A_{(k-1)1},\ldots,A_{(k-1)i_{k-1}};R)\wedge$$
$$V^{+}(A_{k1},\ldots,A_{ki_k};R).$$

□

The set of rules corresponding to a diagram of a prioritized normal default theory of rough default logic (SP rough default logic) is specified employing the usual CAKE methodology.[16]

Provability in a normal default theory and the notion of an extension of a normal default theory are specified as before (see Definitions 10.2.4 and 10.2.5).

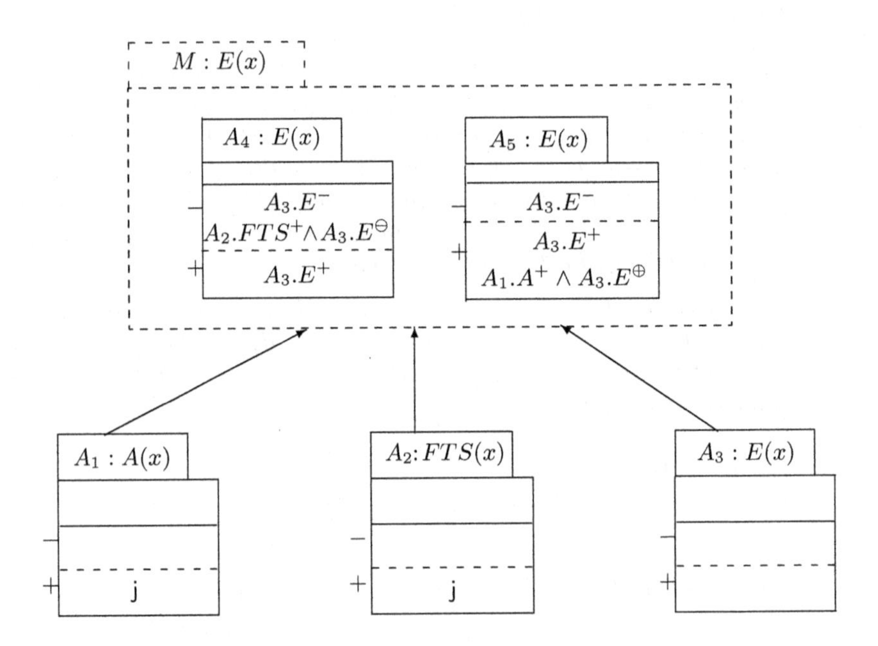

Fig. 10.11. CAKE diagram corresponding to the theory of Example 10.4.5.

Example 10.4.5. Consider the theory

$$W=\{A(\text{j})\wedge FTS(\text{j})\}\qquad D=\left\{\frac{FTS(x):\neg E(x)}{\neg E(x)},\ \frac{A(x):E(x)}{E(x)}\right\}.$$

The following is a non-prioritized default theory for Example 5.2.23 (Chapter 5, Section 5.2.6) with A, FTS, E and j standing for "Adult," "FullTimeStu-

[16] It should be emphasized that the presence of adjudicating granules changes the rules attached to modules (see Chapter 9, Section 9.5.3).

dent,", "Employed" and "John," respectively. The CAKE diagram corresponding to the theory T is provided in Figure 10.11.

Consider the query $E(\mathsf{j})$. It is easily verified that $E(\mathsf{j})$ is not a member of the extension of T. This is because this query is answered FALSE by the granule A_4 and TRUE by the granule A_5. Using the standard voting policy, the final answer is UNKNOWN. As we remarked earlier, see Example 5.2.23, this answer is intuitively problematic. Given that John is both a full time student and an adult the answer to the query should be FALSE. To achieve this effect, we can use a prioritized version of rough default logic with the following prioritization of rules: $\frac{A(x) : E(x)}{E(x)} < \frac{FTS(x) : \neg E(x)}{\neg E(x)}$.

The CAKE diagram for the prioritized theory is similar to that for the non-prioritized version, but in this case, the module M should instead be replaced by that from Figure 10.12. The following set of rules is associated with the diagram for the prioritized theory:

$$
\begin{aligned}
&A_1.A^+(\mathsf{j})\\
&A_2.FTS^+(\mathsf{j})\\
&\{A_3.E^-(x) \lor [A_2.FTS^+(x) \land \neg A_3.E^+(x)]\} \rightarrow A_4.E^-(x)\\
&A_3.E^+(x) \rightarrow A_4.E^+(x)\\
&A_3.E^-(x) \rightarrow A_5.E^-(x)\\
&\{A_3.E^+(x) \lor [A_1.A^+(x) \land \neg A_3.E^-(x)]\} \rightarrow A_5.E^+(x)\\
&\{A_4.E^-(x) \lor [\neg A_4.E^+(x) \land \neg A_4.E^-(x) \land A_5.E^-(x)]\} \rightarrow A_6.E^-(x)\\
&\{A_4.E^+(x) \lor [\neg A_4.E^+(x) \land \neg A_4.E^-(x) \land A_5.E^+(x)]\} \rightarrow A_6.E^+(x)\\
&A_6.E^+(x) \rightarrow M.E^+(x)\\
&A_6.E^-(x) \rightarrow M.E^-(x).
\end{aligned}
$$

The above set is clearly stratified.

The relations occurring in the heads of the above rules are specified by the fixpoint expression given by

$$\mathrm{LFP}\,\overline{X}.\mathcal{B}, \qquad (10.27)$$

where $\overline{X}$ is the tuple $A_1.A^+$, $A_2.FTS^+$, $A_4.E^+$, $A_4.E^-$, $A_5.E^+$, $A_5.E^-$, $A_6.E^+$, $A_6.E^-$, $M.E^+$, $M.E^-$ and $\mathcal{B}$ is the conjunction of all the rules' bodies.

Computing the relations defined by (10.27), results in the following:

$$
\begin{aligned}
&\{\}\\
&\{A_1.A^+(\mathsf{j}), A_2.FTS^+(\mathsf{j})\}\\
&\{A_1.A^+(\mathsf{j}), A_2.FTS^+(\mathsf{j}), A_4.E^-(\mathsf{j}), A_5.E^+(\mathsf{j})\}\\
&\{A_1.A^+(\mathsf{j}), A_2.FTS^+(\mathsf{j}), A_4.E^-(\mathsf{j}), A_5.E^+(\mathsf{j}), A_6.E^-(\mathsf{j})\}\\
&\{A_1.A^+(\mathsf{j}), A_2.FTS^+(\mathsf{j}), A_4.E^-(\mathsf{j}), A_5.E^+(\mathsf{j}), A_6.E^-(\mathsf{j}), M.E^-(\mathsf{j})\}.
\end{aligned}
$$

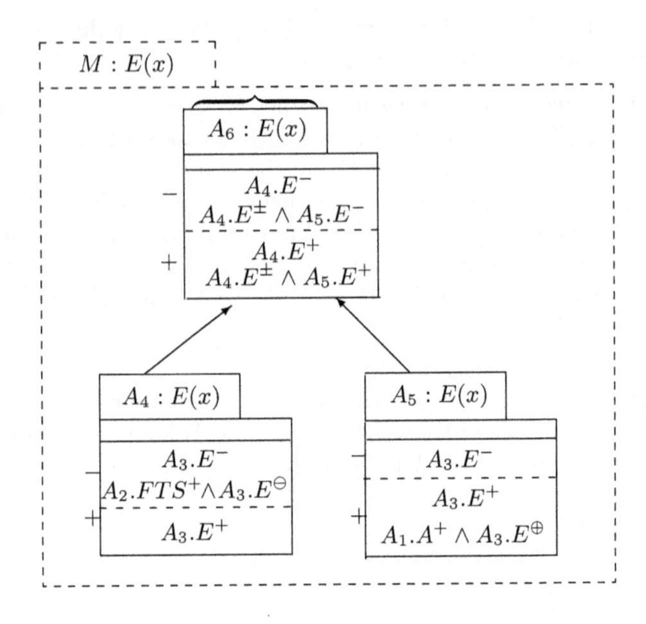

Fig. 10.12. New module corresponding to the prioritized theory of Example 10.4.5.

Since j satisfies the $M.E^-$-coordinate of (10.27), we conclude that $\neg E(\mathrm{j})$ belongs to the extension of the prioritized theory. □

It is obvious that analogous theorems to Theorems 10.2.7 and 10.2.8 hold for prioritized versions of rough default logic and SP rough default logic.

10.5 Semi-Normal Default Theories

In this section, we investigate disjunction-free semi-normal non-prioritized default theories of rough default logic.[17]

Recall that a disjunction-free semi-normal default is any default of the form

$$\frac{l_1 \wedge \cdots \wedge l_n : l'_1 \wedge \cdots \wedge l'_k \wedge l}{l}$$

where $l_1, \ldots, l_n, l'_1, \ldots, l'_k, l$ are literals.

[17] The results in this section can easily be generalized to SP rough default logic and to prioritized versions of both rough default logic and SP rough default logic.

To deal with semi-normal default theories, we proceed analogously as before. The only difference is that additional conjuncts occurring in justifications of semi-normal defaults have to be taken under consideration. The next example illustrates the use of semi-normal defaults.

Example 10.5.1. Consider theory $T = \langle W, D \rangle$, where

$$W = \{D(\mathsf{b})\} \text{ and } D = \left\{ \frac{D(x) : A(x)}{A(x)}, \frac{A(x) : \neg D(x) \wedge E(x)}{E(x)} \right\}.$$

Here A, D, E, and b stand for "Adult," "HighSchoolDropout," "Employed" and "Bill," respectively (see Example 5.2.17, Chapter 5, Section 5.2.4.)

The CAKE diagram corresponding to the theory T is provided in Figure 10.13.

Consider the granule A_5 representing the semi-normal default of our theory. Note that its default method refers to the relation D which occurs in the justification of the default.

The following is the set of rules associated with the above diagram.

$A_2.D^+(\mathsf{b})$
$A_1.A^-(x) \to A_4.A^-(x)$
$\{A_1.A^+(x) \vee [A_2.D^+(x) \wedge \neg A_1.A^-(x)]\} \to A_4.A^+(x)$
$A_3.E^-(x) \to A_5.E^-(x)$
$\{A_3.E^+(x) \vee [A_4.A^+(x) \wedge \neg A_2.D^+(x) \wedge \neg A_3.E^-(x)]\} \to A_5.E^+(x)$.

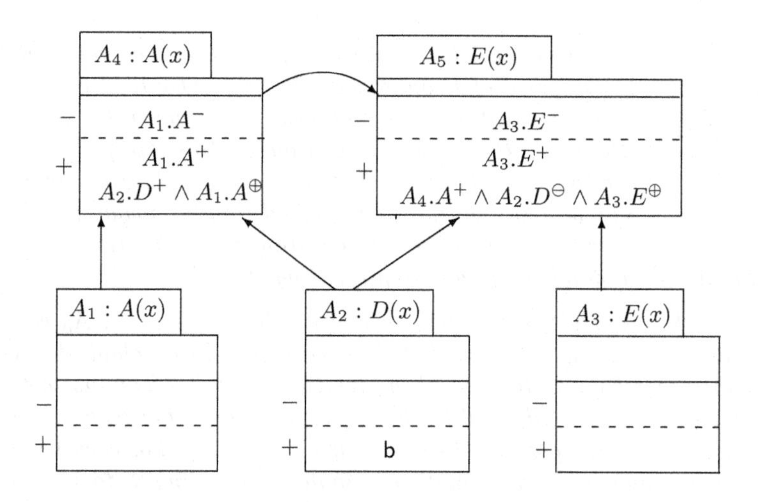

Fig. 10.13. Diagram corresponding to the theory of Example 10.5.1.

The partition P_1, P_2, where P_1 consists of the first rule and P_2 contains the remaining rules, provides a stratification of the above set.

The fixpoint expression specifying the relations occurring in the heads of the rules is given by

$$\text{LFP}\, A_2.D^+, A_4.A^-, A_4.A^+, A_5.E^-, A_5.E^+.\mathcal{B}, \tag{10.28}$$

where $\mathcal{B}$ denotes the conjunction of the rules' bodies.

The resulting computation of the relations specified is provided below. (10.28).

$\{\}$
$\{A_2.D^+(\mathsf{b})\}$
$\{A_2.D^+(\mathsf{b}), A_4.A^+(\mathsf{b})\}.$

Suppose that the query of interest is $E(\mathsf{b})$. It is immediately observed that, according to expectations, b neither satisfies the $A_5.E^+$-coordinate nor the $A_5.E^-$-coordinate of the fixpoint expression (10.28). Consequently, the answer to the query $E(\mathsf{b})$ is UNKNOWN. □

In dealing with semi-normal default theories, the construction of CAKE diagrams has to be adjusted.

Definition 10.5.2. *Let T be a disjunction-free semi-normal default theory. A CAKE diagram corresponding to the theory T is constructed in a manner similar to that from Definition 10.2.1 with the exception that points 4, 5 and 6 are replaced by the following:*

4. *Let A be an independent default granule representing a default d and suppose that the consequent of d contains a relation symbol R. Assume further that $R_1, \ldots, R_k$ are all the relation symbols occurring in the prerequisite of d and let $R_{k+1}, \ldots, R_m$ be all the relation symbols, except R, occurring in the justification of d. For each $1 \leq i \leq m$, we place an arrow from a database granule representing R_i to A and from a module (or independent default granule if it exists) representing R_i to A. An arrow is also placed from a database granule representing R to A.*
5. *Let M be a module representing a relation R. Let $A_1, \ldots, A_p$ be default granules of M and suppose that A_i corresponds to a default d_i. Assume further that $R_1, \ldots, R_k$ are all the relation symbols occurring in the prerequisites of $d_1, \ldots, d_p$ and let $R_{k+1}, \ldots, R_m$ be all the relation symbols, except R, occurring in the justifications of $d_1, \ldots, d_p$. For each $1 \leq i \leq m$, we place an arrow from a database granule representing R_i to M and, from a module (or independent default granule if it exists) representing R_i to M. In addition, we place an arrow from a database granule representing R to M.*

6. *For each default granule (dependent or not) we now assign methods it will employ. Assume that the granule represents a default of the form* $l_1 \wedge \cdots \wedge l_n : l_{n+1} \wedge \cdots \wedge l_m \wedge l/l$, *where* $n \geq 0$, *and suppose that* $R_1, \ldots, R_n, R_{n+1}, \ldots, R_m$, R *are the relation symbols occurring in* $l_1, \ldots, l_n, l_{n+1}, \ldots, l_m, l$, *respectively. First of all, there are two methods, namely* $A_k.R^+$ *and* $A_k.R^-$, *where* A_k *is the database granule responsible for the relation* R. *These methods are placed in the positive and negative parts of the diagram, respectively. In addition, there is one method corresponding to the default represented by the granule. This method is placed in the positive part of the granule's diagram if* l *is a positive literal and in the negative part otherwise. The default method is constructed as follows.*
   a) *Suppose first that none of* $R_1, \ldots, R_n, R_{n+1}, \ldots, R_m$ *is represented either by a default module or by an independent default granule. Then the default method is given by*
   $$A_1.X_1 \wedge \cdots \wedge A_n.X_n \wedge A_{n+1}.X_{n+1} \wedge \cdots \wedge A_m.X_m \wedge A.X \tag{10.29}$$
   where $A_1, \ldots, A_n, A_{n+1}, \ldots, A_m, A$ *are database granules responsible for* $R_1, \ldots, R_n, R_{n+1}, \ldots, R_m, R$, *respectively, each* X_i $(1 \leq i \leq n)$ *is* ${R_i}^+$, *if* l_i *is positive, or* ${R_i}^-$, *if* l_i *is negative, each* X_i $(n+1 \leq i \leq m)$ *is* $R_i^{\oplus}$, *if* l_i *is positive, or* $R_i^{\ominus}$, *if* l_i *is negative. Finally,* X *is* $R^{\oplus}$, *if* l *is positive, or* $R^{\ominus}$, *if* l *is negative.*
   b) *Suppose now that at least one of* $R_1, \ldots, R_n, R_{n+1}, \ldots, R_m$ *is represented by a default module or by an independent default granule. Let* $R_i, \ldots R_j$ *be all relations from* $R_1, \ldots, R_n, R_{n+1}, \ldots, R_m$ *that are represented by default modules (independent default granules) and suppose that* $M_i, \ldots, M_j$ *are the corresponding modules (independent default granules). Then the default method is (10.29) with* $A_i, \ldots, A_j$ *replaced by* $M_i, \ldots, M_j$, *respectively.* □

Since the class of normal default theories is a subset of prioritized normal default theories, it is clear that prioritized normal default theories need not be stratified (see Example 10.2.2 and Section 10.6).

For stratified semi-normal default theories the notion of an extension is defined as usual. The Theorems 10.2.7 and 10.2.8 also hold for stratified semi-normal default theories.

10.6 Non-stratified Default Theories

The notion of an extension of a default theory has been defined for stratified theories only. However, as we have seen (Example 10.2.2), even normal theories need not be stratified. In this section, we briefly describe how non-stratified theories can by approached using the well-founded semantics (see Section 4.5.2).

The process of computing an extension of a default theory T using the well-founded semantics proceeds in three steps.

1. we replace a set of CAKE rules assigned to the diagram of the theory T by a corresponding DATALOG$^{\neg}$ program P_T
2. we compute the well-founded model of P_T
3. using the model obtained in step 2, we compute an extension of T.

The steps 1 and 2 above, have been described in Sections 4.5.2 and 9.5.5, respectively. To describe step 3, we have to reformulate the notion of provability from Definition 10.2.4.

Definition 10.6.1. *Let T be a default theory and suppose that $\mathbf{R}$ is the set of rules corresponding to the diagram of T. Suppose further that I is the well-founded model of the* DATALOG$^{\neg}$ *program corresponding to $\mathbf{R}$. A ground literal of the form $P(\overline{c})$ (respectively $\neg P(\overline{c})$) is said to be* wf-provable *in T if and only if $A.P^{+}$ (respectively $A.P^{-}$) is true in I and A is the name associated with a database granule, an independent default granule or a default module.* □

Definition 10.6.2. *Let T be a default theory. A* wf-extension *of T is the set of all ground literals which are provable in T.* □

Example 10.6.3 (Example 10.2.2 continued). The program corresponding to the rules (10.10)-(10.22) is the following.

$$A_3.Q^{+}(\mathsf{n}) \leftarrow$$
$$A_5.P^{-}(x) \leftarrow A_2.P^{-}(x)$$
$$A_5.P^{-}(x) \leftarrow A_1.R^{+}(x), \neg A_2.P^{+}(x)$$
$$A_5.P^{+}(x) \leftarrow A_2.P^{+}(x)$$
$$A_6.P^{-}(x) \leftarrow A_2.P^{-}(x)$$
$$A_6.P^{+}(x) \leftarrow A_2.P^{+}(x)$$
$$A_6.P^{+}(x) \leftarrow N.Q^{+}(x), \neg A_2.P^{-}(x)$$
$$A_7.Q^{-}(x) \leftarrow A_3.Q^{-}(x)$$
$$A_7.Q^{+}(x) \leftarrow A_3.Q^{+}(x)$$
$$A_7.Q^{+}(x) \leftarrow M.P^{+}(x), \neg A_3.Q^{-}(x)$$
$$A_8.Q^{-}(x) \leftarrow A_3.Q^{-}(x)$$
$$A_8.Q^{+}(x) \leftarrow A_3.Q^{+}(x)$$
$$A_8.Q^{+}(x) \leftarrow A_4.S^{+}(x), \neg A_3.Q^{-}(x)$$
$$M.P^{-}(x) \leftarrow A_5.P^{-}(x), \neg A_5.P^{+}(x), \neg A_6.P^{+}(x)$$
$$M.P^{-}(x) \leftarrow A_6.P^{-}(x), \neg A_5.P^{+}(x), \neg A_6.P^{+}(x)$$

$$M.P^+(x) \leftarrow A_5.P^+(x), \neg A_5.P^-(x), \neg A_6.P^-(x)$$
$$M.P^+(x) \leftarrow A_6.P^+(x), \neg A_5.P^-(x), \neg A_6.P^-(x)$$
$$N.Q^-(x) \leftarrow A_7.Q^-(x), \neg A_7.Q^+(x), \neg A_8.Q^+(x)$$
$$N.Q^-(x) \leftarrow A_8.Q^-(x), \neg A_7.Q^+(x), \neg A_8.Q^+(x)$$
$$N.Q^+(x) \leftarrow A_7.Q^+(x), \neg A_7.Q^-(x), \neg A_8.Q^-(x)$$
$$N.Q^+(x) \leftarrow A_8.Q^+(x), \neg A_7.Q^-(x), \neg A_8.Q^-(x).$$

Using the method described in Section 4.5.2, it is easily checked that the well-founded model of the above program is given by

$$\begin{aligned} \{ \; & N.Q^+(\mathsf{n}), A_3.Q^+(\mathsf{n}), A_7.Q^+(\mathsf{n}), A_8.Q^+(\mathsf{n}), \\ & \neg N.Q^-(\mathsf{n}), \neg A_3 Q^-(\mathsf{n}), \neg A_7.Q^-(\mathsf{n}), \neg A_8.Q^-(\mathsf{n}), \\ & M.P^+(\mathsf{n}), A_6.P^+(\mathsf{n}), \\ & \neg M.P^-(\mathsf{n}), \neg A_2.P^-(\mathsf{n}), \neg A_2.P^+(\mathsf{n}), \\ & \neg A_5.P^-(\mathsf{n}), \neg A_5.P^+(\mathsf{n}), \neg A_6.P^-(\mathsf{n}), \\ & \neg A_1.R^+(\mathsf{n}), \\ & \neg A_4.S^+(\mathsf{n}) \; \}. \end{aligned}$$

Thus, the theory T has a wf-extension given by

$$E = \{P(\mathsf{n}), Q(\mathsf{n})\}.$$

We conclude this section by remarking that for stratified default theories extensions and wf-extensions are identical.

10.7 Conclusions

In this chapter, we have shown that a subset of default logic can easily be formalized using the CAKE methodology. The main advantages of this approach are that the computation of an extension of a default theory is tractable and the extensions are intuitively sound for many of the problematic examples previously considered in the literature. As shown for Reiter's default logic in [98], this problem is computationally hard in the general case, even for propositional disjunction-free theories.

The use of rough set ideas has provided us with a straightforward way to distinguish between "hard" and "soft" facts, when considering the prerequisite of a default. This allows us to properly model various naturally occurring scenarios that are very difficult to model intuitively in traditional versions of default logic.

General non-normal default theories have not been considered in this chapter. In our opinion, they are less interesting from a practical point of view. As we argued elsewhere, all naturally occurring scenarios can be properly dealt with in the framework of semi-normal default theories (for references see Section 10.8).

To compute an extension of a stratified default theory, we construct an appropriate "global" fixpoint expression. In practice, we are generally interested in single queries. For this reason, it would be reasonable to specify a top-down algorithm which performs only necessary calculations. Defining such an algorithm is straightforward, provided that a CAKE diagram corresponding to a given theory does not contain any cycles. To specify such an algorithm in the general case is an interesting research problem.

10.8 Bibliographic Notes

The very first idea of combining rough sets and default reasoning, based on rough truth of Pawlak [149], has been communicated by Skowron. The versions of default logic we have presented in this chapter are based on results reported in [58]. Of course, our considerations have been heavily influenced by the general literature concerning default logic. For references see Section 5.4, Chapter 5.

The idea of distinguishing between weak and strong prerequisites originates from [51] (see also [172]).

The observation that all naturally occurring scenarios can be properly dealt with in the framework of semi-normal default theories has been made in [114].

11

A UAV Scenario: A Case Study

11.1 Introduction

In the current chapter we provide a small case study, based on the WITAS UAV application domain to illustrate various knowledge representation and reasoning techniques presented in the book.

Consider a fully autonomous UAV operating over road and traffic networks. While operating over such an environment, the UAV should be able to navigate autonomously at different altitudes (including autonomous take-off and landing), plan for mission goals such as locating, identifying, tracking and monitoring different vehicle types, and construct internal representations of its focus of attention for use in achieving its mission goals. Additionally, it should be able to identify complex patterns of behavior such as vehicle overtaking, traversing of intersections, parking lot activities, etc.

A UAV has a number of deliberative services such as task planners, trajectory planners, prediction mechanisms, and chronicle recognizers which are dependent on internal qualitative representations of the operational environment over which the UAV operates. The knowledge representation components include a soft-real time database called the dynamic object repository, a standard relational database, a geographic information system containing road and geographic data and a number of front-end query-answering systems which serve as inference engines and may be used by other components in the architecture. In fact, the research described in the current chapter provides a basis for one of these inference engines. In addition to these components, there is an image processing module used for vision tasks and a helicopter control module which is used to position the helicopter and camera dynamically and maintain positions during the execution of task goals which may include highly dynamic tasks such as tracking vehicles through a small village with building obstacles.

P. Doherty et al.: *Knowledge Representation Techniques*, Studfuzz **202**, 213–226 (2006)
www.springerlink.com

Let us examine a particular scenario from the UAV operational environment representative of the use of LCC reasoning in the UAV context. LCC reasoning is described in detail in Chapter 7.

Suppose a UAV receives the following mission goal from its ground control operator:

> Identify and track *all* moving vehicles in region r and log the estimated velocities and positions of *all* small blue vehicles identified, for the duration of their stay in region r, or until the UAV is low on fuel.

Achieving a mission goal such as this in fully autonomous mode is extremely complex and would involve the concurrent use of many of the deliberative and reactive services in the architecture in addition to a great deal of sophisticated reasoning about the operational environment. Both hard and soft real-time constraints must also be taken into consideration, particularly for query-answering during a plan execution phase. In this example, we will focus on a particular type of reasoning capability made possible by the combined use of LCC reasoning and rough knowledge databases.

The first step in achieving the mission goal would be to generate a task plan which would include the following steps:

1. fly to a position that permits viewing region r and possibly an area surrounding the region
2. focus the camera on region r and maintain position, focus and coverage
3. initiate the proper image processing algorithms for identifying moving vehicles in region r
4. use the sensor data gathered in the previous step to produce knowledge as to what is seen or not seen by the UAV in region r
5. use the acquired knowledge to plan for the next series of actions which involve tracking, feature recognition and logging
6. maintain execution of the necessary services and processes until the mission goal is completed.

We will concentrate on steps 3 and 4 whose successful completion is dependent on a combination of the open-world assumption, LCC reasoning and rough knowledge database representations of relations and properties.

Observe that the mission goal above contains two universal statements, the first of which asks to "identify and track *all* moving vehicles in region r". and the second of which asks to "log the estimated velocities and positions of *all* small blue vehicles identified." The meaning of the second universal is naturally dependent on the meaning of the first universal. In order to be able to achieve the mission goal, the inferencing mechanism used by the UAV during plan generation and during plan execution must be able to circumscribe (in the intuitive sense) the meaning of "*all* moving vehicles in region r" and that of "*all* small blue vehicles identified".

In fact, what the UAV *can perceive as moving* given the constraints under which it is operating, the character of the dynamics of its current operational environment, and the capabilities of its sensor and image processing functionalities in this context is not necessarily the same thing as what *is actually moving* in region r. An additional problem of course is that the inferencing mechanism can not appeal to the use of the closed-world assumption. If it could, it would register moving objects in region r and assume via application of the CWA that no other objects are moving. One can not appeal to this mechanism because the open-world assumption is being used. Even if one could, this would be erroneous. Certainly there may be vehicles in region r that are moving but can not be perceived due to limitations associated with the UAV's capabilities and there may also be vehicles outside region r that are moving.

The key to solving this particular representational problem is to note that sensing actions, such as step 3 in the plan sketch above, implicitly generate local or contextual closure information (LCC policies) and that the UAV agent can query the rough knowledge database using the particular contextual closure which exists for the purpose at hand. For example, the sensing action in step 3 above not only generates information about specific moving individuals the UAV can perceive with its current sensor and image processing capabilities, but it also generates knowledge that this is all the UAV can see in the region of interest (ROI), region r. The nature of this information is that it is specific knowledge of what the UAV agent does not see rather than information derived via an assumption such as the CWA.

Of course, one has to (or more specifically, the UAV agent has to) supply the contextual closure information. This will be supplied in terms of one or more integrity constraints and an LCC policy consisting of particular LCC assumptions pertaining to the minimization, maximization, fixing or varying of specific relations. The specific closure context for this situation could be paraphrased as follows:

> After sensing region r with a camera sensor, assume that all moving vehicles in the ROI (r) have been perceived except for those with a signature whose color feature is roadgray.

In the following example, we provide the particulars for representing the scenario above and reasoning about it using the proposed approach.

11.2 A UAV Scenario: Identify, Track and Log

Consider the situation where a UAV observes and classifies cars with different signatures based on color.[1] The considered domains consist of:

- $Cars = \{\mathsf{c1}, \mathsf{c2}, \mathsf{c3}, \mathsf{c4}, \mathsf{c5}, \mathsf{c6}\}$
- $Regions = \{\mathsf{r1}, \mathsf{r2}, \mathsf{r3}, \mathsf{r4}\}$
- $Signatures = \{\mathsf{blue}, \mathsf{roadgray}, \mathsf{green}, \mathsf{yellow}\}$.

The following relations are also defined:[2]

- $Moving(c)$ – the object c is moving
- $InROI(r)$ – the region r is in the region of interest
- $See(c, r)$ – the object c is seen by the UAV in region r
- $In(c, r)$ – the object c is in region r
- $ContainedIn(r, r')$ – region r is contained in region r'
- $Sig(c, s)$ – the object c has signature s.

Suppose the actual situation in the operational environment the UAV is flying over is as depicted in Figure 11.1. For the mission goal, the UAV's initial region of interest, ROI, is region r3.

At mission start, the following facts are in the UAV's on-line EDB:

$$\{ContainedIn^{+}(\mathsf{r1}, \mathsf{r2}), ContainedIn^{+}(\mathsf{r2}, \mathsf{r3})\}.$$

During mission preparation, the ground operator relays the following information to the UAV agent which is placed in the UAV's on-line EDB:

$$\{In^{+}(\mathsf{c2}, \mathsf{r2}), In^{+}(\mathsf{c3}, \mathsf{r3}), Moving^{+}(\mathsf{c2}), \\ Sig^{+}(\mathsf{c2}, \mathsf{roadgray}), Sig^{+}(\mathsf{c3}, \mathsf{green}), Sig^{+}(\mathsf{c6}, \mathsf{roadgray}), InROI^{+}(\mathsf{r3})\}.$$

The intensional database contains the following rules:

$$ContainedIn^{+}(r,s) \leftarrow \exists t.[ContainedIn^{+}(r,t) \land ContainedIn^{+}(t,s)] \quad (11.1)$$

$$InROI^{+}(s) \leftarrow \exists r.[ContainedIn^{+}(s,r) \land InROI^{+}(r)]. \quad (11.2)$$

[1] In an actual scenario, a vehicle signature would be more complex and contain features such as width, height, and length, or vehicle type.

[2] In addition, a number of type properties such as *Car*, *Region*, etc would also be defined.

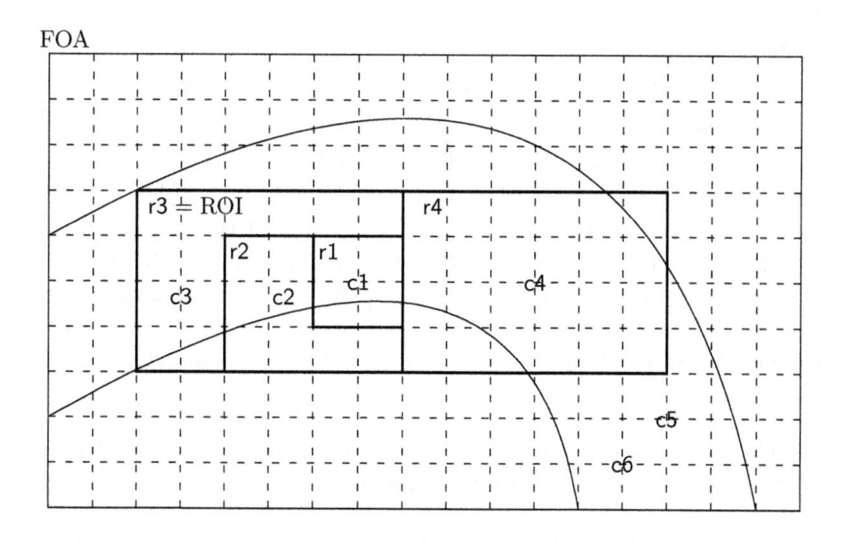

Fig. 11.1. The situation considered in Section 11.2

The current EDB, together with the IDB would allow the UAV agent to infer the following additional facts:

$\{ContainedIn^{+}(\mathsf{r1}, \mathsf{r3}), InROI^{+}(\mathsf{r2}), InROI^{+}(\mathsf{r1})\}.$

Observe that complete information about the *ContainedIn* and *InROI* relations is not yet assumed due to the application of an open-world assumption in the EDB and IDB.

Assume the UAV generates a plan similar to the one above and then executes steps 1-3. Given its sensor capabilities under the current weather conditions, suppose the UAV agent is able to assert the following new set of facts in the EDB generated from its sensor action and image processing facilities (step 3 in the plan):

$\{In^{+}(\mathsf{c1}, \mathsf{r1}), In^{+}(\mathsf{c4}, \mathsf{r4}), Moving^{+}(\mathsf{c1}), Moving^{+}(\mathsf{c4}), Moving^{+}(\mathsf{c5}),$
$Sig^{+}(\mathsf{c1}, \mathsf{blue}), Sig^{+}(\mathsf{c4}, \mathsf{yellow})\}.$

After executing the sensor action in step 3, the UAV's on-line EDB contains the following facts:

$$\{In^{+}(\mathsf{c1}, \mathsf{r1}), In^{+}(\mathsf{c2}, \mathsf{r2}), In^{+}(\mathsf{c3}, \mathsf{r3}), In^{+}(\mathsf{c4}, \mathsf{r4}), \tag{11.3}$$
$$Moving^{+}(\mathsf{c1}), Moving^{+}(\mathsf{c2}), Moving^{+}(\mathsf{c4}), Moving^{+}(\mathsf{c5}), \tag{11.4}$$
$$Sig^{+}(\mathsf{c1}, \mathsf{blue}), Sig(\mathsf{c2}^{+}, \mathsf{roadgray}), Sig^{+}(\mathsf{c3}, \mathsf{green}), \tag{11.5}$$
$$Sig^{+}(\mathsf{c6}, \mathsf{roadgrey}), Sig^{+}(\mathsf{c4}, \mathsf{yellow}), \tag{11.6}$$

$$ContainedIn^{+}(\mathsf{r1},\mathsf{r2}), ContainedIn^{+}(\mathsf{r2},\mathsf{r3}), \tag{11.7}$$
$$InROI^{+}(\mathsf{r3})\}. \tag{11.8}$$

At this point, observe that due to the open-world assumption, it is unknown whether c3 is moving and it is unknown what region c5 is in or what color it is. Additionally, it is unknown what region c6 is in or whether it is moving.

Before proceeding with the execution of the rest of the plan, the UAV must take stock of what it knows about the ROI, r3. In other words, the UAV agent must query the RKDB with a particular policy of contextual closure in order to determine not only what it sees, but all that it sees under the current circumstances. The following closure context discussed above,

> after sensing region r with a camera sensor, assume that all moving vehicles in the ROI r have been perceived except for those with a signature whose color feature is roadgray,

can be represented as the following integrity constraint,

$$\forall x, r, z.[Moving(x) \land In(x,r) \land InROI(r) \land Sig(x,z) \land z \neq \mathsf{roadgray}] \rightarrow See(x,r) \tag{11.9}$$

together with the following LCC policy,

$$lcc_{See} : \text{LCC}[See(x,r), ContainedIn(x,y); Moving] : (11.9). \tag{11.10}$$

This combination states that relations $See(x,r)$ and $ContainedIn(x,y)$ are minimized, the relation $Moving$ is allowed to vary, while all other relations are fixed. In essence, the UAV agent is assuming complete information locally about the $ContainedIn$ and See relations by minimizing them. In addition, new information about moving may also be derived, but the only information about Sig and the other fixed relations that can be inferred is what is already in the EDB. That is the effect of *fixing* relations in this context.

Another way to view this contextual closure is as the equivalent

$$\forall x, r, z.[In(x,r) \land InROI(r) \land Sig(x,z) \land z \neq \mathsf{roadgray} \land \neg See(x,r)] \rightarrow \neg Moving(x) \tag{11.11}$$

which states that "if an object is in the ROI and it has a visible signature relative to the current capabilities of its sensors, if the UAV agents does not see it, then it is not moving." The integrity constraint is intended to represent a strong coupling between moving and seeing due to the character of the sensor capabilities in this context.

The CAKE diagram representing the UAV agent's RKDB (knowledge base) is shown in Figure 11.2.

Given this scenario and its representation as a CAKE diagram, the next section will show how to ask questions about the knowledge represented using techniques already considered in this book.

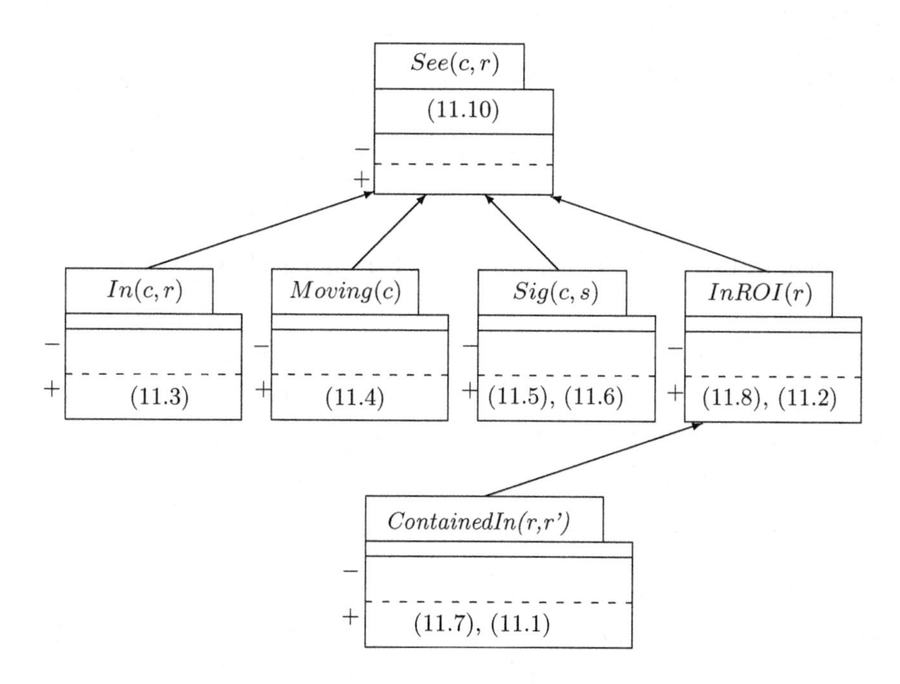

Fig. 11.2. CAKE diagram representing the knowledge database of Section 11.2.

11.3 The Rôle of the Rough Sets Techniques

Rough set techniques are to be used in designing the classifiers for:

- determining whether a given object represents a car
- determining the signature of a given car.

Both classifiers are to be designed via a supervised machine learning technique, which will be described in Chapter 14. In the simple case we deal with, the signature can be determined by means of the RGB values associated with moving objects.[3] In more complicated cases one would rather create a more complex concept of signature based on such concepts as color, length, width, etc.

[3] RGB refers to the standard encoding of colors using Red, Green and Blue.

11.4 Computing the Contextual Closure

In order to obtain the definitions of See_{min} and $Moving_{var}$, we first have to provide the corresponding expansion described in Chapter 7, which is given by the following set of formulas:

$$\{\forall x, r, z.[Moving(x) \wedge In(x,r) \wedge InROI(r) \wedge Sig(x,z) \wedge z \neq \mathsf{roadgray}] \rightarrow See(x,r),$$
$$\forall x, r, z.[\neg See(x,r) \wedge In(x,r) \wedge InROI(r) \wedge Sig(x,z) \wedge z \neq \mathsf{roadgray}] \rightarrow \neg Moving(x,r)\ \}.$$

According to Lemma 6.5.8, the definition of the minimal extension of *See* is given by:

$$See^{+}_{min}(x,r) \equiv See^{+}(x,r) \vee \exists z.[Moving^{+}(x) \wedge In^{+}(x,r) \wedge InROI^{+}(r) \wedge Sig^{+}(x,z) \wedge z \neq \mathsf{roadgrey}] \quad (11.12)$$
$$See^{-}_{min}(x,r) \equiv \neg See^{+}(x,r) \wedge \forall z.[\neg Moving^{+}(x) \vee In^{-}(x,r) \vee InROI^{-}(r) \vee Sig^{-}(x,z) \vee z = \mathsf{roadgrey}]. \quad (11.13)$$

The definition of the varied relation *Moving* is then given by:

$$Moving^{+}_{var}(x) \equiv Moving^{+}(x)$$
$$Moving^{-}_{var}(x) \equiv Moving^{-}(x) \vee \exists r.\exists z.[\neg See^{+}_{min}(x,r) \wedge In^{+}(x,r) \wedge InROI^{+}(r) \wedge Sig^{+}(x,z) \wedge z \neq \mathsf{roadgrey}].$$

The definition of *ContainedIn* is given by:

$$ContainedIn^{+}_{min}(r,s) \equiv ContainedIn^{+}(r,s) \vee \exists t.[ContainedIn^{+}(r,t) \wedge ContainedIn^{+}(t,s)].$$

After applying the LCC policy, now characterized as the relation definitions above, in the CCQ layer which includes integrity constraints, the resulting EDB/IDB combination would contain (explicitly and implicitly) the following facts:

$$\{\ In^{+}(\mathsf{c1},\mathsf{r1}), In^{+}(\mathsf{c2},\mathsf{r2}), In^{+}(\mathsf{c3},\mathsf{r3}), In^{+}(\mathsf{c4},\mathsf{r4}),$$
$$Moving^{+}(\mathsf{c1}), Moving^{+}(\mathsf{c2}), Moving^{+}(\mathsf{c4}), Moving^{+}(\mathsf{c5}),$$
$$Moving^{-}(\mathsf{c3}),$$
$$Sig^{+}(\mathsf{c1},\mathsf{blue}), Sig^{+}(\mathsf{c2},\mathsf{roadgray}), Sig^{+}(\mathsf{c3},\mathsf{green}), Sig^{+}(\mathsf{c4},\mathsf{yellow}),$$
$$ContainedIn^{+}(\mathsf{r1},\mathsf{r2}), ContainedIn^{+}(\mathsf{r2},\mathsf{r3}), ContainedIn^{+}(\mathsf{r1},\mathsf{r3}),$$
$$ContainedIn^{-}(r,r') \text{ for all pairs } r,r' \text{ other than listed above},$$
$$InROI^{+}(\mathsf{r1}), InROI^{+}(\mathsf{r2}), InROI^{+}(\mathsf{r3})\}.$$

It is useful to note the following concerning the UAV agent's knowledge about the ROI resulting from it's sensing action and subsequent reasoning. The UAV agent still has incomplete information about the relations In, Sig, and $InROI$. For example, it is unknown what signature object c5 has or where it is. The UAV agent now knows that object c3 is not moving and it does have complete information about the $ContainedIn$ relation.

What about the relation See which has been minimized? One can now infer the following facts related to the relation See:

- $See(\mathsf{c1}, \mathsf{r3})$
- $\neg See(\mathsf{c2}, \mathsf{r3}), \neg See(\mathsf{c3}, \mathsf{r3}), \neg See(\mathsf{c4}, \mathsf{r3}), \neg See(\mathsf{c6}, \mathsf{r3})$.

Note that c3 is not seen because it is not moving; c4 is not seen because it is not in the ROI, r3; c2 is not seen even though it is moving because of its signature. Most interestingly, it is unknown whether c5 is seen because the UAV agent was not able to discern what region c5 was in, nor what its color signature was. In fact, since the UAV agent could identify c5 as moving, the lack of its discerning a region for c5 could be deduced as the reason for this, due to the tight coupling between moving and seeing. This could provide a reason for focusing on c5 and trying to discern its region. c6 is not seen because of its signature. What is interesting is that the minimization of See does not change the status of c6 w.r.t. $Moving$, i.e., $Moving(\mathsf{c6})$ remains unknown.

The fact that $See(\mathsf{c5}, \mathsf{r3})$ and $Moving(\mathsf{c6})$ remain unknown informs us of the subtlety of the minimization in the context of rough sets. The minimization of a relation in the rough set context does not necessarily create a crisp definition of the relation minimized. What it does do is move tuples in the boundaries of one or more relations into the positive or negative parts of the relation while meeting the conditions of the integrity constraints, while other tuples remain in the boundaries. This is very important because it satisfies the ontological intuition associated with open-world reasoning.

It also worth emphasizing that if the integrity constraint (11.9) was defined as an intensional rule, like the following one

$$\begin{aligned} &See(x,r) \leftarrow \\ &\qquad Moving(x) \wedge In(x,r) \wedge InROI(r) \wedge Sig(x,z) \wedge z \neq \mathsf{roadgray} \end{aligned}$$

then the minimization of See would not result in changes to the relation $Moving$ appearing in the body of the rule.

11.5 Abducing without Contextual Closure

11.5.1 Abducing *See*

Observe that in the example, due to the locality of the contextual closure mechanism, there are several vehicles the UAV agent does not know if it sees or not. Suppose it is interested in determining what sort of information it would have to acquire in order to know whether it sees a specific vehicle in a specific region or not. This is a question of abductive inference. In this section, we show how one can use weakest sufficient and strongest necessary conditions on theories to derive this information.

Assume one wants to abduce $See(\mathsf{c5}, \mathsf{r3})$. In other words, what information would one be required to add to the existing theory in order to infer $See(\mathsf{c5}, \mathsf{r3})$? We are then interested in:[4]

$$\text{WSC}(See(\mathsf{c5}, \mathsf{r3}; \text{EDB} \wedge \text{IDB} \wedge (11.9); -\{See\}). \tag{11.14}$$

According to the techniques described in Chapter 8, (11.14) is equivalent to

$$\forall\, See.[(\text{EDB} \wedge \text{IDB} \wedge (11.9)) \rightarrow See(\mathsf{c5}, \mathsf{r3})]. \tag{11.15}$$

In order to eliminate the second-order quantifier $\forall See$, we simulate the steps of the DLS/DLS* algorithms.[5] We thus first negate (11.15),

$$\exists See.[(\text{EDB} \wedge \text{IDB} \wedge (11.9)) \wedge \neg See(\mathsf{c5}, \mathsf{r3})], \tag{11.16}$$

which is equivalent to

$$\begin{aligned} &\text{EDB} \wedge \text{IDB} \wedge \\ &\exists See.\{\forall x.\forall r.\forall z.[(Moving(x) \wedge In(x,r) \wedge InROI(r) \wedge \\ &\qquad Sig(x,z) \wedge z \neq \mathsf{roadgray}) \rightarrow See(x,r)] \wedge \neg See(\mathsf{c5}, \mathsf{r3})\} \end{aligned} \tag{11.17}$$

and then to

$$\begin{aligned} &\text{EDB} \wedge \text{IDB} \wedge \\ &\exists See.\{\forall x.\forall r.[(Moving(x) \wedge In(x,r) \wedge InROI(r) \wedge \\ &\qquad \exists z.(Sig(x,z) \wedge z \neq \mathsf{roadgray})) \rightarrow See(x,r)] \wedge \neg See(\mathsf{c5}, \mathsf{r3})\}. \end{aligned} \tag{11.18}$$

Quantifier elimination, based on Theorem 2.9.2, results in[6]

[4] Recall that notation "$-\{See\}$" indicates that the target language consists of all considered relation symbols, except of *See* - see Chapter 8.

[5] In fact, in all second-order quantifier eliminations of this chapter we essentially simulate the execution of the DLS/DLS* algorithms.

[6] Observe that there are no fixpoints here, since formula (11.18) is not recursive w.r.t. the relation *See*.

$$\text{EDB} \wedge \text{IDB} \wedge \neg[Moving(\mathsf{c5}) \wedge In(\mathsf{c5}, \mathsf{r3}) \wedge InROI(\mathsf{r3}) \wedge \\ \exists z.(Sig(\mathsf{c5}, z) \wedge z \neq \mathsf{roadgray})].$$

After negating again we obtain the required condition

$$(\text{EDB} \wedge \text{IDB}) \rightarrow [Moving(\mathsf{c5}) \wedge In(\mathsf{c5}, \mathsf{r3}) \wedge InROI(\mathsf{r3}) \wedge \\ \exists z.(Sig(\mathsf{c5}, z) \wedge z \neq \mathsf{roadgray})]. \tag{11.19}$$

Since EDB and IDB entail $Moving(\mathsf{c5})$ and $InROI(\mathsf{r3})$, formula (11.19) reduces to

$$(\text{EDB} \wedge \text{IDB}) \rightarrow [In(\mathsf{c5}, \mathsf{r3}) \wedge \exists z.(Sig(\mathsf{c5}, z) \wedge z \neq \mathsf{roadgray})]. \tag{11.20}$$

In order to be able to see car c5 in region r3, one would then have to make sure that it is in region r3 (in order to satisfy the conjunct $In(\mathsf{c5}, \mathsf{r3})$) and that it has a signature different from roadgray (in order to satisfy the conjunct $\exists z.(Sig(\mathsf{c5}, z) \wedge z \neq \mathsf{roadgray})$). This result could then be used as a strategy for the UAV to isolate and identify the vehicle c5.

11.5.2 Abducing $\neg See$

In the case of abducing $\neg See$, as it was done in the case of abducing See, one gets the following result as the required condition:

$$(\text{EDB} \wedge \text{IDB}) \rightarrow \text{FALSE} \tag{11.21}$$

which, in fact makes perfect sense, since our theory has no means to deduce negative facts about See in the language not containing the relation symbol See. In order to acquire a meaningful result, one has to consider a local closure policy which allows one to infer negative (default) information about See.

11.6 Abducing with Contextual Closure $\mathsf{lcc}_{\mathsf{See}}$

11.6.1 Abducing See_{min}

We want to abduce $See_{min}(\mathsf{c5}, \mathsf{r3})$ w.r.t. EDB, IDB and the integrity constraint expressed by formula (11.9). In the case of contextual closures we can often considerably simplify the necessary computations, as we are given explicit definitions of the suitable relations.

For example, the contextual closure LCC_{See} gives us the explicit definition of See^{+}_{min} due to formula (11.12). According to this definition of See^{+}_{min}, in order to make $See_{min}(\mathsf{c5}, \mathsf{r3})$ true, one has to make true the righthand side of the equivalence given by formula (11.12) with x, r replaced by c5, r3, respectively, i.e., to make the following formula true:

$$See^{+}(\mathsf{c5}, \mathsf{r3}) \vee \exists z.[Moving^{+}(\mathsf{c5}) \wedge In^{+}(\mathsf{c5}, \mathsf{r3}) \wedge \\ InROI^{+}(\mathsf{r3}) \wedge Sig^{+}(\mathsf{c5}, z) \wedge z \neq \mathsf{roadgrey}]$$

which, after moving quantifier $\exists z$ inside the formula, is equivalent to

$$See^{+}(\mathsf{c5}, \mathsf{r3}) \vee [Moving^{+}(\mathsf{c5}) \wedge In^{+}(\mathsf{c5}, \mathsf{r3}) \wedge \\ InROI^{+}(\mathsf{r3}) \wedge \exists z.(Sig^{+}(\mathsf{c5}, z) \wedge z \neq \mathsf{roadgrey})]. \tag{11.22}$$

The current contents of the database, given by literals in (11.3)-(11.8), allows us to simplify the formula (11.22) to

$$In^{+}(\mathsf{c5}, \mathsf{r3}) \wedge \exists z.(Sig^{+}(\mathsf{c5}, z) \wedge z \neq \mathsf{roadgrey}) \tag{11.23}$$

which states that in order to be able to see car c5 in region r3, one has to make sure that c5 is in region r3 and its signature is not roadgrey.

Note that this is similar to abducing without contextual closure (see Section 11.5) but there is a difference when abducing negative information as will be observed in the next section.

11.6.2 Abducing $\neg See$

We want to abduce $\neg See(\mathsf{c5}, \mathsf{r3})$ w.r.t. EDB, IDB and the integrity constraint expressed by formula (11.9). We apply the same methodology as the one presented in Section 11.6.1. We are then interested in satisfying formula $See^{-}_{min}(\mathsf{c5}, \mathsf{r3})$. The contextual closure LCC_{See} provides us with an explicit definition of See^{-}_{min} due to formula (11.13). According to this definition of See^{-}_{min}, in order to make $See^{-}_{min}(\mathsf{c5}, \mathsf{r3})$ true, one has to make true the righthand side of the equivalence given by formula (11.13) with x, r replaced by c5, r3, respectively, i.e., to make the following formula true:

$$\neg See^{+}(\mathsf{c5}, \mathsf{r3}) \wedge \forall z.[\neg Moving^{+}(\mathsf{c5}) \vee In^{-}(\mathsf{c5}, \mathsf{r3}) \vee \\ InROI^{-}(\mathsf{r3}) \vee Sig^{-}(\mathsf{c5}, z) \vee z = \mathsf{roadgrey}].$$

In order to make the formula simpler, we move the quantifier $\forall z$ inside and obtain

$$\neg See^{+}(\mathsf{c5}, \mathsf{r3}) \wedge [\neg Moving^{+}(\mathsf{c5}) \vee In^{-}(\mathsf{c5}, \mathsf{r3}) \vee \\ InROI^{-}(\mathsf{r3}) \vee \forall z.(Sig^{-}(\mathsf{c5}, z) \vee z = \mathsf{roadgrey})]. \tag{11.24}$$

The current contents of the databases, given by literals in (11.3)-(11.8), allows us to simplify the formula (11.24) and obtain

$In^{-}(\mathsf{c5}, \mathsf{r3}) \vee \forall z.(Sig^{-}(\mathsf{c5}, z) \vee z = \mathsf{roadgrey})$

which states that in order not to be able to see car c5 in region r3, one would have to make sure that it is not in the region, or that its signature is roadgrey.

The use of abduction in scenarios such as those considered in this chapter is a very powerful technique for determining what kind of information has to be acquired by the UAV in order to answer certain questions of importance to achieving goals in such scenarios.

11.7 Employing Defaults

Observe that many default rules could be employed in order to better describe the whole situation.

For instance, one could deal with the following default for determining the signature of a car:

> if a car is in the region of interest and it is moving and it is not seen and one cannot exclude that it's signature is roadgrey then assume that it actually is roadgrey.

The default can be expressed as follows:

$$\frac{InROI(r) \wedge In(c, r) \wedge Moving(c) \wedge \neg See(c, r) : Sig(c, \mathsf{roadgrey})}{Sig(c, \mathsf{roadgrey})}.$$

Observe that according to the methodology developed in Chapter 10, the above default is translated into the rule[7]

$$[InROI^{+}(r) \wedge In^{+}(c, r) \wedge Moving^{+}(c) \wedge See^{-}(c, r) \wedge Sig^{\oplus}(c, \mathsf{roadgrey})] \\ \rightarrow Sig^{+}(c, \mathsf{roadgrey}).$$

Consider another default useful in determining whether a car is in a given region:

> if a car entered a region and did not leave the region and one can consistently assume that it is still in the region, then conclude that the car is actually in the region.

In order to express such a default, we need to enrich the language. Consequently, we add the following relations:

[7] We do not indicate granules, as the information sources are obvious here.

- $Enter(c, r)$ meaning that car c has entered the region r
- $Exit(c, r)$ meaning that car c has quit the region r.

The default can then be expressed as follows:

$$\frac{Enter(c, r) \wedge \neg Exit(x, r) : In(x, r)}{In(x, r)}. \tag{11.25}$$

The above default is translated into the following rule:

$$[Enter^{+}(c, r) \wedge Exit^{-}(x, r) \wedge In^{\oplus}(x, r)] \rightarrow In^{+}(x, r).$$

In the case when information about entering and exiting a region is inaccurate, one could replace the default (11.25) by the following one:

$$\frac{Enter(c, r) : \neg Exit(x, r) \wedge In(x, r)}{In(x, r)}$$

whose meaning is:

> if a car entered a region and it is consistent to assume that it did not leave it and there is no information that it is not in the region, then conclude that the car is actually in the region.

The above default is translated into the rule

$$[Enter^{+}(c, r) \wedge Exit^{\ominus}(x, r) \wedge In(x, r)^{\oplus}] \rightarrow In^{+}(x, r).$$

11.8 Bibliographic Notes

The initial version of the UAV scenario considered in this chapter (parts of Sections 11.2 and 11.4) has been considered in [50]. The abduction techniques applied in Section 11.5 have been worked out in [56].

The DLS/DLS* algorithms used for eliminating second-order quantifiers in Section 11.5 are described in [53, 54].

Part III

From Sensors to Relations

12

Information Granules

12.1 Introduction

Solving complex problems by intelligent systems, in such areas as identification of objects by autonomous systems, web mining or sensor fusion, requires techniques for combining information from many different sources with different degrees of quality. Usually, the information is inaccurate and incomplete. One paradigm for dealing with such complex problems is granular computing.

In Chapter 9 we presented the CAKE method. CAKE provides a uniform means for representing complex knowledge structures and reasoning processes. On the other hand, when one wants to base the reasoning on real data, many other techniques can also be applied. For instance, one can use complex classifiers based on analytical tools, difficult or even impossible to be represented in logic in a practical and pragmatic manner. A formalism and methodology which makes the whole software system uniform is then desirable. We show that the paradigm of granular computing suits well to represent both CAKE–based solutions and data analysis tools.

The fundamental concept of the granular computing paradigm is that of an information granule. An information granule is a concise conceptual unit that can be integrated into a larger information infrastructure consisting of other information granules and dependencies between them. Most importantly, the structure described is recursive in nature and each conceptual unit (information granule) is (elaboration) tolerant within the context in which it is used at a particular level of abstraction in the total system infrastructure.

Information granules are building blocks for other, more complex granules and, finally, for the whole information structure used by intelligent systems. One example of an information granule might be a database containing sensory data and making it accessible to other parts of a complex system, another one might be a classifier recognizing and interpreting the gathered data as, e.g., various objects on a road.

P. Doherty et al.: *Knowledge Representation Techniques*, Studfuzz **202**, 229–243 (2006)
www.springerlink.com

In many cases independent information sources provide us with information granules, that must be transformed, analyzed and built into structures that support problem solving, as illustrated in the example below.

Example 12.1.1. Consider the situation shown in Figure 12.1, where:

- ag_1, ag_2, ag are agents involved in producing information granules
- ig_1, ig_2 are information granules provided as input for agents ag_1, ag_2
- og_1, og_2 are information granules output by agents ag_1, ag_2 and provided as input for agent ag
- the hexagonal unit depicts a target concept to be represented. It contains granules $g_1, \ldots, g_4$ which are delivered by agent ag. The granules can be considered as particular cases (clusters, patterns) of the target concept, i.e., subsets of tuples contained (at least to a degree) in the concept or in the complement of the concept.

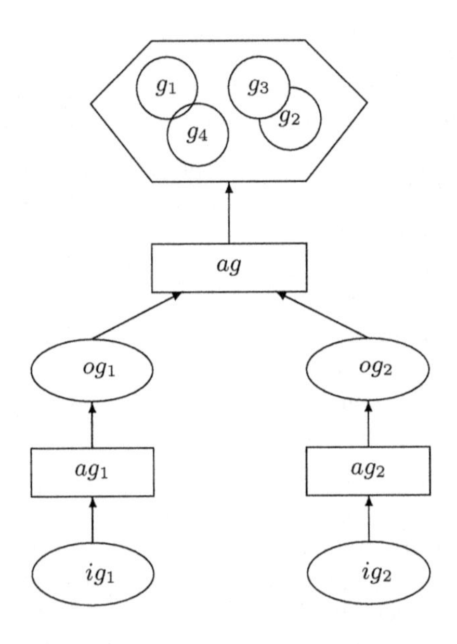

Fig. 12.1. Constructing information granules by agents.

Assume that agent ag_1 collects data (represented by ig_1) from an image processing module of a UAV and produces symbolic information about objects on the road (og_1), e.g., consisting of a vector of values of some relevant attributes for each object. Agent ag_2 collects data (ig_2) from a sensor that measures humidity and outputs symbolic information about the humidity level (og_2). Suppose that the target concept is intended to describe the notion of a "dangerous situation on a road" on the basis of data provided by agents ag_1, ag_2.

The information granules provided by agent ag could then be the following:

- g_1: "a car is moving with a high speed and is close to another car moving on the same lane"
- g_2: "a car is moving with a high speed and is close to a crossroad"
- g_3: "a car is moving with a very high speed and the humidity level is high"
- g_4: "a car is moving with a very high speed and approaches a traffic jam".

Note that the definitions of g_1, g_2, g_3, g_4 are imprecise, as they use vague concepts such as "very high speed" or "high humidity level". In consequence, tools for approximate representation of information granules and approximate reasoning using them are necessary. □

Any calculus of complex information granules should allow

1. to deal with the vagueness of information granules. This can be achieved by adopting fuzzy set theory or rough set theory, either separately or in combination
2. to develop strategies of inducing multi-layered schemes of complex granule construction. This is related to the problem of understanding of reasoning from measurements to perception, to concept approximation learning, as well as to fusion of information from different sources.

In the following sections we concentrate on basic notions concerning information granules, mainly from the knowledge representation perspective.

12.2 Elementary Granules

We start with the notion of a granule. This notion is closely related to the notion of an object in the object-oriented programming paradigm. Granules can be naturally viewed as information objects specialized for distributed information processing and interchange.

Let us start with a general notion of a granule.

Definition 12.2.1. *Let* SIG *be a signature,* P *be a* set of parameters *consisting of variable and constant symbols, and let* N *be a set of* granule names. *By a* granule *we understand a triple* $G = \langle C, I, M \rangle$, *where*

- *C is a set of constant symbols from* SIG.
- *I is an* interface of *G, which is an expression of the form*

  $Name : R_1(a_{11}, \ldots, a_{1m_1}), \ldots, R_k(a_{k1}, \ldots, a_{km_k}),$

  where $Name \in N$, $R_1, \ldots, R_k \in$ SIG *are (rough) relation symbols and $a_{11}, \ldots, a_{1m_1}, \ldots, a_{k1}, \ldots, a_{km_k}$, called* parameters of *G, are members of P.*
- *$M = \{M_i^\sharp : 1 \leq i \leq k$ and $\sharp \in \{+, -\}\}$ is a set of methods. A* method *$M_i^\sharp$, for $1 \leq i \leq k$, is a function that allows one to compute relations $R_i^\sharp(a_{i1}, \ldots, a_{im_i})$ as follows:*[1]

  for a vector of length m_i ($1 \leq i \leq k$) consisting of $0 \leq m \leq m_i$ variables and $(m_i - m)$ constant symbols, $M_i^\sharp$ returns a set of m-ary tuples of constants or truth value UNKNOWN. *If $m = 0$ then the method returns a truth value* TRUE, FALSE *or* UNKNOWN.[2]

  A query to a granule *is any expression of the form $R_i^\sharp(a_{i1}, \ldots, a_{im_i})$, where $1 \leq i \leq k$, $\sharp \in \{+, -, \oplus, \ominus, \pm\}$ and $a_{i1}, \ldots, a_{im_i}$ are constant symbols or variables. An* answer to a query *$R_i^\sharp(a_{i1}, \ldots, a_{im_i})$ is:*

  - *the set computed by method $M_i^\sharp$ with input vector $\langle a_{i1}, \ldots, a_{im_i} \rangle$, when $\sharp \in \{+, -\}$*
  - *the result computed according to definitions of $R_i^\pm, R_i^\oplus, R_i^\ominus$ (see Section 3.2.2), when $\sharp \in \{\oplus, \ominus, \pm\}$.* □

Elementary granules are defined as follows.

Definition 12.2.2. *Let* SIG *be a signature, P be a set of parameters and N be a set of granule names. By an* elementary granule *we understand any granule $G = \langle C, I, M \rangle$, where:*

- *C is a set of constants from* SIG
- *the interface I is of the form $Name : R(\bar{a})$, where $R \in$* SIG *and all parameters in $\bar{a}$ belong to P*
- *M consists of a set of methods that allow one to compute $R^\sharp$, for $\sharp \in \{+, -\}$.* □

We do not require any particular paradigm for defining methods that allow one to compute relations provided by granules. This allows us to abstract from a particular computational mechanism and deal with heterogenous programming environments.

[1] If a superscript $\sharp$ is omitted, it is $+$, by default.

[2] If a method $M_i^\sharp$ is not specified, it is assumed to return always the value UNKNOWN for a query $R_i^\sharp(a_{i1}, \ldots, a_{im_i})$.

Example 12.2.3. Any information system $\mathcal{A}$, with k attributes $a_1, \ldots, a_k$, can be viewed as an elementary granule in which:

Table 12.1. Information table considered in Example 12.2.3.

Object	*Size*	*Color*	*Speed*
1	m	r	l
2	s	b	h
3	m	g	l

- the set of constant symbols is the set of constants appearing in the information table
- the interface consists of the information system name, together with the relation, denoted by $R_{\mathcal{A}}(x, x_{a_1}, \ldots, x_{a_k})$, it represents,[3] where $x, x_{a_1}, \ldots, x_{a_k}$ are new variables corresponding to columns in information table representing $\mathcal{A}$
- the methods it contains correspond to standard deductive database queries of the form $R^{\sharp}_{\mathcal{A}}(w_1, \ldots, w_k)$, where any w_i is a variable or a constant symbol (see Chapter 4).

Consider an information system $\mathcal{A} = \langle U, A \rangle$ shown in Table 12.1, where r, b, g, s, m, l, h stand for "red", "blue", "green", "small", "medium", ""low" and "high", respectively.

It corresponds to an elementary granule, where:

- the set of constants is $\{1, 2, 3, \mathsf{m}, \mathsf{s}, \mathsf{r}, \mathsf{b}, \mathsf{g}, \mathsf{l}, \mathsf{h}\}$
- the interface is

  $\mathcal{A} : R_{\mathcal{A}}(x, x_{Size}, x_{Color}, x_{Speed})$

- the methods correspond to standard deductive database queries with the local closed world assumption applied to $R^{+}_{\mathcal{A}}$ only. For example:
  - query $R_{\mathcal{A}}(x, \mathsf{m}, y, \mathsf{l})$ returns the answer $\{\langle 1, \mathsf{r} \rangle, \langle 3, \mathsf{g} \rangle\}$
  - query $R_{\mathcal{A}}(2, \mathsf{m}, \mathsf{b}, \mathsf{h})$ returns the answer FALSE. □

Example 12.2.4. Any decision rule of the form

$$(a_1 = c_1) \wedge \ldots \wedge (a_n = c_n) \Rightarrow (d = c_d)$$

can be viewed as an elementary granule in which:

[3] Note that any table with k attributes can be viewed as an k-argument relation. This identification is standard in the field of relational databases.

- the set of constant symbols is the set $\{c_1, \ldots, c_n, c_d\}$
- the interface consists of the granule name, together with a relation, denoted by $R(x_{a_1}, \ldots, x_{a_n}, x_d)$
- the methods of the granule return:

$$\begin{cases} \langle c_d \rangle & \text{for query } R^+(c_1, \ldots, c_n, x)\text{, where } x \text{ is a variable,} \\ \text{UNKNOWN} & \text{for all other queries.} \end{cases}$$

□

Example 12.2.5. Any CAKE database granule N with label $Ag : R(x_1, \ldots, x_k)$, can be regarded as an elementary granule, in which:

- the set of constant symbols is the set of constants appearing in the database of N
- the interface is the label of N, where variable symbols $x_1, \ldots, x_k$ serve as parameters
- the methods it contains correspond to the CAKE computing mechanism (see Section 9.5). □

The elementary granules presented in Examples 12.2.3, 12.2.4 and 12.2.5 are *self-contained* in the sense that for computing their methods no information from outside is used. It is also possible to consider elementary granules which are not self-contained, as illustrated in the next example.

Example 12.2.6. Any Boolean descriptor can be viewed as an elementary granule, in which:

- the set of constant symbols is the set of constants appearing in the descriptor
- the interface consists of a granule name, together with a relation, denoted by $R_{\mathcal{A}}(a_1, \ldots, a_k)$, where parameters $a_1, \ldots, a_k$ are all attributes occurring in the descriptor. $R_{\mathcal{A}}$ refers to an information system $\mathcal{A}$ which is to be an input to the granule
- the methods it contains compute answers to queries treating the Boolean descriptor as a standard deductive database query (see Section 4.2). The Boolean descriptor itself always returns the empty set of tuples. This is due to the fact that it contains no information about $R_{\mathcal{A}}$, i.e., from its perspective $R_{\mathcal{A}}$ is empty.[4] □

Example 12.2.7. Any CAKE granule M, not necessarily a database granule, is also an example of an elementary granule, where

[4] In Section 12.3 we provide operations which make such granules useful.

- the set of constant symbols is the set of constants appearing in the rules of M
- the interface is the label of M
- the methods of the granule correspond to the CAKE computing mechanism (see Section 9.5). □

To create an elementary granule G, we introduce a special operation

$MakeEGranule(C, I, M)$,

which constructs the granule $\langle C, I, M \rangle$.

12.3 Information Granule Systems

Information granules are built from elementary granules and previously constructed information granules by means of a number of operations.

Definition 12.3.1. *Let E be a set of elementary granules and let $\mathcal{G}$ be a set of granules such that $E \subseteq \mathcal{G}$. Assume I is a finite set and $\{O_i : i \in I\}$ are k_i-argument operations on $\mathcal{G}$, i.e., for any $i \in I$,*

$$O_i : \underbrace{\mathcal{G} \times \ldots \times \mathcal{G}}_{k_i \ times} \longrightarrow \mathcal{G}.$$

We say that $\langle E, \mathcal{G}, \{O_i : i \in I\}\rangle$ is an information granule system *provided that $\mathcal{G}$ is the least set containing all elementary granules of E and is closed under operations in $\{O_i : i \in I\}$. Any element of $\mathcal{G}$ is called an* information granule. □

12.3.1 Operation *Link*

As we have seen in the previous section, non self-contained granules require some input information from other granules in order to provide meaningful answers to queries. Accordingly, we require a mechanism for linking granules with their inputs. The operation *Link* serves this purpose,[5] where

$$Link : \mathcal{G} \times \underbrace{\mathcal{G} \times \ldots \times \mathcal{G}}_{n \text{ times}} \longrightarrow \mathcal{G}.$$

$Link(G, G_1, \ldots, G_n)$ provides a granule such that:

[5] Observe that, in fact, *Link* is a family of operations, for all $n > 0$.

- the set of constant symbols of the resulting granule is the union of the sets of constant symbols of $G, G_1, \ldots, G_n$
- the interface is the interface of G, except that its name is a fresh name from N
- the methods of the constructed granule compute answers as the methods of G, under the assumption that the results computed by $G_1, \ldots, G_n$ are accessible for methods in G.

Example 12.3.2. As observed in Example 12.2.6, Boolean descriptors are elementary granules. Let G be an elementary granule obtained in the manner described in Example 12.2.6 and corresponding to the Boolean descriptor

$$(Size = \mathsf{m}) \wedge ((Color = \mathsf{r}) \vee (Color = \mathsf{g}))\,.$$

Let $G_{\mathcal{A}}$ be an elementary granule, obtained in the manner described in Example 12.2.3, representing the information table provided in Table 12.1. Consider

Table 12.2. Information table considered in Example 12.3.2.

Object	*Size*	*Color*	*Speed*
1	m	r	l
3	m	g	l

granule $G_L = Link(G, G_{\mathcal{A}})$. Granule G_L represents all objects in Table 12.1, satisfying the considered Boolean descriptor, i.e., the information table shown in Table 12.2. □

12.3.2 Operation Project

Operation *Project* allows one to project out certain columns of a relation.[6]

$$Project : \mathcal{G} \times \underbrace{\omega \times \ldots \times \omega}_{r \text{ times}} \longrightarrow \mathcal{G},$$

where the first argument of *Project* is a granule of the form $G = \langle C, I, M \rangle$ with $I = name\!:\!R(x_1, \ldots, x_k)$ and $0 < r \leq k$.

$Project(G, i_1, \ldots, i_r)$, where $i_1, \ldots, i_r$ are different natural numbers from $\{1, \ldots, k\}$, provides a granule $\langle C_1, I_1, M_1 \rangle$ such that:

- $C_1 = C$

[6] This operation, actually a family of operations, is frequently used in relational and deductive databases.

- I is of the form $name_1 : R_1(x_{i_1}, \ldots, x_{i_r})$, where $name_1$ is a fresh name
- the methods of the constructed granule compute relations $R_1^\sharp$ given by:

$$R_1^\sharp(x_{i_1}, \ldots, x_{i_r}) \stackrel{\text{def}}{\equiv} \exists z_1 \ldots \exists z_p . [R^\sharp(x_1, \ldots, x_k)],$$

where $\{z_1, \ldots, z_p\} = \{x_1, \ldots, x_k\} - \{x_{i_1}, \ldots, x_{i_r}\}$.

12.3.3 Operation *MakeModule*

Operation *MakeModule* allows one to construct a container of granules, responsible for one or more relations. *MakeModule* is defined as:[7]

$$MakeModule : \underbrace{\mathcal{G} \times \ldots \times \mathcal{G}}_{n \text{ times}} \longrightarrow \mathcal{G}.$$

$MakeModule(G_1, \ldots, G_n)$ provides a granule G_M, such that:

- the set of constant symbols of G_M is the union of the sets of constant symbols of $G_1, \ldots, G_n$
- the interface of G_M is of the form $Name : R_1(\bar{x}_1), \ldots, R_m(\bar{x}_m)$ such that $Name$ is a fresh name from N, and $R_1(\bar{x}_1), \ldots, R_m(\bar{x}_m)$ are all the relations occurring in interfaces of $G_1, \ldots, G_n$
- the methods of G_M compute answers using the standard voting policy of CAKE, as defined in Definition 9.3.3.

Example 12.3.3. Consider the following three granules:

- $\langle \{a\}, G_1 : R(x), M_1^+ \rangle$, where the method M_1^+ states that $R^+(a)$ holds
- $\langle \{a, b\}, G_2 : R(x), M_2^+, M_2^- \rangle$, where the methods of G_2 state that $R^-(a)$ and $R^+(b)$ hold
- $\langle \{a\}, G_3 : Q(x), M_3^+ \rangle$, where the method of M_3^+ states that $Q^+(a)$ holds.

Let $G_M = MakeModule(G_1, G_2, G_3)$. G_M is a granule defined as follows.

- the set of constants of G_M is $\{a, b\}$
- the interface of G_M is $name : R(x), Q(x)$, where $name \in N$ is a fresh name
- the methods of G_M provide the answer TRUE, for queries $R^+(b)$ and $Q^+(a)$, and the answer UNKNOWN when asked about $R^+(a)$ and all other queries. □

[7] Observe that, in fact, *MakeModule* is a family of operations, for all $n > 0$.

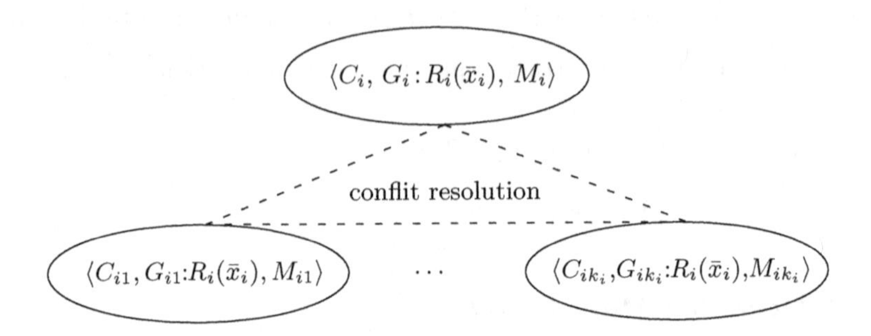

Fig. 12.2. Conflict resolution granules.

The operations *Link* and *MakeModule* can be used to provide more sophisticated voting policies. Suppose methods of the granules G_i, for $1 \leq i \leq n$, resolve conflicts between granules $G_{i1}, \ldots, G_{ik_i}$ (see Figure 12.2) in a non-standard way. In order to implement such voting policies, it suffices to construct a granule

$$MakeModule(Link(G_1, G_{11}, \ldots, G_{1i_1}), \ldots, Link(G_n, G_{n1}, \ldots, G_{ni_n})).$$

In the case of CAKE modules, the corresponding term would be the following:[8]

$$\begin{aligned}&MakeModule[\\ &MakeModule(Link(G_1, G_{11}, \ldots, G_{1i_1}), \ldots, Link(G_n, G_{n1}, \ldots, G_{ni_n})),\\ &\quad G_{11}, \ldots, G_{1i_1}, \ldots, G_{n1}, \ldots, G_{ni_n}].\end{aligned}$$

12.3.4 Operation Hide

Operation *Hide* allows one to hide certain relations from an interface.[9].

$$Hide : \mathcal{G} \times \underbrace{\omega \times \ldots \times \omega}_{r \text{ times}} \longrightarrow \mathcal{G},$$

where the first argument of *Hide* is a granule of the form $G = \langle C, I, M\rangle$ with $I = name{:}R_1(\bar{x}_1), \ldots, R_k(\bar{x}_k)$ and $0 < r \leq k$.

$Hide(G, i_1, \ldots, i_r)$, where $i_1, \ldots, i_r$ are different natural numbers from the set $\{1, \ldots, k\}$, provides a granule $\langle C_1, I_1, M_1\rangle$ such that:

- $C_1 = C$

[8] Observe that, for simplicity, CAKE adjudicating granules are not packed into a separate module.

[9] In fact, *Hide* is a family of operations.

- I_1 is of the form $name_1 : R_{i_1}(\bar{x}_{i_1}), \ldots, R_{i_r}(\bar{x}_{i_r})$, where $name_1$ is a fresh name
- $M_1 = M$.

The sets of constants and methods are the same in G and $Hide(G, i_1, \ldots, i_r)$, as computing relations $R_{i_1}(\bar{x}_{i_1}), \ldots, R_{i_r}(\bar{x}_{i_r})$ might require an access to all constants and relations, including the hidden ones.

12.4 Example: Building Classifiers

Consider a problem of classifying given objects on the basis of some available attributes. For instance, one might want to classify a color on the basis of its RGB attributes.[10] Another, far more complicated example, could depend on classifying traffic situations on a road based on video sequences. Software modules that provide one with such classifications are usually called *classifiers.*

One approach to constructing classifiers is based on providing suitable decision rules (which may be learnt). For instance, one might generate the following decision rules for a classifier recognizing colors, where y, b, dR stand for "yellow", "brown" and "dark red", respectively:

$$(R = 255) \wedge (G = 255) \wedge (B = 51) \Rightarrow (Color = \mathsf{y})$$
$$(R = 153) \wedge (G = 51) \wedge (B = 0) \Rightarrow (Color = \mathsf{b}).$$

The above rules do not lead to any conflicting decisions. However, in general, such conflicts may appear. For instance, if we added the rule

$$(R = 153) \wedge (G = 51) \wedge (B = 0) \Rightarrow (Color = \mathsf{dR}),$$

we would have two different decisions ($Color = \mathsf{b}$ and $Color = \mathsf{dR}$) for attribute values $R = 153, G = 51, B = 0$.

Such classifiers can easily be represented as information granules. The construction process consists of three steps:

1. Construct granules G_j corresponding to each particular decision rule, for $j = 1, \ldots, r$, as done in Example 12.2.4.
2. Construct a granule G solving possible conflicts among decision rules.
3. Construct the term

   $$MakeModule(Link(G, G_1, \ldots, G_r))$$

   which defines the required classifier.

[10] Recall that RGB is a well-known combination of red, green and blue.

Observe that there may be a number of different strategies for solving conflicts. The conflict resolution granule can be defined as $\langle C,\, G\!:\!R(d),\, M\rangle$, where:

- C is the set of constants occurring in $G_1, \ldots, G_r$
- R is a one-argument relation returning one-argument tuples identified with decisions
- M is a method which can be defined in many ways, e.g.:

$$d \text{ represents all decisions supported by a maximal number of rules} \tag{12.1}$$

$$d \text{ represents all decisions supported by at least one rule} \tag{12.2}$$

$$d \text{ represents a unique decision supported by some rules or UNKNOWN in the case when there is a conflict or no rule supports a decision.} \tag{12.3}$$

In classifier construction one has also to consider what attributes are required in order to compute the decision. In the case when the query does not contain all the required attributes, the classifier's method should return the value UNKNOWN or a partial matching method is to be applied in such situations.

Observe that in this chapter we consider simple rules which are precise. In real-world applications it is often necessary to use approximate rules whose lefthand sides may be satisfied for objects having attributes close to the respective values appearing in the rules. This topic is considered in Chapter 13.

Example 12.4.1. An example of classifier construction is illustrated in Figure 12.3, where three sets of decision rules are presented for the decision values $1, 2, 3$, respectively. Consider the following cases and corresponding conflict resolution strategies given as (12.1)-(12.3) above:

- the input granule matches Boolean descriptors α_1 and α_2; then all the strategies (12.1)-(12.3) return $\{\langle 1\rangle\}$
- the input granule matches Boolean descriptors α_1, α_2 and β_1; then strategy (12.1) returns $\{\langle 1\rangle\}$, strategy (12.2) returns $\{\langle 1\rangle, \langle 2\rangle\}$ and strategy (12.3) returns UNKNOWN. □

From conceptual point of view the notion of a classifier is then relatively simple. However, constructing a classifier is a complex task. The main difficulties lie in providing suitable decision rules and resolving conflicts.

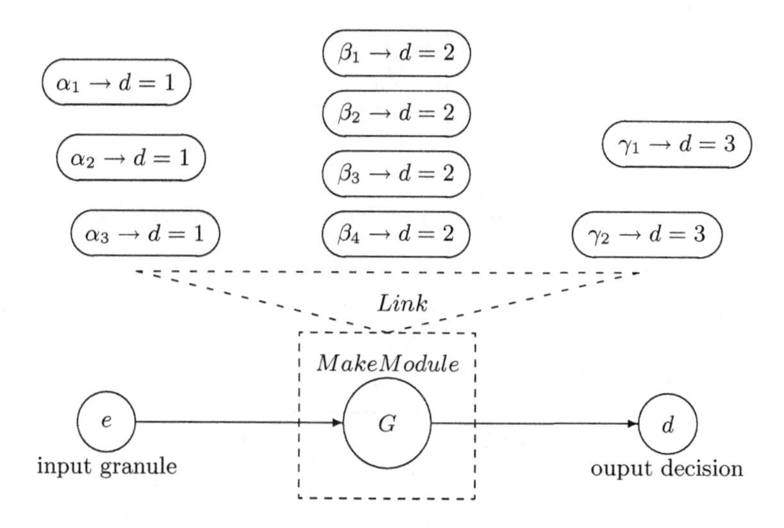

Fig. 12.3. Classifiers as information granules.

12.5 Extending CAKE to Deal with Information Granules

In this section we define a simple extension of the CAKE method in order to provide diagrams for representing general information granules. This allows one to graphically represent the whole information structure of an intelligent system in a uniform manner. In this case, however, a semantics and computation mechanism is to be attached separately, since granules can be constructed using various tools, or even be bought as "black boxes" for which only specification of the methods is provided.

The diagrams for representing granules are similar to those of CAKE, except that in order to indicate that a given diagram represents a non CAKE granule we draw a line on the righthand side of the diagram and additionally indicate the set of constants in the resulting box (see Figure 12.4, where $G : R(\bar{x})$ is the granule's interface, C represents its set of constants and M_R^-, M_R^+ represent its methods).

Operations *Link* and *MakeModule* have a very natural graphical representation.

Let, for $1 \leq i \leq n$, $G_i = \langle C_i, N_i : \bar{R}_i(\bar{x}_i), \bar{M}_i \rangle$ and $G = \langle C, G : \bar{R}(\bar{x}), \bar{M} \rangle$ be information granules. Operation $Link(G, G_1, \ldots, G_n)$ amounts to drawing arrows leading from G and $G_1, \ldots, G_n$ to the resulting granule G_L (see Figure 12.5). In order to indicate the special rôle of argument G, we draw a thick

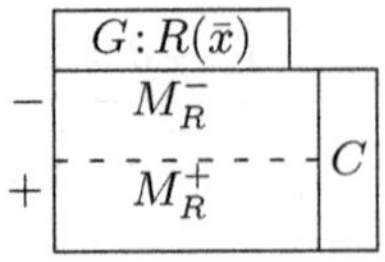

Fig. 12.4. Diagram representing an information granule.

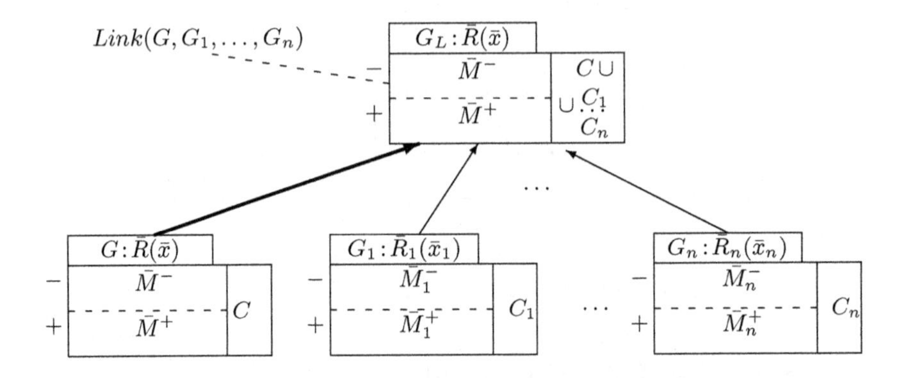

Fig. 12.5. Diagram representing operation *Link*.

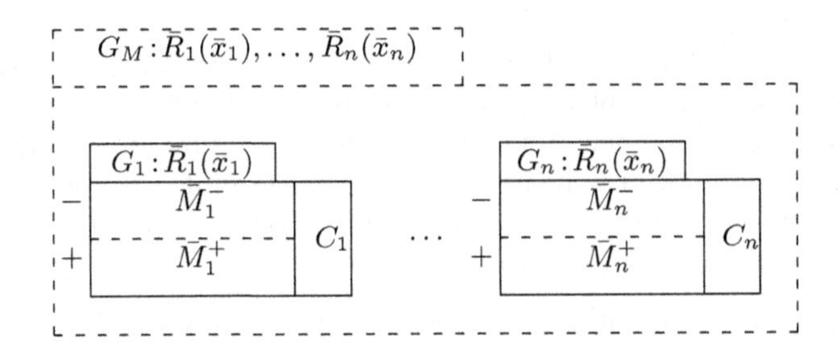

Fig. 12.6. Diagram representing operation $MakeModule$.

arrow between G and G_L. For operation $MakeModule(G_1, \ldots, G_n)$ we use the CAKE–like representation of modules (see Figure 12.6).

Example 12.5.1. Consider the classifier constructed in Example 12.4.1. The diagram visualizing the structure of classifier's granules is shown in Figure 12.7, where

- granules $G_1, \ldots, G_9$ represent the decision rules $\alpha_i \to d = 1$ (for $i = 1, 2, 3$), $\beta_j \to d = 2$ $(j = 1, 2, 3, 4)$ and $\gamma_k \to d = 3$ $(k = 1, 2)$, respectively

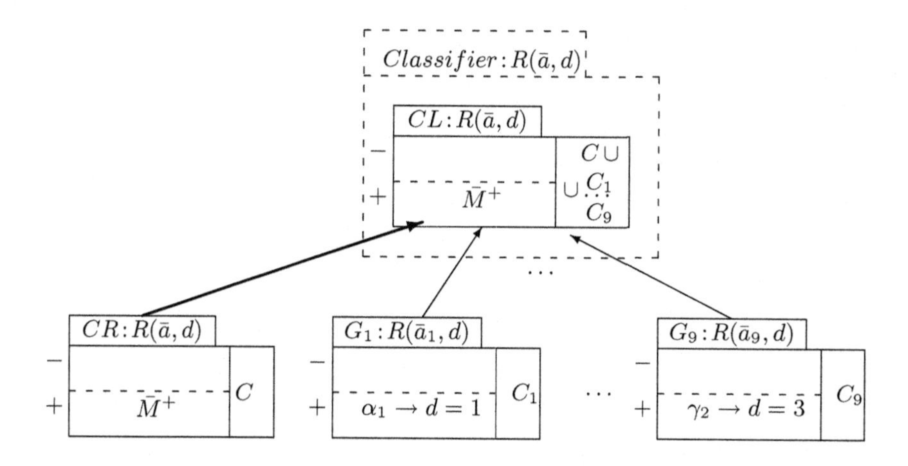

Fig. 12.7. Diagram representing the classifier considered in Example 12.4.1.

- granule CR serves the purpose of conflict resolution. It is linked with granules $G_1, \ldots, G_9$; granule CL is the result of $Link(G, G_1, \ldots, G_9)$
- the module $Classifier$ returns the set of tuples representing the decision d when values of input attributes $\bar{a}$ are given. □

12.6 Bibliographical Notes

The granular computing paradigm was proposed by Zadeh in [236, 237].

The approach to information granules presented in this chapter is novel. It was reported in [59]. It is related to many other sources in the field, including [110, 137, 164, 166, 167, 184] and, in particular, [192, 193].

The information granule calculi aimed at synthesizing solutions to problems posed under uncertainty are presented in [159, 164, 167]. The approach is strongly related to concept approximation learning in layered machine learning [209] as well as to fusion of information from different sources (see., e.g., [29, 151, 153, 236, 237, 238]).

Structures and examples of information granules have been studied in a number of papers (see, e.g., [110, 151, 153, 154, 164, 167, 184, 191, 192, 207]).

One can distinguish several classes of operations on information granules [128, 192]. Yet another important class of operations, namely, operations defined by generalized terms defined on concepts, are discussed in [192, 198].

Bibliographical Notes

13

Tolerance Spaces

13.1 Introduction

In traditional approaches to knowledge representation, notions such as tolerance measures on data, distance between objects or individuals, and similarity measures between primitive and complex data structures such as properties and relations, elementary and complex descriptors, decision rules, information systems, and relational databases, are rarely considered. This is unfortunate because many complex systems which have knowledge representation components receive and process data which is incomplete, noisy, and uncertain. There is often a need to use tolerance and similarity measures in processes of data and knowledge abstraction. This is a particular problem in the area of cognitive robotics where data input by sensors has to be fused, filtered and integrated with more traditional qualitative knowledge structures. A great many levels of knowledge abstraction and data reduction must be used as one tries to integrate newly acquired data with existing data which has previously been abstracted and represented implicitly in the form of more qualitative data and knowledge structures.

One important aspect of this process is that one would like the uncertainty or approximativeness of lower level data to be somehow retained and represented in higher levels of knowledge abstraction in the systems in question. In this manner, notions of similarity and tolerance measures naturally apply to more complex data and knowledge structures such as sets of decision rules, or information systems, or relational databases, because the uncertainty and approximativeness of the primitive data and knowledge structures used to construct the more complex structures is induced through the many levels of data and knowledge abstraction.

One of the important ontological choices made in this book is the manner in which individuals or elements in a universe are viewed. It is generally the case that a set of attributes and values for these attributes are associated with

P. Doherty et al.: *Knowledge Representation Techniques*, Studfuzz **202**, 245–276 (2006)
www.springerlink.com

each of these individuals. Attribute/value pairs are used to define elementary descriptors which are then used to define decision rules. A set of decision rules may be viewed as a decision system. A set of values for each attribute associated with an individual may be viewed as a tuple. A set of k-tuples is a k-argument relation, sets of relations are associated with relational structures. Sets of k-tuples represent information systems. It is clear that complex data structures and representational systems are constructed recursively from the primitive notions of individual, attribute and value. Consequently, notions of tolerance and similarity can be induced through these structures via the tolerance and similarity measures placed on primitive data or value sets.

The strategy that will be used is to define tolerance and distance measures on the value sets associated with attributes or primitive data domains associated with particular applications. These tolerance and distance measures will be induced through the different levels of data and knowledge abstraction in complex representational structures. The representational structures will in some sense inherit the tolerance measures from the primitive data domains and value sets used in these structures at lower levels of abstraction. This point of view is particularly relevant when considering concept formation and machine learning, two aspects which will be thoroughly investigated in Chapter 14. It is important to note that both primitive and complex representational structures may be viewed as information granules. This generic view of representational structures can be used as a framework for comparing similarity of information granules of whatever degree of complexity in a straightforward and uniform manner.

Let us begin with the notion of tolerance and tolerance measures. Webster's dictionary defines tolerance as,

> "the amount of variation allowed from a standard, accuracy, etc."

For example, suppose a system receives data about an attribute a from two sources, where source one asserts that that $a = 1.04$ and source two asserts that $a = 0.98$. Depending on the context, the system might want to consider the values 1.04 and 0.98 as the same relative to some tolerance measure since their distance is only 0.06. In another application this difference may have serious repercussions on system safety, so it is important to make sure that tolerance measures are contextual and can be tuned either automatically or manually relative to the application and context at hand.

By defining parameterized measures of tolerance via distance measurements on values sets and primitive domains, one can cluster sets of values into tolerance neighborhoods and view the clusters as individual elements. Similarly, individuals whose identities are dependent on sets of attribute/value pairs can also be clustered into tolerance neighborhoods and viewed as indiscernible entities to a particular degree of tolerance when used in other data structures. One can recursively apply this idea to increasingly more complex representational structures such as decision rules, relations, and information systems and consider notions of tolerance between such structures and of tolerance neighborhoods for each of these structures. These notions can then in turn be used as a basis for measuring similarity between quite complex representational structures.

Tolerance measures play an important rôle when comparing two representational structures for similarity. What does it mean for two pieces of fruit to be similar? Are two apples of different type similar to each other? Are an apple and a pear similar to each other? Is an apple and a pear more similar than an apple and a banana? Similar questions can be asked about color, taste and acoustic domains. When framing these questions in terms of similarity among concepts, properties or relations, we are in familiar territory and a great deal of research has been directed toward these issues. Less so though in areas of traditional knowledge representation.

On the other hand, what about more complex representational structures? When can we say that two representational structures such as two information systems or relational databases are similar? What does it mean to state that two sets of decision rules are similar, or two ontological structures are similar? Much less research has been directed in this area, but it is of fundamental importance when dealing with complex distributed information systems, where comparisons, sharing, merging and fusing between data and knowledge structures are necessary for many applications. In both cases, similarity measures are necessarily contextual. Similarity depends on what attributes associated with the particular structures are being used in the comparison and what measures of tolerance are supplied with the primitive data domains or value sets associated with the structures.

Webster's dictionary defines *similarity* as

> "like; resembling; having a general resemblance, but not exactly the same."

Webster's dictionary also defines *discern* as

> "to see or understand the difference; to make distinction."

It is clear that these notions are obviously contextual and in the eye of the beholder or based on the beholder's purpose in discerning or comparing. Quite often, the level of abstraction at which the comparison is being made and the individual domains used in defining the structures being compared, play a central rôle in the comparison. Similarity is in some sense, a generalization of the mathematical notion of equivalence, so notions of generalized inclusion that take tolerance measures into account will be introduced. Here again, there is some room for choice and tuning just as there is for associating tolerance measures with primitive data domains and value sets. In our approach, the particular definition of inclusion used can be parameterized if necessary in a highly modular manner.

The basic primitive in the ideas presented is that of a tolerance function. Let's begin with a value set V and two elements $x, y \in V$. A tolerance function τ provides us with a distance measure between x and y normalized to the real interval $[0, 1]$ where the higher the value, the closer in tolerance the two elements are. Given a parameter $p \in [0, 1]$, a tolerance relation τ^p is then introduced among individuals with a threshold p which tunes the tolerance to be within a certain degree. If $\tau(x, y) \geq p$ then $\langle x, y \rangle$ is in the relation τ^p. Both the tolerance function and the parameter p must be provided by a knowledge engineer or must be machine learned. One can continually refine these values.

Once this is done for individual value sets or primitive data domains, it can be generalized to tuples of values and tolerance can be measured between two tuples $\langle x_1, \ldots, x_k \rangle$ and $\langle y_1, \ldots, y_k \rangle$ using pairwise comparison of associated tolerance relations.

Given a value set V with associated tolerance measures, we can then take subsets $V_1, V_2 \subseteq V$ and induce tolerance measures and neighborhood functions on the subsets. Likewise, given a set T of k-tuples with associated tolerance measures, we can then take subsets $T_1, T_2 \subseteq T$ and induce tolerance measures and neighborhood functions on the subsets. Subsets of V can be viewed as properties or concepts and subsets of T can be viewed as k-argument relations.

We can then define a similarity measure on sets of individuals or sets of tuples, by defining an inclusion operation based on comparing the neighborhoods for individuals or tuples in the sets being compared.

These ideas can be generalized further to sets of sets and sets of tuples, where the tolerance and similarity measures between these structures is induced from the primitive tolerance measures in the base value sets. Once the tolerance and similarity measures are in place, an important structuring generalization can be made where the idea of a *tolerance space* is introduced. A tolerance space $TS = \langle U, \tau, p \rangle$, consists of a universe U, a tolerance function $\tau : U \longrightarrow [0, 1]$ and a parameter $p \in [0, 1]$ that allows us to define a tolerance relation τ^p. These are used as a basis for automatically generating tolerance relations, neighborhood functions and similarity measures on the structures in U, which can be complex representational structures such as information systems or relational databases.

It is important to emphasize that the initial tolerance functions and parameters on individuals must be provided and often tuned as complex knowledge structures are specified and used. Tuning may be done explicitly by knowledge engineers familiar with the various primitive data domains and value sets or via various machine learning and parameter tuning techniques. Once these are in place though, the induction process through the complex representational structures is automatic and grounded in the base tolerances associated with the individual data domains.

Given a universe U of objects in a tolerance space with the associated tolerance measures, we can provide a generalization of the notions of upper and lower approximations on sets to subsets of U. The lower and upper approximations will again be induced from the particular tolerance measures provided by the tolerance space in question. Rather than using equivalence classes of individuals constructed from subsets of attributes in information systems as is done in the traditional rough set approach, one would work instead with equivalence classes generated from neighborhood functions of individuals.

There is an interesting connection between the idea of tolerance spaces proposed in this chapter and the work of Gärdenfors with conceptual spaces. Gärdenfors observes that within cognitive science there are two major approaches to modeling representations. The first is the symbolic approach where representation and use of representations is based on the use of symbol structures, manipulation of those structures and various forms of inference on those structures. The other is the subsymbolic approach which is based on ideas from associationism and their instantiations such as connectionism and neural network technologies. In order to bridge the gap between the two approaches, Gärdenfors proposes the level of conceptual spaces. Conceptual spaces are built up using multi-dimensional spaces of quality dimensions (attributes) and providing geometric constraints between these dimensions in order to model distance measures and similarity.

There is an interesting issue as to what relationship there is between conceptual spaces and tolerance spaces. Tolerance spaces contribute to a generalization of conceptual spaces in the sense that concepts can be generalized to approximate concepts based on tolerance measures and the geometric constraints used are less rigid than with conceptual spaces. In fact, we use the notion of semi-distances rather than of distances. In order to place tolerance spaces in the proper context with conceptual spaces, we define a simple version of conceptual spaces and show how tolerance spaces may be integrated in this framework.

Information granulation refers to the approximate treatment of objects and information granules. Information granulation can be viewed from the following two levels of abstraction:

- the object level, where one deals with clusters of similar objects, and
- the information granule level, where one deals with clusters of similar granules. Note that the representational structures encapsulated in these granules can be arbitrarily complex.

As an example of information granulation at the object level, consider a classifier consisting of a set of precise decision rules for classifying objects. Suppose

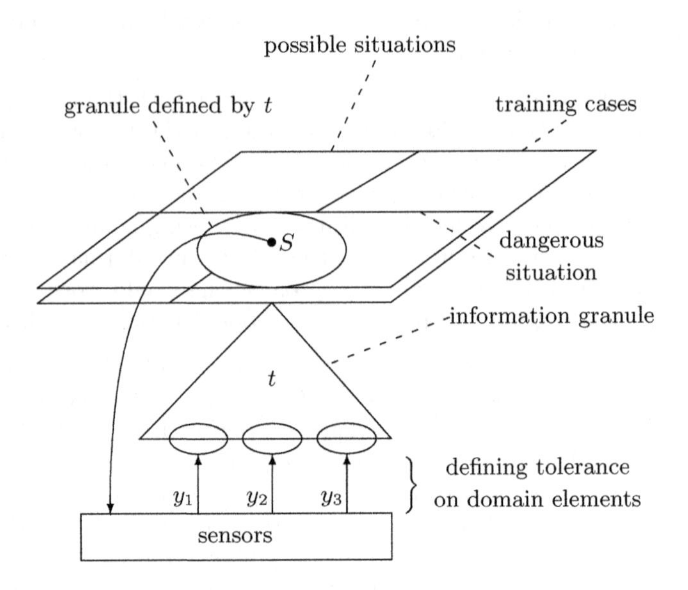

Fig. 13.1. Granulation of information using tolerance.

a system which uses this classifier acquires an object that should be analyzed. It is often the case that the attribute values associated with the object do not satisfy any of the rules making up the classifier. On the other hand, there may have been another object which has been classified properly. If that object is similar enough to the current object being classified, for example it is in the object's tolerance neighborhood or it is evaluated as similar, then we can assert that the new object has been classified properly by the classifier or at least assert that it is reasonable to accept such a classification.

Suppose that we cannot find a similar object and classify using such a technique. In this case, we might lift the question to the next level of abstraction and ask if there are any classifiers similar to the one we have been using. If so, we might try to classify the object in question with one of these similar classifiers. This illustrates a use of information granulation at the granule level.

Before proceeding to details, let us take a closer look at an application of the ideas using a scenario and show how the technical details would fit in to this scenario. Suppose an autonomous UAV is flying over a specific area and its mission is to identify dangerous situations on the road.

Consider the situation illustrated in Figure 13.1. Assume that y_1, y_2, y_3 represent tuples of attributes related to a ground camera, a camera on the UAV platform and a humidity sensor, respectively. Tolerance measures are defined on the value sets associated with each of the attributes.

A classifier representing dangerous situations on the road is defined as an information granule. That information granule in turn is defined by other information granules representing specific dangerous situations. In fact these would be sub-regions in a multi-dimensional attribute space or several subregions in different attribute spaces related via sets of logical constraints. Particular subregions might be "a car moving with high speed on the same lane as another car moving with low speed", or "a car moving with very high speed with a very high level of humidity in the atmosphere". Some of the concepts and relations involved in defining such situations could possibly have been acquired through machine learning techniques and the use of training samples acquired from previous video and sensor information from the ground system and the UAV. In most cases, the information granules used in defining the target concept of dangerous situations on the road will be approximate, inheriting the combined tolerance measures associated with the attributes and value sets from which they are defined. Some of these information granules may be defined in terms of attributes from y_1, y_2, and y_3.

Suppose that the UAV gathers information via its sensors about one or more vehicles and the humidity level in terms of a combination of attributes which include those in y_1, y_2 and y_3. In the figure, the information is gathered as an information granule t. Granule t, because it is defined in terms of attributes from y_1, y_2 and y_3, would inherit the tolerances associated with these attributes.

In order to classify the current situation represented as t, the system would try and match t with any of the information granules composing the target concept. One such situation is shown in the figure as S. The matching procedure would be done in terms of checking for a certain degree of similarity between t and the existing information granules in the target concept. In most cases, the match would not be exact due to the complexity and ambiguity in the definition of a dangerous situation and the fact that the information granules used in defining the target concept are approximate. Using these techniques, small deviations in the sensory data gathered would still lead to certain situations being classified properly as leading to dangerous situations. Consequently, the tolerance measures and similarity measures based on them make good sense in applications such as these.

Let us summarize the methodology we propose:

- we start with a quantitative representation of the similarity of considered concepts given by semi-distance or tolerance functions
- the definition of tolerance spaces and neighborhoods allows us to transform the quantitative representation of the similarity into a qualitative representation of the concepts. Such a transformation can also be applied to complex representational structures. Tolerance parameters allow us to tune the similarities to fit particular application domains

- the approximations we define allow us to isolate objects that surely satisfy a given property and that might satisfy the property. In consequence, we also obtain a characterization of objects that surely do not satisfy the property
- finally one can apply various deduction mechanisms to reason about the considered concepts.

13.2 Conceptual Spaces

For the purpose of our approach we consider finite conceptual spaces only as defined below.

Definition 13.2.1. *Let U be a nonempty finite set of objects. By a* quality dimension over U *we understand any semi-metric space $\langle U, \delta \rangle$. By a* conceptual space over U *we mean any pair $\langle U, Q \rangle$, where Q is a finite set of quality dimensions over U.* □

For example, if one measures colors of objects, quality dimensions can correspond to hue, chromaticity and brightness. The concept "fruit" may have dimensions corresponding to weight, taste, color, etc.

Usually, with any quality dimension one associates a relational structure representing a domain of values corresponding to the quality dimension, together with functions and relations allowing one to calculate (semi-)distances.

For instance, with the quality dimension "weight" one can associate a relational structure defining arithmetic on the real numbers.

Any information system $\mathcal{A} = \langle U, A \rangle$ provides a basis for defining a conceptual space over U in which quality dimensions correspond to the attributes A. However, distances on value sets of attributes have to be additionally specified. In our approach this is done by using tolerance functions.

13.3 Tolerance and Inclusion Functions

We begin by defining a tolerance function on individuals. From this a parameterized tolerance relation follows naturally.

Definition 13.3.1. *By a* tolerance function *on a set U we mean any function $\tau : U \times U \longrightarrow [0, 1]$ such that for all $x, y \in U$,*

$$\tau(x, x) = 1 \quad \textit{and} \quad \tau(x, y) = \tau(y, x).$$

□

Given a conceptual space $\langle U, Q\rangle$ and a quality dimension $\langle U, \delta\rangle \in Q$, a *tolerance function* τ*, based on the quality dimension* can be defined as follows:

$$\tau(u, u') \stackrel{\text{def}}{=} 1 - \frac{\delta(u, u')}{\max\{\delta(x, y) : x, y \in U\}}.$$

Of course, the same approach could be used for an attribute a and its value set V_a in an information system $\mathcal{A}$, provided δ is given.

Definition 13.3.2. *For* $p \in [0, 1]$ *by a* tolerance relation to a degree at least p based on τ, *we mean the relation* τ^p *given by*

$$\tau^p \stackrel{\text{def}}{=} \{\langle x, y\rangle \mid \tau(x, y) \geq p\}.$$

The relation τ^p *is also called the* parameterized tolerance relation. □

In what follows, $\tau^p(x, y)$ is used to denote the characteristic function for the relation τ^p.

Intuitively, $\tau(x, y)$ provides a degree of similarity between x and y, whereas $\tau^p(x, y)$ states that the degree of similarity between x and y is at least p. Note that $\tau^{1.0}(x, y)$ states that there is the largest possible similarity between x and y.

In what follows we limit ourselves to tolerance relations where it is assumed that the parameter p has been provided and is tuned to fit particular applications.[1]

Often one considers objects to be similar if a given distance between them is not greater than a given threshold, say d. Given a quality dimension $\langle U, \delta\rangle$ and a threshold $d \geq 0$, one can define the parameter p from Definition 13.3.2 to be

$$p \stackrel{\text{def}}{=} 1 - \frac{d}{\max\{\delta(x, y) : x, y \in U\}}. \tag{13.1}$$

A parameterized tolerance relation is used to construct tolerance neighborhoods for individuals.

Definition 13.3.3. *By a* neighborhood function w.r.t. τ^p,

$$n^{\tau^p} : U \longrightarrow \text{Pow}(U),$$

we mean a function given by $n^{\tau^p}(u) \stackrel{\text{def}}{=} \{u' \in U \mid \tau^p(u, u') \text{ holds}\}$. *By a* neighborhood of u w.r.t. τ^p *we mean the value of* $n^{\tau^p}(u)$. □

[1] The tuning can be done by an expert or by applying various machine learning techniques - see Chapter 14.

Example 13.3.4. Let $U = \{\mathsf{lR}, \mathsf{mR}, \mathsf{dR}, \mathsf{lG}, \mathsf{g}, \mathsf{b}\}$, where lR, mR, dR, lG, g, b stand for "light red", "medium red", "dark red", "light green", "green" and "blue", respectively. Assume τ^p is given by

$$\{\langle \mathsf{lR}, \mathsf{lR}\rangle, \langle \mathsf{mR}, \mathsf{mR}\rangle, \langle \mathsf{dR}, \mathsf{dR}\rangle, \langle \mathsf{lR}, \mathsf{mR}\rangle, \langle \mathsf{mR}, \mathsf{dR}\rangle,$$
$$\langle \mathsf{lG}, \mathsf{g}\rangle, \langle \mathsf{mR}, \mathsf{lR}\rangle, \langle \mathsf{dR}, \mathsf{mR}\rangle, \langle \mathsf{g}, \mathsf{lG}\rangle,$$
$$\langle \mathsf{lG}, \mathsf{lG}\rangle, \langle \mathsf{g}, \mathsf{g}\rangle, \langle \mathsf{b}, \mathsf{b}\rangle\}.$$

The neighborhood function induced by τ^p is the following:

$$n^{\tau^p}(\mathsf{lR}) = \{\mathsf{lR}, \mathsf{mR}\},\ n^{\tau^p}(\mathsf{mR}) = \{\mathsf{lR}, \mathsf{mR}, \mathsf{dR}\},\ n^{\tau^p}(\mathsf{dR}) = \{\mathsf{mR}, \mathsf{dR}\}$$
$$n^{\tau^p}(\mathsf{lG}) = \{\mathsf{lG}, \mathsf{g}\},\ n^{\tau^p}(\mathsf{g}) = \{\mathsf{g}, \mathsf{lG}\},\ n^{\tau^p}(\mathsf{b}) = \{\mathsf{b}\}.$$

□

13.4 Tolerance Spaces

The concept of tolerance spaces plays a fundamental rôle in our approach to granulation of information.

Definition 13.4.1. *A* tolerance space *is defined as the tuple* $TS = \langle U, \tau, p\rangle$, *which consists of*

- *a nonempty set* U*, called the* domain *of* TS
- *a* tolerance function τ
- *a* tolerance parameter $p \in [0, 1]$.

The parameterized tolerance relation τ^p *is defined as in Definition 13.3.2. In the sequel we write* $n_{TS}^{\tau^p}$ *to denote* n^{τ^p}*, where* τ^p *is a tolerance relation induced from* TS.[2] *We call* $n_{TS}^{\tau^p}$ *a* neighborhood function induced by TS. □

In standard rough set theory, given a universe U of individuals, a set of attributes A and a set $X \subseteq U$, the definition of the lower and upper approximation to X is defined in terms of a partitioning of the universe U in indiscernibility classes relative to a subset of the attributes A. Given a tolerance space $TS = \langle U, \tau, p\rangle$, rather than considering an individuals indiscernibility class as a basis for defining the lower and upper approximation of a subset $X \subseteq U$, we can instead use the neighborhood of an individual induced by the tolerance function/parameter pair(s) provided by the tolerance space. In addition, we

[2] We often drop the superscripts and subscripts when the tolerance spaces and relations are known from context.

can tune our definition of upper approximation via a parameter p which determines how much of a neighborhood must be part of X in order for it to be included in the upper approximation. In order to define the approximations we shall need a standard inclusion function.

Definition 13.4.2. *Let $U_1, U_2 \subseteq U$. By the* standard inclusion function *we mean the function given by*

$$\mu(U_1, U_2) \stackrel{\text{def}}{=} \begin{cases} \dfrac{|U_1 \cap U_2|}{|U_1|} & \text{if } U_1 \neq \emptyset \\ 1 & \text{otherwise.} \end{cases}$$

□

The definition of approximations follows.

Definition 13.4.3. *Let $TS = \langle U, \tau, p \rangle$ be a tolerance space and $X \subseteq U$. The* lower and upper approximations of X w.r.t. TS to a degree $q \in [0,1]$, *X^q_{TS+} and $X^q_{TS\oplus}$, are defined by*

$$X^q_{TS+} = \{u \in U : \mu(n^{\tau^p}(u), X) = 1\},$$
$$X^q_{TS\oplus} = \{u \in U : \mu(n^{\tau^p}(u), X) > q\},$$

where μ is the standard inclusion function.

The approximations X^0_{TS+}, $X^0_{TS\oplus}$, where $q = 0$, are called the lower and upper approximations of X w.r.t. TS *and are often denoted by X_{TS+}, $X_{TS\oplus}$, respectively.* □

In the following examples, we will use the data in Table 13.1 to exemplify the definition and use of tolerance spaces. The table provides information about product brands of insulation sheeting to cover sensitive material on the ground during the winter months. Our primary use of this table will be to demonstrate the use of tolerance measures on the value sets associated with the *temperature, weight* and *material* attributes.

Example 13.4.4. We define a tolerance space for the integer value domain V_T of the attribute *Temperature* in Table 13.1. We use a threshold of 7°C with a distance function defined as $\delta(x, y) = | x - y |$.

Let $TS_T = \langle V_T, \tau_1, p_1 \rangle$ where

- $V_T = \{x \mid -16 \leq x \leq 16\}$
- $\tau_1(x, y) = 1 - \dfrac{\delta(x, y)}{\delta(-16, 16)}$
- $p_1 = 1 - \dfrac{7}{\delta(-16, 16)} = 0.78125.$

Table 13.1. Information table considered in Examples 13.4.4 to 13.4.6.

Object	*Temperature*(°C)	*Weight*(g/m^2)	*Cost*(*Euros*)	*Material*
ob1	7	940	75	m1
ob2	3	1880	70	m2
ob3	0	1280	125	m3
ob4	0	1750	90	m2
ob5	0	1900	120	m4
ob6	-3	1490	150	m3
ob7	-3	1550	150	m1
ob8	-7	1450	220	m5
ob9	-7	2060	140	m6
ob10	-7	1850	175	m3
ob11	-15	2100	200	m7
ob12	3	970	190	m8
ob13	3	800	175	m8
ob14	-3	1690	165	m9
ob15	-3	1200	185	m8
ob16	-3	1500	220	m8
ob17	-7	1380	175	m8
ob18	-7	1460	225	m9
ob19	-10	1820	275	m9
ob20	-10	1390	335	m8
ob21	-15	1800	275	m8

The following tolerance relation is generated

$$\tau_1^{p_1} = \{\langle x, y\rangle \mid \tau_1(x, y) \geq 0.78125 \text{ and } x, y \in V_T\}$$

which is equivalent to

$$\tau_1^{p_1} = \{\langle x, y\rangle \mid \mid x - y \mid\leq 7 \text{ and } x, y \in V_T\}.$$

The following pairs from the table are in $\tau_1^{p_1}$,

$$\{(-15,-10), (-10,-7), (-10,-3), (-7,-3), (-7,0), \\ (-3,0), (-3,3), (0,3), (0,7), (3,7)\}$$

in addition to those pairs that follow from symmetry and reflexivity.

The following neighborhood function is generated

$$n_1^{\tau_1^{p_1}}(x) = \{y \in V_T \mid \tau_1^{p_1}(x, y)\}$$

which is equivalent to

$$n_1^{\tau_1^{p_1}}(x) = \{y \in V_T \mid \mid x - y \mid\leq 7\}.$$

The neighborhood function generates the following neighborhoods for the values in the table,

$$n_1^{\tau_1^{p_1}}(-15) = \{-15, -10\}, \quad n_1^{\tau_1^{p_1}}(-10) = \{-15, -10, -7, -3\},$$
$$n_1^{\tau_1^{p_1}}(-7) = \{-10, -7, -3, 0\}, \quad n_1^{\tau_1^{p_1}}(-3) = \{-10, -7, -3, 0, 3\},$$
$$n_1^{\tau_1^{p_1}}(0) = \{-7, -3, 0, 3, 7\}, \quad n_1^{\tau_1^{p_1}}(3) = \{-3, 0, 3, 7\},$$
$$n_1^{\tau_1^{p_1}}(7) = \{0, 3, 7\}.$$

Let $X = \{-15, -10, -7, -3\}$. We compute the upper and lower approximations of X to degree 0.5. We can do this using Definition 13.4.3, where the standard inclusion function μ is used to compare neighborhoods of individuals in V_T with X.

$$\mu(n_1^{\tau_1^{p_1}}(-15), X) = \tfrac{2}{2} = 1.0, \quad \mu(n_1^{\tau_1^{p_1}}(-10), X) = \tfrac{4}{4} = 1.0,$$
$$\mu(n_1^{\tau_1^{p_1}}(-7), X) = \tfrac{3}{4} = 0.75, \quad \mu(n_1^{\tau_1^{p_1}}(-3), X) = \tfrac{3}{5} = 0.6,$$
$$\mu(n_1^{\tau_1^{p_1}}(0), X) = \tfrac{2}{5} = 0.4, \quad \mu(n_1^{\tau_1^{p_1}}(3), X) = \tfrac{1}{4} = 0.25,$$
$$\mu(n_1^{\tau_1^{p_1}}(7), X) = \tfrac{0}{3} = 0.0\,.$$

Consequently, the lower and upper approximations of X to degree 0.5 are,

$$X^{0.5}_{TS_T^+} = \{-15, -10\},$$
$$X^{0.5}_{TS_T^\oplus} = \{-15, -10, -7, -3\}.$$

□

Example 13.4.5. We define a tolerance space for the integer value domain V_W of the attribute $Weight$ in Table 13.1. We use a threshold of 400 grams with a distance function defined as $\delta(x, y) = \mid x - y \mid$.

Let $TS_W = \langle V_W, \tau_2, p_2 \rangle$ where

- $V_W = \{x \mid 800 \leq x \leq 2100\}$
- $\tau_2(x, y) = 1 - \dfrac{\delta(x, y)}{\delta(800, 2100)}$
- $p_2 = 1 - \dfrac{400}{\delta(800, 2100)} = 0.692307.$

The following tolerance relation is generated

$$\tau_2^{p_2} = \{\langle x, y \rangle \mid \tau_2(x, y) \geq 0.692307 \text{ and } x, y \in V_W\}$$

which is equivalent to

$$\tau_2^{p_2} = \{\langle x, y \rangle \mid \ \mid x - y \mid \leq 400 \text{ and } x, y \in V_W\}.$$

The following neighborhood function is generated

$$n_2^{\tau_2^{p2}}(x) = \{y \in V_W \mid \tau_2^{p_2}(x, y)\}$$

which is equivalent to

$$n_2^{\tau_2^{p2}}(x) = \{y \in V_W \mid \mid x - y \mid \leq 400\}.$$

As an example, the neighborhood function generates the following neighborhoods for the following values in the table,

$$n_2^{\tau_2^{p2}}(1380) = \{y \in V_W \mid 1200 \leq y \leq 1750\}$$

$$n_2^{\tau_2^{p2}}(1850) = \{y \in V_W \mid 1450 \leq y \leq 2100\}.$$

□

Example 13.4.6. We define a tolerance space for the symbol value domain V_M of the attribute *Material* in Table 13.1.

Let $TS_M = \langle V_M, \tau_3, p_3 \rangle$ where

- $V_M = \{\mathsf{m1}, \mathsf{m2}, \mathsf{m3}, \mathsf{m4}, \mathsf{m5}, \mathsf{m6}, \mathsf{m7}, \mathsf{m8}, \mathsf{m9}\}$
- τ_3 is a tolerance function defined by

$$\tau_3(x, y) = \begin{cases} 1 \text{ for } x, y \in \{\mathsf{m1}, \mathsf{m2}, \mathsf{m6}, \mathsf{m7}\} \text{ or} \\ \qquad x, y \in \{\mathsf{m2}, \mathsf{m4}, \mathsf{m5}\} \text{ or } x, y \in \{\mathsf{m8}, \mathsf{m9}\} \\ 0 \text{ otherwise} \end{cases}$$

- $p_3 = 1.0$.

$\tau_3^{1.0}$ is the tolerance relation based on τ_3, where

$$\tau_3^{1.0} = \{\langle x, y \rangle \mid x, y \in V_M \text{ and } \\ x, y \in \{\mathsf{m1}, \mathsf{m2}, \mathsf{m6}, \mathsf{m7}\} \text{ or } x, y \in \{\mathsf{m2}, \mathsf{m4}, \mathsf{m5}\} \text{ or } \\ x, y \in \{\mathsf{m8}, \mathsf{m9}\} \ \}.$$

□

Example 13.4.7. Let $\mathcal{A} = \langle U, A \rangle$ be an information system whose associated information table is shown in Table 13.2. The values m, mR, dR, lR stand for "medium", "medium red", "dark red" and "light red", respectively.

Let $TS = \langle V_C, \tau, p \rangle$ be a tolerance space for the value set V_C associated with the attribute *Color* in $\mathcal{A}$, where

Table 13.2. Information system considered in Example 13.4.7.

Object	*Size*	*Color*
1	m	mR
2	m	dR
3	m	lR

- $V_C = \{\text{lR}, \text{mr}, \text{dR}\}$
- $\tau(x, y) = \begin{cases} 1 \text{ for } x, y \in \{\text{lR}, \text{mR}\} \text{ or } x, y \in \{\text{mR}, \text{dR}\} \text{ or } x = y \\ 0 \text{ otherwise} \end{cases}$
- $p = 1.0$.

The neighborhood function $n_{TS}^{\tau^{1.0}}$ induced by $\tau^{1.0}$ is given by

$$n_{TS}(\text{lR}) = \{\text{lR}, \text{mR}\},\ n_{TS}(\text{mR}) = \{\text{lR}, \text{mR}, \text{dR}\},\ n_{TS}(\text{dR}) = \{\text{mR}, \text{dR}\}.$$

Let $X = \{\text{mR}, \text{lR}\}$. We compute the lower and upper approximations of X to degree 0 and to degree 0.6, using Definition 13.4.3 where the standard inclusion function μ is used to compare neighborhoods of individuals in V_C computed above, with X:

$$\mu(n_{TS}(\text{lR}), X) = \mu(\{\text{lR}, \text{mR}\}, \{\text{mR}, \text{lR}\}) = \frac{2}{2} = 1.0$$

$$\mu(n_{TS}(\text{mR}), X) = \mu(\{\text{lR}, \text{mR}, \text{dR}\}, \{\text{mR}, \text{lR}\}) = \frac{2}{3}$$

$$\mu(n_{TS_C}(\text{dR}), X) = \mu(\{\text{mR}, \text{dR}\}, \{\text{mR}, \text{lR}\}) = \frac{1}{2} = 0.5.$$

It then follows that

- $X^0_{TS^+} = \{\text{lR}\}$ and $X^0_{TS^\oplus_C} = \{\text{dR}, \text{lR}, \text{mR}\}$
- $X^{0.6}_{TS^+} = \{\text{lR}\}$ and $X^{0.6}_{TS^\oplus} = \{\text{mR}, \text{lR}\}$. □

13.5 Defining Tolerance on Elementary Granules

In this section we provide definitions of tolerance relations for typical elementary granules on the basis of a tolerance relation defined on domain elements.

13.5.1 Defining Tolerance on Sets

Consider a tolerance space $TS = \langle U, \tau, p \rangle$. First, we would like to extend the tolerance and neighborhood functions induced by TS to deal with subsets of U. We shall need a notion of generalized inclusion function ν^{τ^p} which will be used as a basis for measuring similarity between complex information structures.

One of the important motivations behind the definition provided is that we require a generalized inclusion function to coincide with the standard inclusion function in the case of a trivial tolerance space (identifying equal elements and distinguishing elements that are not equal).[3]

Definition 13.5.1. *Let* $U_1, U_2 \subseteq U$. *By the* generalized inclusion function induced by τ^p *we mean the function given by*

$$\nu^{\tau^p}(U_1, U_2) \stackrel{\text{def}}{=} \begin{cases} \frac{|\{u_1 \in U_1 : \exists u_2 \in U_2[u_1 \in n^{\tau^p}(u_2)]\}|}{|U_1|} & \textit{if } U_1 \neq \emptyset \\ 1 & \textit{otherwise.} \end{cases}$$

For $q \in [0,1]$, *we say that* U_1 is included in U_2 to a degree at least q w.r.t. ν^{τ^p} *if and only if* $\nu^{\tau^p}(U_1, U_2) \geq q$. □

In the sequel we write $\nu^{\tau p}_{TS}$ and $n^{\tau p}_{TS}$, respectively, to denote $\nu^{\tau p}$ and n^{τ^p}, where τ^p is a tolerance relation induced from a tolerance space TS.[4]

Definition 13.5.2. *Let* $TS = \langle U, \tau, p\rangle$ *be a tolerance space. By a* power tolerance space induced by TS *we mean* $T^{TS} = \langle U^{TS}, \tau^{TS}, s\rangle$, *where*

- $U^{TS} \stackrel{\text{def}}{=} \text{Pow}(U)$, *is the set of all subsets of* U
- *for* $U_1, U_2 \in U^{TS}$, $\tau^{TS}(U_1, U_2) \stackrel{\text{def}}{=} \min\{\nu^{\tau^p}(U_1, U_2), \nu^{\tau^p}(U_2, U_1)\}$
- $s \in [0,1]$ *is a tolerance parameter.* □

Remark 13.5.3. It should be emphasized that the task of specifying a tolerance relation on sets on the basis of a tolerance relation on domain elements has no unique solution. Therefore, the Definition 13.5.2 should be regarded as a possible approach to the problem. Although it works properly in many cases, there are also scenarios where it is problematic.[5] To see this, consider a set $U = \{m_1, \ldots, m_{100}\}$ of mushrooms. Suppose that m_{100} is the only poisonous one. Assume further that we do not distinguish between non-poisonous mushrooms, i.e., $\tau(m_i, m_j) = 1$, for $1 \leq i, j \leq 99$, whereas $\tau(m_i, m_{100}) = 0$, for

[3] We also require such "continuity" in other definitions. Namely, the trivial tolerance space should always lead to standard notions that are accepted when tolerance is not considered.

[4] We often drop the superscripts and subscripts when the tolerance spaces and relations are known from the context.

[5] The same remark concerns various definitions of tolerance relations for typical elementary granules provided in the rest of this section.

$1 \leq i \leq 99$. Let $U_1 = U$ and $U_2 = U - \{m_{100}\}$. According to Definition 13.5.2 (with $p = 1$), $\tau(U_1, U_2) = 0.99$. On the other hand, since U_1 contains the only poisonous mushroom, whereas U_2 does not, we intuitively feel that U_1 and U_2 are very different. To achieve this goal, Definition 13.5.2 should be changed. A possible solution is to define $\tau^{TS}(U_1, U_2) \stackrel{\text{def}}{=} \min\{\tau(m_i, m_j)\}$, for $m_i \in U_1, m_j \in U_2$. □

13.5.2 Defining Tolerance on Tuples

The following is a definition for tolerance spaces on sets of tuples. We consider tuples to be ordered sets of the same cardinality.

Let us, however, remark that tolerance on tuples can be defined in many different ways and the choice of a definition suitable for a particular application should be based on the characteristics of the application.

Definition 13.5.4. *Let $TS_1 = \langle U_1, \tau_1, p_1 \rangle, \ldots, TS_k = \langle U_k, \tau_k, p_k \rangle$ be tolerance spaces defined on the sets $U_1, \ldots, U_k$, respectively. By the* generalized inclusion function over tuples *from $U_1 \times \ldots \times U_k$ we mean the function given by*

$$\nu_o^{\tau^{p_1} \ldots \tau^{p_k}}(\langle u_1, \ldots, u_k \rangle, \langle u'_1, \ldots, u'_k \rangle) \stackrel{\text{def}}{=} \begin{cases} \dfrac{|\{u_i : 1 \leq i \leq k \text{ and } u_i \in n^{\tau^{p_i}}(u'_i)\}|}{k} & \text{if } k \neq 0 \\ 1 & \text{otherwise.} \end{cases}$$

□

Definition 13.5.5. *Let $TS_1 = \langle U_1, \tau_1, p_1 \rangle, \ldots, TS_k = \langle U_k, \tau_k, p_k \rangle$ be tolerance spaces defined on the sets $U_1, \ldots, U_k$, respectively. By a k-*tuple tolerance space induced by *$TS_1, \ldots, TS_k$ we mean the tolerance space $\langle U, \tau^k, p \rangle$, where*

- *$U \stackrel{\text{def}}{=} U_1 \times \ldots \times U_k$*
- *for $\bar{u}_1, \bar{u}_2 \in U$, $\tau^k(\bar{u}_1, \bar{u}_2) \stackrel{\text{def}}{=} \nu_o^{\tau^{p_1} \ldots \tau^{p_k}}(\bar{u}_1, \bar{u}_2) = \nu_o^{\tau^{p_1} \ldots \tau^{p_k}}(\bar{u}_2, \bar{u}_1)$ (the equality between $\nu_o^{\tau^p}(\bar{u}_1, \bar{u}_2)$ and $\nu_o^{\tau^p}(\bar{u}_2, \bar{u}_1)$ follows from the symmetry of τ)*
- *$p \in [0, 1]$ is a tolerance parameter.* □

Example 13.5.6. Consider a part of the information table shown in Table 13.1, consisting of columns corresponding to attributes $Temperature$, $Weight$ and $Material$. Assume tolerance spaces on those attributes, TS_T, TS_W, TS_M, are defined as in Examples 13.4.4, 13.4.5 and 13.4.4, respectively. The 3-tuple tolerance space induced by TS_T, TS_W, TS_M is $\langle U, \tau^3, p \rangle$, where

- $U = V_T \times V_W \times V_M$
- $\tau^3(\langle t, w, m\rangle, \langle t', w', m'\rangle) = \nu_o^{p_1 p_2 p_3}(\langle t, w, m\rangle, \langle t', w', m'\rangle)$
- $p \in [0, 1]$ is a tolerance parameter.

For example,

$$\tau^3(\langle 7, 940, \mathsf{m1}\rangle, \langle 3, 970, \mathsf{m8}\rangle) = \frac{2}{3}$$

since $7 \in n^{\tau^{p_1}}(3)$, $940 \in n^{\tau^{p_2}}(970)$ and $\mathsf{m1} \notin n^{\tau^{p_3}}(\mathsf{m8})$,

$$\tau^3(\langle 7, 940, \mathsf{m1}\rangle, \langle -15, 2100, \mathsf{m7}\rangle) = \frac{1}{3}$$

since $7 \notin n^{\tau^{p_1}}(-15)$, $940 \notin n^{\tau^{p_2}}(2100)$ and $\mathsf{m1} \in n^{\tau^{p_3}}(\mathsf{m7})$. □

13.5.3 Defining Tolerance on Elementary Descriptors

An elementary descriptor $a = v$ consists of an attribute $a \in A$ and a value $v \in V_a$ from the appropriate value set. Any elementary descriptor can be represented as a tuple $\langle a, v\rangle$.

Definition 13.5.7. *Let $A = \{a_1, \ldots, a_k\}$ be a set of attributes and $V_{a_1}, \ldots, V_{a_k}$ be their associated value sets. Let, for $1 \leq i \leq k$, $TS_i = \langle V_{a_i}, \tau_i, p_i\rangle$ be a tolerance space for the value set V_{a_i}.*

$TS_d = \langle U_d, \tau_d, p_d\rangle$ is defined to be a tolerance space for elementary descriptors over A where,

- *$U_d \stackrel{\text{def}}{=} \{\langle a_i, v\rangle \mid a_i \in a \text{ and } v \in V_{a_i}\}$ is the set of all elementary descriptors over A*
- $\tau_d(\langle a_i, v\rangle, \langle a_j, v'\rangle) \stackrel{\text{def}}{=} \begin{cases} \tau_i(v, v') & \text{when } i = j \\ 0 & \text{otherwise.} \end{cases}$
- *$p_d \in [0, 1]$ is a tolerance parameter.* □

Example 13.5.8. Assume tolerance spaces TS_T, TS_W, TS_M on attributes *Temperature*, *Weight* and *Material* are defined as in Examples 13.4.4, 13.4.5 and 13.4.4, respectively. The tolerance space for elementary descriptors over $\{Temperature, Weight, Material\}$, $\langle U_d, \tau_d, p_d\rangle$, is given by

- $U_d = \{\langle Temperature, t\rangle, \langle Weight, w\rangle, \langle Material, m\rangle \mid t \in V_T, w \in V_W \text{ and } m \in V_M\}$

- $$\tau_d(\langle a, v\rangle, \langle b, v'\rangle) \stackrel{\text{def}}{=} \begin{cases} \tau_1(v, v') \text{ when } a{=}b{=}Temperature \text{ and } v, v' {\in} V_T \\ \tau_2(v, v') \text{ when } a{=}b{=}Weight \text{ and } v, v' \in V_W \\ \tau_3(v, v') \text{ when } a{=}b{=}Material \text{ and } v, v' \in V_M \\ 0 \quad\quad\quad \text{otherwise.} \end{cases}$$

□

13.5.4 Defining Tolerance on Templates

Recall that a template is a conjunction of elementary descriptors. A template can be represented as a set of tuples. For example, the template

$(Size = \mathsf{m}) \wedge (Speed = \mathsf{l})$

can be represented as the set of tuples

$\{\langle Size, \mathsf{m}\rangle, \langle Speed, \mathsf{l}\rangle\}$.

For the sake of simplicity we assume that the set of attributes is linearly ordered and consider templates to be tuples, where conjuncts appear in order specified by the ordering. For example, if $Speed < Size$ then the above template is represented as tuple

$\langle\langle Speed, \mathsf{l}\rangle, \langle Size, \mathsf{m}\rangle\rangle$.

Thus tolerance on templates is defined as tolerance on tuples of elementary descriptors.

Definition 13.5.9. *Let $A = \{a_1, \ldots, a_k\}$ be a set of attributes such that $a_1 < a_2 < \ldots < a_k$, where $<$ is a linear order on the set of attributes. Let $V_{a_1}, \ldots, V_{a_k}$ be value sets of $a_1, \ldots, a_k$ and let $TS_1 = \langle V_{a_1}, \tau_1, p_1\rangle, \ldots, TS_k = \langle V_{a_k}, \tau_k, p_k\rangle$ be tolerance spaces for the value sets $V_{a_1}, \ldots, V_{a_k}$, respectively.*

$TS_t = \langle U_t, \tau_t, p_t\rangle$ is defined to be the tolerance space for templates *over A where,*

- *U_t is the set of templates over A*
- $\tau_t(\langle t_1, \ldots, t_n\rangle, \langle t'_1, \ldots, t'_m\rangle) \stackrel{\text{def}}{=}$

$$\begin{cases} \tau^n(\langle t_1, \ldots, t_n\rangle, \langle t'_1, \ldots, t'_m\rangle) & \text{when } m = n \\ 0 & \text{otherwise,} \end{cases}$$

  where τ^n is the n-tuple tolerance function (see Definition 13.5.5) induced by n copies of the tolerance space TS_d of elementary descriptors over A (see Definition 13.5.7)
- *$p_t \in [0, 1]$ is a tolerance parameter.*

□

Example 13.5.10. Consider the tolerance space for elementary descriptors constructed in Example 13.5.8. Let t_1, t_2, t_3 be templates specified by

$$t_1 = Temperature = 7 \wedge Weight = \mathsf{1000} \wedge Material = \mathsf{m2}$$
$$t_2 = Temperature = 12 \wedge Weight = \mathsf{1390} \wedge Material = \mathsf{m7}$$
$$t_3 = Temperature = 15 \wedge Weight = \mathsf{1800} \wedge Material = \mathsf{m6}.$$

It is easily observed that $\tau_t(t_1, t_2) = 1$, $\tau_t(t_1, t_3) = \frac{1}{3}$ and $\tau_t(t_2, t_3) = \frac{2}{3}$. □

13.5.5 Defining Tolerance on Decision Rules

A decision rule $a_1 = v_1 \wedge \ldots \wedge a_n = v_n \rightarrow a_{n+1} = v_{n+1}$ consists of a set of elementary descriptors $a_j = v_j$ where $a_1, \ldots, a_n$, $n \geq 1$, are conditional attributes and a_{n+1} is a decision attribute. Any decision rule can be represented as a set of tuples $\{\langle a_1, v_1\rangle, \ldots, \langle a_n, v_n\rangle, \langle a_{n+1}, v_{n+1}\rangle\}$.

The conditional part of a rule is a template. We then have the following definition.

Definition 13.5.11. *Let $A = \{a_1, \ldots, a_k\}$ be a set of conditional attributes such that $a_1 < a_2 < \ldots < a_k$, where $<$ is a linear order on A. Let a_{k+1} be a decision attribute. Let $V_{a_1}, \ldots, V_{a_{k+1}}$ be value sets of $a_1, \ldots, a_k, d$ and let $TS_1 = \langle V_{a_1}, \tau_1, p_1\rangle, \ldots, TS_k = \langle V_{a_k}, \tau_{k+1}, p_{k+1}\rangle$ be tolerance spaces for the value sets $V_{a_1}, \ldots, V_{a_{k+1}}$, respectively.*

$TS_{dr} = \langle U_{dr}, \tau_{dr}, p_{dr}\rangle$ is defined to be the tolerance space for decision rules *over A where,*

- *U_{dr} is the set of decision rules over $A \cup \{d\}$*
- $\tau_{dr}(\langle t_1, \ldots, t_n\rangle, d_1, \langle t'_1, \ldots, t'_m\rangle, d_2) \stackrel{\text{def}}{=}$

$$\begin{cases} \min\{\tau^n(\langle t_1, \ldots, t_n\rangle, \langle t'_1, \ldots, t'_m\rangle), \tau_{k+1}(d_1, d_2)\} & \textit{when } m = n \\ 0 & \textit{otherwise,} \end{cases}$$

  where τ^n is the n-tuple tolerance function induced by $TS_1, \ldots, TS_n$
- *$p_t \in [0, 1]$ is a tolerance parameter.* □

Example 13.5.12. Let $A = \{H, S, D\}$ be a set of attributes, where H, S and D stand for humidity, speed and danger, respectively. Let V_H, V_S and V_D be the value sets associated with A, $V_H = \{\mathsf{80}, \mathsf{90}, \mathsf{98}\}$, where

- 80, 90, and 98, stand for 80%, 90%, and 98% humidity
- $V_S = \{\mathsf{l}, \mathsf{m}, \mathsf{h}\}$, where l, m, and h stand for low, medium and high speed

- $V_D = \{\mathsf{nd}, \mathsf{hd}, \mathsf{vhd}\}$, where nd, vd and vhd stand for normal danger, high danger and very high danger.

The following tolerance spaces are defined over the value sets V_H, V_S and V_D,

- $TS_H = \langle V_H, \tau_h, 0.9 \rangle$, where $\tau_h(h_1, h_2) = 1 - \frac{\mid h_1 - h_2 \mid}{100}$
- $TS_S = \langle V_S, \tau_s, 1.0 \rangle$, where $\tau_s(s_1, s_2) = \begin{cases} 1 \text{ when } s_1 = s_2 \\ 0 \text{ otherwise} \end{cases}$
- $TS_D = \langle V_D, \tau_d, 1.0 \rangle$, where

  $\tau_d(\mathsf{nd}, \mathsf{hd}) = 0.5, \tau_d(\mathsf{nd}, \mathsf{vhd}) = 0.2, \tau_d(\mathsf{hd}, \mathsf{vhd}) = 0.8.$

Consider the decision rules:

$$(H = \mathsf{90}) \wedge (S = \mathsf{h}) \Rightarrow (D = \mathsf{hd}) \quad (13.2)$$
$$(H = \mathsf{80}) \wedge (S = \mathsf{m}) \Rightarrow (D = \mathsf{nd}) \quad (13.3)$$
$$(H = \mathsf{98}) \wedge (S = \mathsf{h}) \Rightarrow (D = \mathsf{vhd}). \quad (13.4)$$

Consider the tolerance space $TS = \langle U_{dr}, \tau_{dr}, p_{dr} \rangle$ for decision rules. According to Definition 13.5.11 we have that:

$$\tau_{dr}((13.2), (13.3)) = 0.5$$
$$\tau_{dr}((13.2), (13.4)) = 0.8$$
$$\tau_{dr}((13.3), (13.4)) = 0.0.$$

Intuitively, rules (13.2), (13.4) are most similar (tolerance equal to 0.8) since their conditional parts are similar in degree 1.0 and decisions are similar in degree 0.8. Rules (13.3) and (13.4) are not similar at all, since for their conditional parts tolerance is 0.0. □

13.6 Defining Tolerance on Arbitrary Information Granules

In this section we show how to induce a tolerance function on arbitrary granules on the basis of a tolerance function defined on domain elements.

The methodology we apply here is founded on a simple observation that more complex granules can essentially be formed as sets or tuples of simpler granules.

For example an information table is a set of tuples corresponding to rows in the table. A decision table can be viewed as a set of decision rules. A relational database consists of a set of relations which are again sets of tuples. A rule-based classifier is a collection of rules together with a conflict solving relation which itself can be viewed as a collection of tuples or rules.

Let us consider an example illustrating this idea.

Example 13.6.1. Consider rules (13.2), (13.3) and (13.4) provided in Example 13.5.12. Let C_1 and C_2 be two very simple classifiers, the first one consisting of rules (13.2), (13.3) and the second one of rules (13.3), (13.4). More formally, $C_1 = \{(13.2), (13.3)\}$ and $C_2 = \{(13.3), (13.4)\}$ In order to compare C_1 with C_2 we use the power tolerance space over the tolerance space TS of decision rules defined in Example 13.5.12.

Now, for example, if $p_{dr} = 0.75$ then we can easily calculate that

$$\tau^{TS}(C_1, C_2) = \frac{2}{2} = 1.0.$$

If $p_{dr} = 0.9$ then $\tau^{TS}(C_1, C_2) = \frac{1}{2} = 0.5$. □

13.7 Sensor Models and Tolerance Spaces

In this section, we provide a simple sensor model and one method for modeling uncertainty in sensor data which integrates well with tolerance spaces. We also discuss the construction of virtual sensors from combinations of actual and other virtual sensors.

The point to this is an assumption that basic or primitive properties and relations in many domains will be derived through various aggregations of sensor data. Sensor data involves individual readings of values with uncertainty due to noise, etc. In essence, individual data values have neighborhoods of indiscernibility. Tolerance spaces can be used to represent such indiscernibility or similarity. Using tolerance spaces on data readings as building blocks, primitive approximate relations can be defined which inherit the approximate readings of data values.

A sensor is used to measure one or more physical attributes in an environment E. The value sets associated with a physical attribute might be the real numbers, as in the case of measurement of the temperature or velocity of an object; Boolean values, as in the measurement of the presence or absence of an object such as a red car; integer values, as in the case of measurement of the number of vehicles in a particular intersection; or scalar values, such as the specific color of a vehicle.

An environment E can be viewed as an abstract entity containing a collection of physical attributes that are measurable. Vectors or n-dimensional arrays of attribute/value pairs could be used to represent a particular environment. One may want to add a temporal argument to E, so that the current state of the environment is dynamic and changes with time.

Any attribute a can be viewed as a function of the environment E and time point t, i.e., $a : E \times \text{TIME} \longrightarrow V_a$, where TIME is the set of considered time points and V_a is the set of possible values of a.

We consider a sensor S_i as a pair of functions of the environment E and time point t, $S_i(E,t) = \langle V_i(E,t), \epsilon_i(E,t) \rangle$. Depending on the type of sensor being modeled, $V_i(E,t)$ can be a function that returns the values of the physical attributes associated with the sensor, as sensed at time t in environment E. V_i might return a single value, as in the case of a single temperature sensor, or a vector or array of values for more complex sensors. For any physical attribute measured, explicit accuracy bounds are supplied in the form of $\epsilon_i(E,t)$. The temporal and environment arguments are supplied since the accuracy of a sensor may vary with time and change of environment. As in the case of V_i, ϵ_i might return a single accuracy bound or a vector or array of accuracy bounds.

For example, suppose S_{temp} is a sensor measuring the temperature. Let a_{temp} be the physical attribute associated with temperature in the environment, where the actual temperature is $a_{temp}(E,t)$ and the value returned by the sensor is $V_{temp}(E,t)$. The following constraint holds:

$$a_{temp}(E,t) \in [V_{temp}(E,t) - \epsilon_{temp}(E,t), V_{temp}(E,t) + \epsilon_{temp}(E,t)].$$

By using tolerance spaces, accuracy bounds for a physical attribute can be represented equivalently as parameterized tolerance relations on the value set for the attribute. In this manner, we can use neighborhood functions to reason about the tolerance or accuracy neighborhoods around individual sensor readings and combine these into neighborhoods for more complex virtual sensors.

In the following, we will drop the environment and temporal argument for ϵ_{ik} and assume the accuracy bounds for attributes do not change with environment and time.

Let $\bar{a}_i = \langle a_{i1}, \ldots, a_{ik} \rangle$ be the tuple of attributes measured by sensor S_i and let $TS_{ik} = \langle V_{ik}, \tau_{ik}, p_{ik} \rangle$ be a tolerance space for the kth physical attribute, a_{ik}, associated with the sensor S_i, where,

- $V_{ik} \stackrel{\text{def}}{=} \{x \in D_{ik} \mid lb_{ik} \leq x \leq ub_{ik}\}$, where D_{ik} is a value domain such as the reals or integers. It is assumed that the legal values for a physical attribute have a lower and upper bound, lb_{ik}, ub_{ik}. We associate the same distance measurement $\delta(x,y) \stackrel{\text{def}}{=} |x - y|$ with all value sets V_{ik}
- both the tolerance function τ_{ik}, and the tolerance parameter p_{ik} are defined by

$$\tau_{ik}(x,y) \stackrel{\text{def}}{=} 1 - \frac{\delta(x,y)}{\delta(lb_{ik}, ub_{ik})}, \quad p_{ik} \stackrel{\text{def}}{=} 1 - \frac{\epsilon_{ik}}{\delta(lb_{ik}, ub_{ik})}.$$

The neighborhood function can be used to compute the possible actual values of a physical attribute in the environment, given a sensor reading, under the assumption that the accuracy bounds have been generated correctly for a particular sensor and the sensor remains calibrated. For example, if $V_{temp}(E,t)$ is the current value measured by the sensor S_{temp}, then we would know that $a_{temp}(E,t) \in n^{p_{temp}}(V_{temp}(E,t))$. So, the tolerance neighborhood around a sensor reading always contains the actual value of the physical attribute in the environment E and it would be correct to reason with the neighborhoods of sensor values, rather than the sensor values themselves.

Example 13.7.1. Let S_R, S_G and S_B be sensors detecting values of R, G, B color attributes.[6] The universe of values is restricted in those cases to integers in interval $[0, 255]$. Assume that all sensors have the same accuracy, say 5. Then the tolerance space for all three cases is $\langle [0, 255], \tau, p \rangle$, where:

$$\tau(x,y) = 1 - \frac{|x-y|}{255}, \quad p = 1 - \frac{5}{255} \approx 0.9804.$$

In this case an agent using sensor data from S_R, S_G, S_B is unable to distinguish between color values where values of τ on R, G, B attributes are greater than or equal to 0.9804. These physical attributes and their associated tolerance spaces can be used to construct more complex attributes and knowledge structures in terms of these. These new attributes and knowledge structures would inherit the accuracy (inaccuracy) of the primitive sensor data used in their construction. □

13.8 Tolerance-Based Decision Rules

When decision rules are built over continuous attribute domains, e.g., over the set of reals, the main problem is that it is unlikely that new objects to be classified by such rules match conditions exactly. For instance, consider the following rule

$$Temperature = 96.4 \rightarrow Danger = \mathsf{yes} \tag{13.5}$$

derived from a decision table. The rule is too specific. In order to characterize all dangerous situations one would need an impractical number of decision

[6] Of course, there are many techniques for dealing with noise and uncertainty associated with color, more sophisticated than R, G, B attributes. On the other hand, this domain provides a simple and intuitive vehicle to present our ideas.

rules, each for any possible dangerous attribute value (up to some measurement accuracy).[7] On the other hand, given tolerance spaces on attributes *Pressure* and *Temperature* one could identify similar templates and make the above rule applicable to all objects matching the template on the lefthand side of the decision rule.

Consider the tolerance space for attribute *Temperature*,

$$TS_T = \langle V_T, \tau_T, 0.95 \rangle,$$

where $V_T = [0.0, 100.0]$ and $\tau_T(t_1, t_2) \stackrel{\text{def}}{=} 1 - \frac{|t_1 - t_2|}{100}$. In such a case, $n^{0.95}(x) = \{y| \ |x - y| \leq 5.0\}$, i.e., TS_T identifies temperatures which differ from each other not more than by 5.0, assuming they are from the domain. Thus rule (13.5) can be reformulated to cover an interval rather than a particular real value:

$$91.4 \leq Temperature \leq 100.0 \rightarrow Danger = \text{yes}.$$

In the case of more complicated decision rules one would have to apply tolerance space for templates which have been defined in Section 13.5.4.

From methodological point of view the use of tolerance spaces for matching objects with conditional parts of decision rules allows one to deal with clusters of similar objects in a well-controlled manner.

In the machine learning literature one can find a number of other methods solving this problem (for instance the discretization process described in Section 14.4.1). These methods frequently lead to good results. However, these methods typically define intervals for given values independently of a particular application domain. This sometimes may decrease the quality of resulting classifiers, especially when the tolerance function is not linear or there are huge gaps between attribute values in the training data table.

13.9 Tolerance Agents

From the point of view advocated in this book, agents are similar to granules. The main difference is that granules are passive, i.e., provide information about relations when it is required. On the other hand, agents are active. They use information granules and other agents as information sources that allow them to make decisions and act in a given environment. In addition, it is assumed that active agents have a number of additional functionalities

[7] The number of such rules exponentially grows when the number of attributes is increased.

which use passive knowledge sources such as planing and prediction modules. Active agents are also autonomous, or at least semi-autonomous.

According to one of the many definitions in the literature,

> an *autonomous agent* is a system situated within and a part of an environment that senses that environment and acts on it, over time, in pursuit of its own agenda and so as to effect what it senses in the future.

Consequently, the following definition will be used to characterize agents.[8]

Definition 13.9.1. *Let* SIG *be a signature, P be a* set of parameters *consisting of variable and constant symbols, and let N be a set of* agent names. *By an* agent *we understand a tuple $Ag = \langle C, I, M, A\rangle$, where*

- *C is a set of constant symbols from* SIG
- *I is an* interface of *Ag, which is an expression of the form*

  $$Name : R_1(a_{11}, \ldots, a_{1m_1}), \ldots, R_k(a_{k1}, \ldots, a_{km_k}),$$

  where $Name \in N$, $R_1, \ldots, R_k \in$ SIG are (rough) relation symbols and $a_{11}, \ldots, a_{1m_1}, \ldots, a_{k1}, \ldots, a_{km_k}$, called parameters of *$Ag$, are members of P*
- *$\{M_i^\sharp : 1 \leq i \leq k$ and $\sharp \in \{+, -\}\ \}$ is a set of methods.* Methods *and* queries *of agents are defined as in Definition 12.2.1 except that methods may refer to other agents as well*
- *A is an active component of Ag, which represents agent activities that may refer to and use other information granules and agent activities that may involve different types of communication and joint cooperation with other agents.*[9] □

13.9.1 The Definition of Tolerance Agents

In this chapter rough sets are identified with pairs of sets representing lower and upper approximations of a set. More precisely, by a rough set Z we shall understand a pair $Z = \langle X, Y\rangle$, where $X \subseteq Y$. The set X is interpreted as the lower approximation of Z and Y as the upper approximation of Z. We also use the usual notation Z^+ and $Z^\oplus$ to denote X and Y, respectively.

---

[8] Note the similarity with Definition 12.2.1.

[9] This book focuses primarily on the knowledge representation aspects of an active agent, so the precise character of the active component is not necessary to pin down. It may contain executable code, actions performed by humans, diverse functionalities such as planners, etc.

The following definition provides a basis for the semantics of methods of tolerance agents.

Definition 13.9.2. *Let $TS = \langle U, \tau, p \rangle$ be a tolerance space. By a* lower *and* upper approximation of a rough set $Z = \langle X, Y \rangle$ w.r.t. TS *we mean*

$$Z^{\tau^p}_{TS+} \stackrel{\text{def}}{=} \{u \in U : n^{\tau^p}_{TS}(u) \subseteq X\}$$
$$Z^{\tau^p}_{TS\oplus} \stackrel{\text{def}}{=} \{u \in U : n^{\tau^p}_{TS}(u) \cap Y \neq \emptyset\}.$$

In consequence, $Z^{\tau^p}_{TS-}$ is defined as $-Z^{\tau^p}_{TS\oplus}$. □

Remark 13.9.3. Note that when it is clear from context which τ^p is current, then $Z^{\tau^p}_{TS+}$, $Z^{\tau^p}_{TS\oplus}$ and $n^{\tau^p}_{TS}$ can be written as Z_{TS+}, $Z_{TS\oplus}$ and n_{TS}, respectively. □

Example 13.9.4. Let $TS = \langle I, \tau, 1.0 \rangle$ be a tolerance space such that I is the set of integers and $\tau(x,y) = \begin{cases} 1 \text{ when } |x-y| \leq 1 \\ 0 \text{ otherwise.} \end{cases}$

Observe that for any $i \in I$, $n_{TS}(i) = \{i-1, i, i+1\}$.

Let Z be the rough set $\langle \{2,3,4\}, \{1,2,3,4,5\} \rangle$. The lower and upper approximations of Z w.r.t. TS and $\tau^{1.0}$ are defined by

$$Z_{TS+} = \{3\},\ Z_{TS\oplus} = \{0,1,2,3,4,5,6\}.$$

□

As argued in previous chapters, tolerance spaces are fundamentally important for intelligent systems which process inaccurate data and use information granules. How then can one view an agent acting in environments characterized by incompleteness and uncertainty? First observe that each agent can have its own view of the environment due to the use of different sensor suites, methods, knowledge structures, reasoning processes, etc. Consequently, agents can have different understandings of the underlying concepts and can measure objects and phenomena with different accuracy. According to Definition 13.9.1, agents are equipped with their own methods and knowledge structures and can also reason differently about the same phenomena. How about the difference in accuracy measurements used by agents? In order to address this problem, it is assumed that a tolerance agent is additionally equipped with its own tolerance space. The following important issues then have to be addressed:

- what is the intended semantics of an agent's methods given that objects and phenomena are perceived and identified in the context of inaccurate measurements and use of inaccurate data?

- how can agents with different knowledge structures and tolerance spaces understand each other and effect meaningful communication?

The notion of a tolerance agent is formally defined as the following.

Definition 13.9.5. *By a* tolerance agent *we shall understand any pair* $\langle Ag, TS\rangle$*, where* Ag *is an agent and* TS *is a tolerance space.* □

Observe that tolerance spaces can be implemented as information granules. Consequently, agents, tolerance agents and information granules can be implemented using a uniform framework. This introduces an interesting perspective as far as knowledge structures are concerned. Rather than viewing knowledge structures as passive entities which are processed by active entities such as agents, one can view knowledge structures as dynamic entities with intelligent capabilities of their own.

A semantics for methods used by tolerance agents still has to be provided. Informally, suppose that a tolerance agent has to determine whether, say $R(\mathsf{a})$, holds, where R is a relation symbol and a is a constant symbol. Due to its limited perceptive capabilities, one can assume that the agent may not recognize the difference between a and other objects in the neighborhood of a. Thus, the agent can only be sure that $R(\mathsf{a})$ holds only if all elements in the neighborhood of a satisfy R. The agent also cannot exclude the possibility that $R(\mathsf{a})$ holds if there is at least one element in the neighborhood of a satisfying R. Consequently, it is clear that R can be viewed as a rough set such that:

- its lower approximation only contains elements that, together with all elements in their neighborhood, satisfy R
- its upper approximation contains elements for which there is at least one element in their neighborhood that satisfies R.

The following example illustrates this approach.

Example 13.9.6. Assume that the universe consists of the following: Mary, lR, mR, dR and that a tolerance agent, say TA, only knows the following three facts:

$$Likes(\mathsf{Mary}, \mathsf{lR}),\ Likes(\mathsf{Mary}, \mathsf{mR}),$$
$$\neg Likes(\mathsf{Mary}, \mathsf{dR}),$$

where lR, mR and dR denote "light red", "medium red" and "dark red", respectively. Assume further that the single tolerance relation associated with the tolerance space of TA identifies lR with mR and mR with dR.

In addition, suppose that the agent TA is given the task of verifying whether Mary likes a color it senses as being lR. Based on the agent's tolerance relation, its sensors are not capable of recognizing the difference between lR and mR. However, Mary likes both colors, so *TA* can be sure that she likes the color sensed by TA with certainty.

If TA classified the sensed color as mR then it could not be sure whether Mary likes this color or not, since it does not see any difference between mR and dR. Consequently, the sensed color might actually be dR which Mary does not like. On the other hand, *TA* could not exclude the alternative that Mary likes this color, as it could equally well be mR.

In summary, lR is in the lower approximation of a (unary) relation $Likes(\mathsf{Mary}, x)$ and mR, and dR are in the upper approximation of the relation. The agent *TA* uses this rough relation together with its knowledge and tolerance space when answering questions about Mary's likes or dislikes. □

These intuitions are formalized in the following definition.

Definition 13.9.7. *Let* TA $= \langle Ag, TS \rangle$ *be a tolerance agent. Then the semantics of a relation R w.r.t.* TA *is given by:*

$$R_{TA^+} \stackrel{\text{def}}{=} R_{TS^+},\ R_{TA^\oplus} \stackrel{\text{def}}{=} R_{TS^\oplus} \text{ and } R_{TA^-} \stackrel{\text{def}}{=} R_{TS^-},$$

where R_{TS^+}, $R_{TS^\oplus}$ and R_{TA^-} are as defined in Definition 13.9.2. □

Remark 13.9.8. It is important to note that Definition 13.9.7 refers to an arbitrary relation, not necessarily to a relation in the agent's interface. Since any first-order or fixpoint query returns a relation as its result, the definition also provides us with the semantics of queries asked to and answered by tolerance agents. □

Example 13.9.9. Consider the tolerance agent *TA* again, with the same tolerance space and facts given in Example 13.9.6. The answer returned by agent *TA* to the sample query, $Likes(\mathsf{Mary}, x)$, will be computed using Definition 13.9.7.

According to Example 13.9.6, $Likes = \{\langle \mathsf{Mary}, \mathsf{lR}\rangle, \langle \mathsf{Mary}, \mathsf{mR}\rangle\}$. Consequently, *Likes* is approximated by *TA* as follows:

$$\begin{aligned}
&Likes_{TA^+} = Likes_{TS^+} = \\
&\quad \{u \mid u \in \{\langle \mathsf{Mary}, \mathsf{lR}\rangle, \langle \mathsf{Mary}, \mathsf{mR}\rangle, \langle \mathsf{Mary}, \mathsf{dR}\rangle\} \wedge n_{TS}(u) \subseteq Likes\} = \\
&\quad \{\langle \mathsf{Mary}, \mathsf{lR}\rangle\} \\
&Likes_{TA^\oplus} \stackrel{\text{def}}{=} Likes_{TS^\oplus} = \\
&\quad \{u \mid u \in \{\langle \mathsf{Mary}, \mathsf{lR}\rangle, \langle \mathsf{Mary}, \mathsf{mR}\rangle, \langle \mathsf{Mary}, \mathsf{dR}\rangle\} \wedge n_{TS}(u) \cap Likes \neq \emptyset\} = \\
&\quad \{\langle \mathsf{Mary}, \mathsf{lR}\rangle, \langle \mathsf{Mary}, \mathsf{mR}\rangle, \langle \mathsf{Mary}, \mathsf{dR}\rangle\}.
\end{aligned}$$

Thus, the following facts hold:

$$Likes(\mathsf{Mary}, \mathsf{lR})_{TA^+},\ Likes(\mathsf{Mary}, \mathsf{lR})_{TA^\oplus},$$
$$Likes(\mathsf{Mary}, \mathsf{mR})_{TA^\oplus},\ Likes(\mathsf{Mary}, \mathsf{dR})_{TA^\oplus}.$$

These results reflect the intuitions described in Example 13.9.6. Namely, lR is the only color perceived by *TA* as surely satisfying the query $Likes(\mathsf{Mary}, x)$. Color dR is in the upper approximation of the answer, since *TA* is unable to distinguish it from mR, and mR satisfies the query $Likes(\mathsf{Mary}, x)$. The color classified by *TA* to be mR might actually be dR, and *TA* knows that $\neg Likes(\mathsf{Mary}, \mathsf{dR})$ holds. Thus the *TA* can not be sure whether the color it perceives as mR does or does not satisfy the query. □

Example 13.9.10. Let $TS = \langle I, \tau, 1.0\rangle$ be the tolerance space considered in Example 13.9.4. Assume that a tolerance agent $TA = \langle Ag, TS\rangle$ knows only that the following facts about a (crisp) relation, say R, hold:

$R(1)$, $R(2)$, $R(3)$, $R(4)$.

In other words, $R = \{1, 2, 3, 4\}$ and it is a crisp set. Consequently,

$$R_{TA^+} = R_{TS^+} = \{2, 3\}$$
$$R_{TA^\oplus} = R_{TS^\oplus} = \{0, 1, 2, 3, 4, 5\}.$$

Now it is easily observed that $R_{TA^\oplus}(0)$ and $R_{TA^+}(2)$ hold, but $R_{TA^+}(0)$ and $R_{TA^+}(1)$ do not hold. □

13.9.2 Mutual Understanding among Tolerance Agents

Given that two tolerance agents have different tolerance spaces it becomes necessary to define the meaning of queries and answers relative to the two agents. As advocated before, a tolerance agent, when asked about a relation, answers by using the rough set obtained by approximating the relation w.r.t. its tolerance space. On the other hand, the agent that asked the query has to understand the answer provided by the other agent w.r.t. to its own tolerance space. The dialog between agents, say TA_1 (query agent) and TA_2 (answer agent), conforms then to the following schema:

1. TA_1 asks a query Q to TA_2
2. TA_2 computes the answer approximating it according to its tolerance space and returns as an answer the rough set $QA = \langle Q_{TA_2^+}, Q_{TA_2^\oplus}\rangle$
3. TA_1 receives QA as input and approximates it according to its own tolerance space. The resulting rough set is the answer to the query, as understood by TA_1.

This schema works properly under the assumption that the two agents operate with a common vocabulary when communicating. This does not imply that the agents have to have the same vocabulary, simply that there is some overlap. The following is a tentative attempt at formalizing one version of common vocabulary for the purposes of discussing the central ideas involved.

- the universes of agents' tolerance spaces are the same
- the same constant symbols used by different agents have the same meaning
- the same relation symbols used by different agents have the same meaning.

In cases where one agent has a one or more relation symbols in its vocabulary that the other does not, one can apply the techniques described in Section 8.5.1 which approximate any query so it can be understood in the questioning agent's own vocabulary.

The definition describing this interaction now follows.

Definition 13.9.11. *Let TA_1, TA_2 be tolerance agents and let Q be a query, expressed in a logic, which is asked by TA_1 and answered by TA_2. Then the* meaning of the query *is the rough set*

$$\langle\langle Q_{TA_2^+}, Q_{TA_2^\oplus}\rangle_{TA_1^+}, \langle Q_{TA_2^+}, Q_{TA_2^\oplus}\rangle_{TA_1^\oplus}\rangle. \tag{13.6}$$

□

Remark 13.9.12. If the tolerance relations are computable in deterministic time polynomial in the size of agents' databases then the process of computing answers to queries can also be implemented in deterministic polynomial time. This follows from the fact that, in addition to the standard querying mechanism, one has to compute at most $O(n)$ neighborhoods of size not greater than n, where n is the number of all objects stored in the databases. □

13.10 Bibliographic Notes

The research on similarity-based reasoning has a long history. In many application domains objects can be aggregated using, among others, their indiscernibility, proximity or similarity. Approaches based on rough sets, tolerance (similarity) rough sets, and rough mereology are to be found, e.g., in [34, 60, 61, 62, 64, 82, 102, 152, 153, 154, 157, 160, 164, 165, 166, 168, 182, 184, 186, 189, 191, 192, 193, 195, 196, 199, 200, 207]). In this chapter, we use an approach introduced in [60, 61, 62] based on tolerance spaces. A generalization of tolerance spaces used in this chapter to similarity spaces can be found in [64].

An application of tolerance spaces in agent communication is presented in [61, 62]. Other applications of approximated reasoning in multi-agent systems are presented, e.g., in [182, 183].

The notion of conceptual spaces is due to Gärdenfors [77]. However, we use the notion of semi-distances rather than of distances.

A different approach, based on a notion of approximation spaces, based on object neighborhoods and rough inclusion, has been introduced in [191]. It can be used to extend the rough set approach to the case when the indiscernibility is not an equivalence relation but a tolerance or similarity relation [81, 82, 102, 165, 191, 197].

Tuning parameters of tolerance spaces is important for applications. To do this one can adopt strategies known from literature (see [100, 128, 165]).

A number of inclusion functions were considered in the literature. An example of inclusion, known in knowledge discovery [3], for the case of association rules is defined by means of two thresholds called support and confidence. Measures of closeness of rules are discussed, e.g., in [65, 217]. For other measures see [233].

Similarity has also been intensively studied in machine learning (see, e.g., [87]).

14

A Rough Set Approach to Machine Learning

14.1 Introduction

This chapter is primarily devoted to a rough set methodology for *supervised machine learning.*

The rôle of machine learning techniques in the approach we pursue in this book is substantial in application domains, where precise definitions, in the form of logical theories, of required concepts would be too time consuming or even impossible to be provided by experts. Machine learned rules for classifying objects to concepts can be used, e.g., in constructing approximation transducers and trees (see Chapter 7) or CAKE diagrams (see Chapter 9). Rough sets are applied here to model classifiers as information granules providing rough relations as well as a substantial technique for automated generation of classifiers' rules. In fact, we we provide a number of rough set methods that can be used in constructing classifiers. A larger example of the process used in such constructions is presented in Chapter 15.

In the supervised machine learning paradigm, a learning algorithm is given a training data set, usually in the form of a decision system $\mathcal{A} = \langle U, A, d \rangle$, prepared by an expert. Each such decision system classifies elements from U into decision classes. The purpose of the algorithm is to return a set of decision rules together with a matching procedure and conflict resolution strategy, called a classifier, which classifies objects not described in the original decision table.

Most of the techniques discussed here are based on computing prime implicants.[1] Accordingly, they are computationally hard. However, many heuristics have been developed which turned out to be very promising. The results of experiments on many data sets, reported in the literature on these heuristics

[1] See Definition 2.5.3.

P. Doherty et al.: *Knowledge Representation Techniques*, Studfuzz **202**, 277–309 (2006)
www.springerlink.com

(see Section 14.9), show a very good quality of classification of objects outside the training set.

14.2 Reducts in Information and Decision Systems

A crucial concept in the rough set approach to machine learning is that of a reduct. In fact, the term "reduct" corresponds to a wide class of concepts. What typifies all of them is that they are used to reduce information by removing redundant attributes. In this section, we consider three kinds of reducts which will be used in the remainder of this chapter.

Definition 14.2.1. *Given an information system* $\mathcal{A} = \langle U, A \rangle$, *a* reduct *is a minimal set (w.r.t. inclusion) of attributes* $B \subseteq A$ *such that* $\text{IND}_{\mathcal{A}}(B) = \text{IND}_{\mathcal{A}}(A)$. *The intersection of all reducts is called a* core. □

Intuitively, a reduct is a minimal set of attributes from A that preserves the original classification defined by A. Reducts are extremely valuable in applications. Unfortunately, finding a minimal reduct in the general case is NPTIME-hard. One can also show that, for any m, there is an information system with m attributes having an exponential (w.r.t. m) number of reducts. Fortunately, there are reasonably good heuristics which allow one to compute sufficiently many reducts in an acceptable amount of time.

To provide a general method for computing reducts, we will use the following constructs.

Definition 14.2.2. *Let* $\mathcal{A} = \langle U, A \rangle$ *be an information system with* n *objects. The* discernibility matrix *of* $\mathcal{A}$ *is an* $n \times n$ *matrix with elements* c_{ij} *consisting of the set of attributes from* A *on which objects* x_i *and* x_j *differ, i.e.,*

$$c_{ij} = \{a \in A \mid a(x_i) \neq a(x_j)\}, \text{ for } i, j = 1, ..., n. \tag{14.1}$$

A discernibility function $f_{\mathcal{A}}$ *for* $\mathcal{A}$ *is a propositional formula of* m *Boolean variables,* $a_1^*, ..., a_m^*$, *corresponding to the attributes* $a_1, ..., a_m$, *defined by*

$$f_{\mathcal{A}}(a_1^*, ..., a_m^*) = \bigwedge_{1 \leq j < i \leq m} \; \bigvee_{c \in c_{ij}^*, c_{ij} \neq \emptyset} c, \tag{14.2}$$

where $c_{ij}^* = \{a^* \mid a \in c_{ij}\}$. *In the sequel, we write* a_i *instead of* a_i^*. □

The discernibility function $f_{\mathcal{A}}$ describes constraints which must hold if one would like to preserve discernibility between all pairs of discernible objects from $\mathcal{A}$. In particular, it requires keeping at least one attribute from each non-empty element of the discernibility matrix corresponding to any pair of discernible objects.

Table 14.1. Information table considered in Example 14.2.4.

Object	*Speed*	*Color*	*Humidity*
car1	medium	green	large
car2	medium	yellow	small
car3	large	blue	large

Table 14.2. The discernibility matrix for the information table provided in Table 14.1.

$\mathcal{M}(\mathcal{A})$	car1	car2	car3
car1		c, h	s, c
car2	c, h		s, c, h
car3	s, c	s, c, h	

Theorem 14.2.3. *Let $\mathcal{A} = \langle U, A \rangle$ be an information system. The set of all prime implicants of $f_{\mathcal{A}}$ determines the set of all reducts of $\mathcal{A}$.* □

Example 14.2.4. Consider the information system $\mathcal{A}$ whose associated information table is provided in Table 14.1.

The discernibility matrix for $\mathcal{A}$ is presented in Table 14.2. (The letters s, c and h stand for *Speed*, *Color* and *Humidity*, respectively.) The discernibility function for the information system $\mathcal{A}$ is then given by

$$f_{\mathcal{A}}(s, c, h) \equiv (c \vee h) \wedge (s \vee c) \wedge (s \vee c \vee h).$$

The prime implicants of $f_{\mathcal{A}}(s, c, h)$ can be computed in order to derive the reducts for $\mathcal{A}$:

$$\begin{aligned} f_{\mathcal{A}}(s, c, h) &\equiv (c \vee h) \wedge (s \vee c) \wedge (s \vee c \vee h) \\ &\equiv (c \vee h) \wedge (s \vee c) \\ &\equiv c \vee (h \wedge s). \end{aligned}$$

The prime implicants of $f_{\mathcal{A}}(s, c, h)$ are c and $h \wedge s$. Accordingly, there are two reducts of $\mathcal{A}$, namely $\{Color\}$ and $\{Humidity, Speed\}$. □

The second type of reduct used in this chapter are the *decision-relative reducts* for decision systems.

Definition 14.2.5. *Let $\mathcal{A} = \langle U, A, d \rangle$ be a decision system. The* generalized decision *function for $\mathcal{A}$ is the function $\partial_A : U \longrightarrow \textsc{Pow}(V_d)$ defined by*

$$\partial_A(x) = \Big\{ i \mid \exists x' \in U. [\, \langle x', x \rangle \in \textsc{Ind}_{\mathcal{A}}(A) \text{ and } d(x') = i] \Big\}.$$

$\mathcal{A}$ is called consistent (deterministic), *if* $|\partial_A(x)| = 1$, *for any* $x \in U$. *Otherwise $\mathcal{A}$ is said to be* inconsistent (non-deterministic). *Any set consisting of all objects with the same generalized decision value is called a* generalized decision class. □

In terms of decision tables, $\partial_A(x)$ specifies all rows in the table whose conditional attribute values are the same as x and then collects the decision values from each row.

Definition 14.2.6. *Let $\mathcal{A} = \langle U, A, d \rangle$ be a decision system. A* decision-relative reduct *of $\mathcal{A}$ is a minimal (w.r.t. inclusion) non-empty set of attributes $B \subseteq A$ such that $\partial_B = \partial_A$.* □

Intuitively, the definition states that B allows us to classify exactly the same objects, as belonging to equivalence classes $U/\{d\}$, as A. In terms of decision tables, the columns associated with the attributes $A - B$ may be removed without affecting the classification power of the original table.

To compute decision-relative reducts, we extend the definitions of discernibility matrix and discernibility function in the following straightforward manner.

Definition 14.2.7. *Let $\mathcal{A} = \langle U, A, d \rangle$ be a consistent decision system. The discernibility matrix for $\mathcal{A}$ is the discernibility matrix of the information system $\langle U, A \rangle$.* □

Definition 14.2.8. *Let $\mathcal{A} = \langle U, A, d \rangle$ be a consistent decision system and suppose that $\mathcal{M}(\mathcal{A}) = [c_{ij}]$ is a discernibility matrix of $\mathcal{A}$. We construct a new matrix, $\mathcal{M}'(\mathcal{A}) = [c'_{ij}]$, where*

$$c'_{ij} = \begin{cases} \emptyset & \text{if and only if} \quad d(x_i) = d(x_j) \\ c_{ij} & \text{otherwise.} \end{cases}$$

The matrix $\mathcal{M}'(\mathcal{A})$ is called the decision-relative discernibility matrix *of $\mathcal{A}$. The* decision-relative discernibility function $f^r_{\mathcal{A}}$ *for $\mathcal{A}$ is constructed from the decision-relative discernibility matrix for $\mathcal{A}$ in the same way as a discernibility function is constructed from a discernibility matrix.* □

Theorem 14.2.9. *Let $\mathcal{A} = \langle U, A, d \rangle$ be a consistent decision system. The set of all prime implicants of $f^r_{\mathcal{A}}$ determines the set of all decision-relative reducts of $\mathcal{A}$.* □

Example 14.2.10. Consider the decision table associated with a decision system $\mathcal{A}$ as represented in Table 14.3.

Table 14.3. Decision table considered in Example 14.2.10.

Object	*Speed*	*Color*	*Humidity*	*Dangerous*
car1	medium	green	large	no
car2	medium	yellow	small	no
car3	large	blue	large	yes

Table 14.4. The decision-relative discernibility matrix corresponding to decision system shown in Table 14.3.

$\mathcal{M}^r(\mathcal{A})$	car1	car2	car3
car1			s, c
car2			s, c, h
car3	s, c	s, c, h	

The discernibility matrix for $\mathcal{A}$ is the same as the one given in Table 14.2, and the decision-relative discernibility matrix for $\mathcal{A}$ is provided in Table 14.4.

Using the decision-relative discernibility matrix, we can compute the decision-relative discernibility function for $\mathcal{A}$:

$$f^r_{\mathcal{A}}(s, c, h) \equiv (s \vee c) \wedge (s \vee c \vee h) \equiv (s \vee c).$$

The set of all prime implicants of $f^r_{\mathcal{A}}(s, c, h)$ is $\{s, c\}$. Therefore, there are two decision-relative reducts of $\mathcal{A}$, namely $\{Speed\}$ and $\{Color\}$. □

To each decision-relative reduct B of a decision system $\mathcal{A}$, we assign a new decision system, called the B-*reduction* of $\mathcal{A}$. The details are as follows.

Definition 14.2.11. *Let $\mathcal{A} = \langle U, A, d\rangle$ be a consistent decision system and suppose that B is a decision-relative reduct of $\mathcal{A}$. A B-*reduction *of $\mathcal{A}$ is a decision system $\mathcal{A}^* = \langle V, B, d\rangle$, where:*[2]

- $V = \{[x]_B : x \in U\}$
- $a([x]_B) = a(x)$, *for each* $a \in B$ *and each* $[x]_B \in V$
- $d([x]_B) = d(x)$, *for each* $[x]_B \in V$. □

Example 14.2.12 (Example 14.2.10 continued). Let $\mathcal{A}^*$ be the $\{Speed\}$-reduction of the decision system $\mathcal{A}$. The decision table associated with $\mathcal{A}^*$ is provided in Table 14.5. □

The third type of reduct considered in this chapter is used in applications where approximations to reducts are preferable to using standard reducts.

[2] Recall that $[x]_B$, where $x \in U$, denotes an equivalence class of the relation $\mathrm{IND}_{\mathcal{A}}(B)$ containing x (see Chapter 3, Section 3.3).

Table 14.5. $\{Speed\}$-reduction of the decision system $\mathcal{A}$.

Objects	*Speed*	*Dangerous*
car1, car2	medium	no
car3	large	yes

Such reducts are called α-reducts, where α is a real number from the interval $[0, 1]$.

Definition 14.2.13. *Let $\mathcal{A} = \langle U, A, d\rangle$ be a decision system and let $\mathcal{M}(\mathcal{A})$ be the discernibility matrix of $\mathcal{A}$. Assume further that n is the number of non-empty sets in $\mathcal{M}(\mathcal{A})$. A set of attributes $B \subseteq A$ is called an* α-reduct *if and only if $\frac{m}{n} \geq \alpha$, where m is the number of sets that have a non-empty intersection with B.* □

14.3 Attribute Selection

In the supervised machine learning approach, a learning algorithm is provided with training data. In the context of rough set machine learning techniques, training data is provided in the form of training decision systems, or their equivalent representations as decision tables.

Since the conditional attributes of a specific decision table are typically extracted from large sets of unstructured data, it is often the case that some of the attributes are irrelevant for the purpose of classification. Such attributes should be removed from the table if possible. The *attribute selection problem* is the problem of choosing a relevant subset of attributes, while removing the irrelevant ones.

A natural solution for the attribute selection problem is to assume that the intersection of the decision-relative reducts of a training decision table is the source of the relevant attributes. Unfortunately, there are two problems with this solution. Firstly, the intersection can be empty. Secondly, the number of attributes contained in all decision-relative reducts is typically small. Consequently, although these attributes perfectly characterize the training decision table, they are in general inadequate for providing a satisfactory classification of new objects not part of the training data.

To deal with the attribute selection problem, it is often reasonable to use various approximations of decision-relative reducts.

Definition 14.3.1. *Let $\mathcal{A} = \langle U, A, d\rangle$ be a consistent decision system. Any subset B of A is called an* approximate reduct *of $\mathcal{A}$. The number*

$$\varepsilon_{A,\{d\}}(B) = \frac{(\gamma(A, \{d\}) - \gamma(B, \{d\}))}{\gamma(A, \{d\})} = 1 - \frac{\gamma(B, \{d\})}{\gamma(A, \{d\})} \tag{14.3}$$

is called an error of reduct approximation.[3] □

The error of reduct approximation expresses exactly how the set of attributes B approximates the set of condition attributes A with respect to determining d. Note that $\varepsilon_{A,\{d\}}(B) \in [0, 1]$, where 0 indicates no error and the closer $\varepsilon_{A,\{d\}}(B)$ is to 1, the greater the error.

There are two general approaches to attribute selection: an *open-loop approach* and a *closed-loop approach.* Methods based on the open-loop approach are characterized by the fact that they do not use any feedback information about classifier quality for attribute selection. In contrast, the methods based on the closed-loop approach use feedback information as criteria for attribute selection.

A number of attribute selection algorithms have been proposed in the machine learning literature, but they will not be considered here since our focus is on rough set based techniques. Rough set techniques which attempt to solve the attribute selection problem are typically based on the closed-loop approach and consist of the following basic steps:[4]

1. decision-relative reducts are extracted from a training decision table. The attributes contained in these reducts (or in their intersection) are viewed as potentially relevant
2. using the specific machine learning algorithm, a classifier based on the chosen attributes is constructed
3. the classifier is then tested on a new set of training data. If its performance is unsatisfactory (w.r.t. some measure), a new set of attributes is constructed by extracting approximate additional reducts for the initial training table, and the process is repeated.

Reducts need not be the only source of information used in the selection of attributes. The rough set approach offers another interesting possibility. The main idea is to generalize the notion of attribute reduction by introducing the concept of *significance of attributes.* This measure enables attributes to be evaluated using a multi-valued scale which assigns a real number from the interval [0,1] to an attribute. This number, expressing the importance of an attribute in a decision system, is evaluated by measuring the effect of removing the attribute from the table.

Definition 14.3.2. *Let* $\mathcal{A} = \langle U, A, d \rangle$ *be a decision system and* $a \in A$. *The* significance of an attribute a *in* $\mathcal{A}$ *is defined by*

[3] Recall that the coefficient $\gamma(X, Y)$ expresses the degree of dependency between sets of attributes X and Y (see Definition 3.8.1).

[4] There are public domain software packages, for instance the RSES system (for references see Section 14.9), which offer software that may be used to solve the attribute selection problem.

$$\sigma_{A,\{d\}}(a) = \frac{(\gamma(A,\{d\}) - \gamma(A - \{a\},\{d\}))}{\gamma(A,\{d\})} = 1 - \frac{\gamma(A - \{a\},\{d\})}{\gamma(A,\{d\})}. \quad (14.4)$$

Assume that $B \subseteq A$. The significance coefficient can be extended to sets of attributes as follows,

$$\sigma_{(A,\{d\})}(B) = \frac{(\gamma(A,\{d\}) - \gamma(A - B,\{d\}))}{\gamma(A,\{d\})} = 1 - \frac{\gamma(A - B,\{d\})}{\gamma(A,\{d\})}. \quad (14.5)$$

□

The coefficient $\sigma_{A,\{d\}}(B)$, can be understood as a classification error which occurs when the attributes $a \in B$ are removed from the decision system. Note that $\sigma_{A,\{d\}}(B) \in [0,1]$, where 0 indicates removing attributes in B causes no error and the closer $\sigma_{A,\{d\}}(B)$ is to 1, the greater the error is.

Remark 14.3.3. In this section we have mainly concentrated on the case where the attributes are selected from the set of attributes of the input decision system. In some cases it might be useful to replace some attributes by a new one.

For example, if one considers a concept of a safe distance between vehicles, then attributes, say VS standing for "vehicle speed" and SL standing for "speed limit", can be replaced by an attribute DIF representing the difference $SL - VS$. In fact, the new attribute better corresponds to the concept of safe distance than the pair $\langle VS, SL\rangle$.

The use of this technique, also called *feature extraction*, is illustrated in Chapter 15. □

14.4 Value Set Reduction

Consider a decision system with a large number of attribute values. There is a very low probability that a new object will be properly recognized by matching its attribute value vector with any of the rows in the decision table associated with the decision system. So, in order to construct a high quality classifier, it is often necessary to reduce the cardinality of the value sets of specific attributes in a training decision table. The task of reducing the cardinality of value sets is referred to as the *value set reduction problem.*

In this section, two methods of value set reduction are considered:

1. discretization, used for real value attributes, and
2. symbolic attribute value grouping, used for symbolic attributes.

Table 14.6. The discretization process: (a) The original decision system $\mathcal{A}$ considered in Example 14.4.2. (b) The C-discretization of $\mathcal{A}$ considered in Example 14.4.4.

$\mathcal{A}$	a	b	d
u_1	0.8	2.0	1
u_2	1.0	0.5	0
u_3	1.3	3.0	0
u_4	1.4	1.0	1
u_5	1.4	2.0	0
u_6	1.6	3.0	1
u_7	1.3	1.0	1

(a) $\Rightarrow$

$\mathcal{A}^C$	a^C	b^C	d
u_1	0	2	1
u_2	1	0	0
u_3	2	3	0
u_4	3	1	1
u_5	3	2	0
u_6	4	3	1
u_7	2	1	1

(b)

14.4.1 Discretization

A discretization replaces value sets of conditional real-valued attributes with intervals. The replacement ensures that a consistent decision system is obtained (assuming a given consistent decision system) by substituting any object's original value in the decision table by the unique name of the interval in which it is contained. This substantially reduces the size of the value sets of real-valued attributes.

The use of discretization is not specific to the rough set approach to machine learning. In fact, a majority of rule or tree induction algorithms require it for a good performance.

Definition 14.4.1. *Let $\mathcal{A} = \langle U, A, d\rangle$ be a consistent decision system. Assume $V_a = [l_a, r_a) \subset \mathcal{R}$,*[5] *for any $a \in A$ and $l_a < r_a$. A pair $\langle a, c\rangle$, where $a \in A$ and $c \in V_a$, is called a* cut *on V_a.*

Any attribute $a \in A$ defines a sequence of real numbers $v_1^a < v_2^a < \cdots < v_{k_a}^a$, where $\{v_1^a, v_2^a, \ldots, v_{k_a}^a\} = \{a(x) : x \in U\}$. The set of basic cuts *on a, written B_a, is specified by*

$$B_a = \{\langle a, (v_1^a + v_2^a)/2\rangle, \langle a, (v_2^a + v_3^a)/2\rangle, \ldots, \langle a, (v_{k_a-1}^a + v_{k_a}^a)/2\rangle\}.$$

The set $\bigcup_{a\in A} B_a$ is called the set of basic cuts *on $\mathcal{A}$.* □

Example 14.4.2. Consider a consistent decision system $\mathcal{A}$ and the associated decision table presented in Table 14.6(a).

We assume that the initial value domains for the attributes a and b are:

[5] Recall that $\mathcal{R}$ denotes the set of real numbers.

$V_a = [0, 2) ; V_b = [0, 4)$.

The sets of values of a and b for objects from U are:

$$a(U) = \{0.8,\ 1.0,\ 1.3,\ 1.4,\ 1.6\}$$
$$b(U) = \{0.5,\ 1.0,\ 2.0,\ 3.0\}.$$

By definition, the sets of basic cuts for a and b are

$$B_a = \{\langle a, 0.9\rangle,\ \langle a, 1.15\rangle,\ \langle a, 1.35\rangle,\ \langle a, 1.5\rangle\}$$
$$B_b = \{\langle b, 0.75\rangle;\ \langle b, 1.5\rangle;\ \langle b, 2.5\rangle\}.$$

□

Using the idea of cuts, decision systems with real-valued attributes can be discretized.

Definition 14.4.3. *Let $\mathcal{A} = \langle U, A, d\rangle$ be a decision system. For $a \in A$, let $C_a = \{\langle a, c_1^a\rangle, \langle a, c_2^a\rangle, \ldots, \langle a, c_k^a\rangle\}$ be any set of cuts of a. Assume that $c_1^a < c_2^a < \cdots < c_k^a$. The set of cuts $C = \bigcup_{a\in A} C_a$ defines a new decision system $\mathcal{A}^C = \langle U, A^C, d\rangle$, called the C-*discretization *of $\mathcal{A}$, where*

- $A^C = \{a^C : a \in A\}$
- $a^C(x) = \begin{cases} 0 & \text{if and only if } a(x) < c_1^a \\ i & \text{if and only if } a(x) \in [c_i^a, c_{i+1}^a),\ \text{for } i \in \{1, \ldots, k-1\} \\ k+1 & \text{if and only if } a(x) > c_k^a. \end{cases}$

□

Example 14.4.4 (Example 14.4.2 continued). Let $C = B_a \cup B_b$. It is easily checked that the C-discretization of $\mathcal{A}$ is the decision system whose decision table is provided in Table 14.6 (b). □

Since a decision system can be discretized in many ways, a natural question arises as to how to evaluate various possible discretizations.

Definition 14.4.5. *A set of cuts C is called $\mathcal{A}$-*consistent, *if $\partial_{\mathcal{A}} = \partial_{\mathcal{A}^C}$, where $\partial_{\mathcal{A}}$ and $\partial_{\mathcal{A}^C}$ are generalized decision functions for $\mathcal{A}$ and $\mathcal{A}^C$, respectively. An $\mathcal{A}$-consistent set of cuts C is $\mathcal{A}$-*irreducible *if C' is not $\mathcal{A}$-consistent for any $C' \subset C$. The $\mathcal{A}$-consistent set of cuts C is $\mathcal{A}$-*optimal, *if $|C| \le |C'|$, for any $\mathcal{A}$-consistent set of cuts C'.* □

It is easily observed that the set of cuts considered in Example 14.4.4 is $\mathcal{A}$-consistent. However, as we shall see (Example 14.4.8), it is neither optimal nor irreducible.

Since the purpose of the discretization process is to reduce the size of individual value sets of attributes, we are primarily interested in optimal sets of cuts. These are extracted from the basic sets of cuts for a given decision system. The details follow.

Let $\mathcal{A} = \langle U, A, d \rangle$ be a consistent decision system where $U = \{u_1, \ldots, u_n\}$. Recall that any attribute $a \in A$ defines a sequence $v_1^a < v_2^a < \cdots < v_{k_a}^a$, where $\{v_1^a, v_2^a, \ldots, v_{k_a}^a\} = \{a(x) : x \in U\}$. Let $ID(\mathcal{A})$ be a set of pairs $\langle i, j \rangle$ such that $i < j$ and $d(u_i) \neq d(u_j)$. We now construct a propositional formula, called the *discernibility formula* of $\mathcal{A}$, as follows:

1. to each interval of the form $\left[v_k^a, v_{k+1}^a\right)$, $a \in A$ and $k \in \{1, \ldots, n_a - 1\}$, we assign a Boolean variable denoted by p_k^a. The set of all these variables is denoted by $V(\mathcal{A})$
2. we first construct a family of formulas

   $$\{B(a, i, j) : a \in A \text{ and } \langle i, j \rangle \in ID(\mathcal{A})\},$$

   where $B(a, i, j)$ is a disjunction of all elements from the set

   $$\left\{p_k^a : \left[v_k^a, v_{k+1}^a\right) \subseteq \left[\min\{a(u_i), a(u_j)\}, \max\{a(u_i), a(u_j)\}\right) \right\}$$

3. next, we construct a family of formulas

   $$\{C(i, j) : i, j \in \{1, \ldots, n\}, i < j \text{ and } \langle i, j \rangle \in ID(\mathcal{A})\},$$

   where $C(i, j) = \bigvee_{a \in A} B(a, i, j)$
4. finally, the discernibility formula for $\mathcal{A}$, $D(\mathcal{A})$, is defined as

   $$D(\mathcal{A}) = \bigwedge C(i, j),$$

   where $i < j$ and $\langle i, j \rangle \in ID(\mathcal{A})$ and $C(i, j) \not\equiv$ FALSE.

Any non empty set $S = \{p_{k_1}^{a_1}, \ldots, p_{k_r}^{a_r}\}$ of Boolean variables from $V(\mathcal{A})$ uniquely defines a set of cuts, $C(S)$, given by

$$C(S) = \{\langle a_1, (v_{k_1}^{a_1} + v_{k_1+1}^{a_1})/2 \rangle, \cdots, \langle a_1, (v_{k_r}^{a_r} + v_{k_r+1}^{a_r})/2 \rangle\}.$$

Theorem 14.4.6. *Let $\mathcal{A} = \langle U, A, d \rangle$ be a consistent decision system. For any non-empty set $S \subseteq V(\mathcal{A})$ of Boolean variables, the following two conditions are equivalent:*

1. *The conjunction of variables from S is a prime implicant of the discernibility formula for $\mathcal{A}$.*
2. *$C(S)$ is an $\mathcal{A}$-irreducible set of cuts on $\mathcal{A}$.* □

Corollary 14.4.7. *Let $\mathcal{A} = \langle U, A, d\rangle$ be a consistent decision system. For any non-empty set $S \subseteq V(\mathcal{A})$ of Boolean variables, the following two conditions are equivalent:*

1. *The conjunction of variables from S is a minimal (w.r.t. to length) prime implicant of the discernibility formula for $\mathcal{A}$.*
2. *$C(S)$ is an $\mathcal{A}$-optimal set of cuts on $\mathcal{A}$.* □

Example 14.4.8 (Example 14.4.4 continued).

$$ID(\mathcal{A}) = \{\langle 1,2\rangle,\ \langle 1,3\rangle,\ \langle 1,5\rangle,\ \langle 2,4\rangle,\ \langle 2,6\rangle,\ \langle 2,7\rangle\ \langle 3,4\rangle,\ \langle 3,6\rangle,\ \langle 3,7\rangle,\ \langle 4,5\rangle,\ \langle 5,6\rangle,\ \langle 5,7\rangle\}.$$

1. We introduce four Boolean variables, p_1^a, p_2^a, p_3^a, p_4^a, corresponding respectively to the intervals

   $$[0.8, 1.0)\,,\ [1.0, 1.3)\,,\ [1.3, 1.4)\,,\ [1.4, 1.6)\,,$$

   of the attribute a, and three Boolean variables, p_1^b, p_2^b, p_3^b, corresponding respectively to the intervals

   $$[0.5, 1.0)\,,\ [1.0, 2.0)\,,\ [2, 3.0)\,,$$

   of the attribute b
2. the following are the formulas $B(a,i,j)$ and $B(b,i,j)$, where $i < j$ and $\langle i,j\rangle \in ID(\mathcal{A})$:

$B(a,1,2) \equiv p_1^a$	$B(b,1,2) \equiv p_1^b \vee p_2^b$
$B(a,1,3) \equiv p_1^a \vee p_2^a$	$B(b,1,3) \equiv p_3^b$
$B(a,1,5) \equiv p_1^a \vee p_2^a \vee p_3^a$	$B(b,1,5) \equiv \textsc{False}$
$B(a,2,4) \equiv p_2^a \vee p_3^a$	$B(b,2,4) \equiv p_1^b$
$B(a,2,6) \equiv p_2^a \vee p_3^a \vee p_4^a$	$B(b,2,6) \equiv p_1^b \vee p_2^b \vee p_3^b$
$B(a,2,7) \equiv p_2^a$	$B(b,2,7) \equiv p_1^b$
$B(a,3,4) \equiv p_3^a$	$B(b,3,4) \equiv p_2^b \vee p_3^b$
$B(a,3,6) \equiv p_3^a \vee p_4^a$	$B(b,3,6) \equiv \textsc{False}$
$B(a,3,7) \equiv \textsc{False}$	$B(b,3,7) \equiv p_2^b \vee p_3^b$
$B(a,4,5) \equiv \textsc{False}$	$B(b,4,5) \equiv p_2^b$
$B(a,5,6) \equiv p_4^a$	$B(b,5,6) \equiv p_3^b$
$B(a,5,7) \equiv p_3^a$	$B(b,5,7) \equiv p_2^b$

3. the following are the formulas $C(i,j)$, where $i < j$ and $\langle i,j\rangle \in ID(\mathcal{A})$:

Table 14.7. The C-discretization considered in Example 14.4.8

$\mathcal{A}^C$	a^C	b^C	d
u_1	0	2	1
u_2	1	0	0
u_3	1	2	0
u_4	1	1	1
u_5	1	2	0
u_6	2	2	1
u_7	1	1	1

$C(1,2) \equiv p_1^a \vee p_1^b \vee p_2^b$
$C(1,3) \equiv p_1^a \vee p_2^a \vee p_3^b$
$C(1,5) \equiv p_1^a \vee p_2^a \vee p_3^a$
$C(2,4) \equiv p_2^a \vee p_3^a \vee p_1^b$
$C(2,6) \equiv p_2^a \vee p_3^a \vee p_4^a \vee p_1^b \vee p_2^b \vee p_3^b$
$C(2,7) \equiv p_2^a \vee p_1^b$
$C(3,4) \equiv p_3^a \vee p_2^b \vee p_3^b$
$C(3,6) \equiv p_3^a \vee p_4^a$
$C(3,7) \equiv p_2^b \vee p_3^b$
$C(4,5) \equiv p_2^b$
$C(5,6) \equiv p_4^a \vee p_3^b$
$C(5,7) \equiv p_3^a \vee p_2^b$

4. the discernibility formula for $\mathcal{A}$ is then given by

$$\begin{aligned} D(\mathcal{A}) \equiv\ & (p_1^a \vee p_1^b \vee p_2^b) \wedge (p_1^a \vee p_2^a \vee p_3^b) \wedge \\ & (p_1^a \vee p_2^a \vee p_3^a) \wedge (p_2^a \vee p_3^a \vee p_1^b) \wedge \\ & (p_2^a \vee p_3^a \vee p_4^a \vee p_1^b \vee p_2^b \vee p_3^b) \wedge (p_2^a \vee p_1^b) \wedge \\ & (p_3^a \vee p_2^b \vee p_3^b) \wedge (p_3^a \vee p_4^a) \wedge (p_2^b \vee p_3^b) \wedge \\ & p_2^b \wedge (p_4^a \vee p_3^b) \wedge (p_3^a \vee p_2^b). \end{aligned}$$

The prime implicants of the formula $D(\mathcal{A})$ are

$p_2^a \wedge p_4^a \wedge p_2^b$
$p_2^a \wedge p_3^a \wedge p_2^b \wedge p_3^b$
$p_3^a \wedge p_1^b \wedge p_2^b \wedge p_3^b$
$p_1^a \wedge p_4^a \wedge p_1^b \wedge p_2^b.$

Suppose we take the the prime implicant $p_1^a \wedge p_4^a \wedge p_1^b \wedge p_2^b$. Its corresponding set of cuts is

$C = \{\langle a, 0.9\rangle, \langle a, 1.5\rangle, \langle b, 0.75\rangle, \langle b, 1.5\rangle\}.$

The decision table for the C-discretization of $\mathcal{A}$ is provided in Table 14.7.

Observe that the set of cuts corresponding to the prime implicant $p_2^a \wedge p_4^a \wedge p_2^b$ is $\{\langle a, 1.15\rangle, \langle a, 1.5\rangle, \langle b, 1.5\rangle\}$. Thus C is not an optimal set of cuts. □

The problem of searching for an optimal set of cuts P in a given decision system $\mathcal{A}$ is NPTIME-hard. However, it is possible to devise efficient heuristics which, in general, return reasonable sets of cuts. One of them, called MD-heuristics, is presented below.

We say that a cut $\langle a, c\rangle$ discerns objects x and y if and only if $a(x) < c \leq a(y)$ or $a(y) < c \leq a(x)$.

Algorithm 14.4.9 (MD-heuristics).

INPUT: a decision system $\mathcal{A} = \langle U, A, d\rangle$
OUTPUT: a set of cuts $\mathcal{C}$.

1. *Set $\mathcal{C}$ to $\emptyset$.*
2. *Let $\bigcup_{a\in A} C_a$ be the set of basic cuts on $\mathcal{A}$ (see Definition 14.4.1).*
3. *Construct an information table $\mathcal{A}^* = \langle U^*, A^*\rangle$ such that*
   - *U^* is the set of pairs $\langle u_i, u_j\rangle$ of objects discerned by d (in $\mathcal{A}$) such that $i < j$;*
   - *$A^* = \bigcup_{a\in A} C_a$, where, for each $c \in A^*$,*

   $$c(\langle x, y\rangle) = \begin{cases} 1 \text{ if and only if } c \text{ discerns } x \text{ and } y \text{ (in } \mathcal{A}) \\ 0 \text{ otherwise} \end{cases}$$
4. *Choose a column from $\mathcal{A}^*$ with the maximal number of occurrences of 1's; add the cut corresponding to this column to $\mathcal{C}$; delete the column from $\mathcal{A}^*$, together with all rows marked with 1 in it;*
5. *If A^* is non-empty, then go to step 4 else stop.* □

Let n be the number of objects and let k be the number of attributes of a decision system $\mathcal{A}$. It can be shown that the best cut can be found in $O(kn)$ steps using $O(kn)$ space only.

Example 14.4.10. Consider the associated decision table for a decision system $\mathcal{A}$ provided in Table 14.6 from Example 14.4.2. The associated information table for the information system $\mathcal{A}^*$ is presented in Table 14.8.

Under the assumption that columns with maximal number of 1's are chosen from left to right (if many such columns exist in a given step), the set of cuts returned by the algorithm is $\{\langle a, 1.35\rangle, \langle b, 1.5\rangle, \langle a, 1.15\rangle, \langle a, 1.5\rangle\}$. However, as shown in Example 14.4.8, it is not an optimal set of cuts. □

14.4.2 Symbolic Attribute Value Grouping

Symbolic attribute value grouping is a technique for reducing the cardinality of value sets of symbolic attributes.

Table 14.8. The C-discretization considered in Example 14.4.8

$\mathcal{A}^*$	$\langle a, 0.9\rangle$	$\langle a, 1.15\rangle$	$\langle a, 1.35\rangle$	$\langle a, 1.5\rangle$	$\langle b, 0.75\rangle$	$\langle b, 1.5\rangle$	$\langle b, 2.5\rangle$
$\langle u_1, u_2\rangle$	1	0	0	0	1	1	0
$\langle u_1, u_3\rangle$	1	1	0	0	0	0	1
$\langle u_1, u_5\rangle$	1	1	1	0	0	0	0
$\langle u_2, u_4\rangle$	0	1	1	0	1	0	0
$\langle u_2, u_6\rangle$	0	1	1	1	1	1	1
$\langle u_2, u_7\rangle$	0	1	0	0	1	0	0
$\langle u_3, u_4\rangle$	0	0	1	0	0	1	1
$\langle u_3, u_6\rangle$	0	0	1	1	0	0	0
$\langle u_3, u_7\rangle$	0	0	0	0	0	1	1
$\langle u_4, u_5\rangle$	0	0	0	0	0	1	0
$\langle u_5, u_6\rangle$	0	0	0	1	0	0	1
$\langle u_5, u_7\rangle$	0	0	1	0	0	1	0

Definition 14.4.11. *Let* $\mathcal{A} = \langle U, A, d\rangle$ *be a decision system. Any function* $c_a : V_a \to \{1, \ldots, m\}$ *(where* $m \leq |V_a|$*) is called a* clustering function *for* V_a*. The* rank of c_a*, denoted by* $rank\,(c_a)$*, is the value* $|\{c_a(x) \mid x \in V_a\}|$.

For $B \subseteq A$*, a family of clustering functions* $\{c_a\}_{a\in B}$ *is* B-consistent *if and only if*

$$\forall a \in B\ \ [c_a(a(u)) = c_a(a(u'))]$$

implies

$$\langle u, u'\rangle \in \text{IND}_{\mathcal{A}}(B) \cup \text{IND}_{\mathcal{A}}(\{d\}),\ \textit{for any pair}\ \langle u, u'\rangle \in U.$$

□

The notion of B-consistency has the following intuitive interpretation: if two objects are indiscernible w.r.t. clustering functions for value sets of attributes from B, then they are indiscernible either by the attributes from B or by the decision attribute.

We consider the following problem, called the *symbolic value partition grouping problem*:

> Given a decision system $\mathcal{A} = \langle U, A, d\rangle$, where $U = \{u_1, \ldots, u_k\}$, and a set of attributes $B \subseteq A$, search for a B-consistent family $\{c_a\}_{a\in B}$ of clustering functions such that $\sum_{a\in B} rank\,(c_a)$ is minimal.

In order to solve this problem, we apply the following steps:

1. Introduce a set of new Boolean variables:[6]

[6] The introduced variables serve to discern between pairs of objects w.r.t. an attribute a.

$\{a_v^{v'} \mid a \in B \text{ and } v, v' \in V_a \text{ and } v \neq v'\}$.

We extract a subset S of this set such that $a_v^{v'} \in S$ implies that $v' < v$ w.r.t. some arbitrary linear order $<$ on the considered domain

2. Construct matrix $\mathcal{M} = [c_{ij}]_{i,j=1,\ldots,k}$ as follows:

$$c_{ij} = \{a_v^{v'} \in S \mid v' = a(u_i) \text{ and } v = a(u_j) \text{ and } d(u_i) \neq d(u_j)\}.$$

It is easily seen that in the case of a binary decision, the matrix can be reduced by placing objects corresponding to the first decision in rows and those corresponding to the second decision in columns. Such a matrix we call a *reduced discernibility matrix*

3. Using the reduced matrix, $\mathcal{M}'$, obtained in the previous step, construct the function

$$\bigwedge_{c_{ij} \in \mathcal{M}'} \left(\bigvee_{c \in c_{ij}, c_{ij} \neq \emptyset} c \right)$$

4. Compute the shortest prime implicant I of the constructed function
5. Using I, construct, for each attribute $a \in B$, an undirected graph $\Gamma_a = \langle V_a^{\Gamma}, E_a^{\Gamma} \rangle$, where
   - $V_a^{\Gamma} = \{a_v \mid v \in V_a\}$
   - $E_a^{\Gamma} = \{\langle a_x, a_y \rangle \mid x, y \in U \text{ and } a(x) \neq a(y)\}$.

   Note that using I one can construct E_a^{Γ} due to the equality

   $E_a^{\Gamma} = \{\langle a_v, a_{v'} \rangle : a_v^{v'} \text{ occurs in } I\}$
6. Find a minimal coloring of vertices for Γ_a.[7] The coloring defines a partition of V_a^{Γ} by assuming that all vertices of the same color belong to the same partition and no partition contains vertices with different colors. Partitions are named using successive natural numbers.

   The clustering function for V_a^{Γ} is $c_a(a_v) = i$, provided that a_v is a member of the i-th partition.

Remark 14.4.12. In practical implementations usually one does not construct the matrix $\mathcal{M}$ explicitly, as required in Steps (2)-(3) above. Instead, prime implicants are directly extracted from the original decision system.

It should be emphasized that in Step (4) above, there can be many different shortest prime implicants and in Step (6) there can be many different

[7] The colorability problem is solvable in polynomial time for $k = 2$, but remains NPTIME-complete for all $k \geq 3$. But, similarly to discretization, one can apply some efficient search heuristics for generating (sub-) optimal partitions.

colorings of the obtained graphs. Accordingly, one can obtain many substantially different families of clustering functions resulting in different classifiers. In practice, one often generates a number of families of clustering functions, tests them against data and chooses the best one. □

Using the construction above to generate a family of partitions, it is usually possible to obtain a substantially smaller decision table, according to the following definition.

Definition 14.4.13. *Let $\mathcal{A} = \langle U, A, d \rangle$ be a decision system and $B \subseteq A$. Any family of clustering functions $c = \{c_a\}_{a \in B}$ specifies a new decision system $\mathcal{A}^c = \langle U, A^c, d \rangle$ called the* c-reduction *of $\mathcal{A}$ w.r.t. B, where $A^c = \{a^c \mid a \in B\}$ and $a^c(x) = c_a(a(x))$.* □

Example 14.4.14. Consider the decision table provided in Table 14.9. The goal is to solve the symbolic value partition problem for $B = A$.

One then has to perform the following steps:

1. Introduce new Boolean variables a_v^u, b_x^w, for all $u, v \in V_a, u < v$ and $w, x \in V_b, w < x$
2. The reduced matrix $\mathcal{M}'$ is presented in Table 14.10
3. The required Boolean function is given by

$$\begin{aligned}
&b_{b_4}^{b_1} \wedge b_{b_4}^{b_2} \wedge (a_{a_2}^{a_1} \vee b_{b_4}^{b_3}) \wedge (a_{a_3}^{a_1} \vee b_{b_4}^{b_1}) \wedge \\
&(a_{a_2}^{a_1} \vee b_{b_2}^{b_1}) \wedge a_{a_2}^{a_1} \wedge b_{b_3}^{b_2} \wedge (a_{a_3}^{a_2} \vee b_{b_2}^{b_1}) \wedge \\
&a_{a_2}^{a_1} \wedge (a_{a_2}^{a_1} \vee b_{b_2}^{b_1}) \wedge b_{b_3}^{b_1} \wedge a_{a_3}^{a_2} \wedge \\
&(a_{a_4}^{a_1} \vee b_{b_2}^{b_1}) \wedge a_{a_4}^{a_1} \wedge (a_{a_4}^{a_2} \vee b_{b_3}^{b_2}) \wedge (a_{a_4}^{a_3} \vee b_{b_2}^{b_1}) \wedge \\
&(a_{a_3}^{a_1} \vee b_{b_4}^{b_1}) \wedge (a_{a_3}^{a_1} \vee b_{b_4}^{b_2}) \wedge (a_{a_3}^{a_2} \vee b_{b_4}^{b_3}) \wedge b_{b_4}^{b_1} \wedge \\
&(a_{a_2}^{a_1} \vee b_{b_5}^{b_1}) \wedge (a_{a_2}^{a_1} \vee b_{b_5}^{b_2}) \wedge b_{b_5}^{b_3} \wedge (a_{a_3}^{a_2} \vee b_{b_5}^{b_1})
\end{aligned}$$

4. The shortest prime implicant for the function is:

$$I \equiv a_{a_2}^{a_1} \wedge a_{a_3}^{a_2} \wedge a_{a_4}^{a_1} \wedge a_{a_4}^{a_3} \wedge b_{b_4}^{b_1} \wedge b_{b_4}^{b_2} \wedge b_{b_3}^{b_2} \wedge b_{b_3}^{b_1} \wedge b_{b_5}^{b_3}$$

5. The graphs corresponding to a and b are shown in Figure 14.1
6. The graphs are 2-colored, as shown in Figure 14.1, where nodes marked by $\otimes$ are colored black and the other nodes are colored white. These colorings generate the following clustering functions:

$$\begin{aligned}
&c_a(a_1) = c_a(a_3) = 1 \\
&c_a(a_2) = c_a(a_4) = 2
\end{aligned}$$

Table 14.9. The decision table considered in Example 14.4.14

$\mathcal{A}$	a	b	d
u_1	a_1	b_1	0
u_2	a_1	b_2	0
u_3	a_2	b_3	0
u_4	a_3	b_1	0
u_5	a_1	b_4	1
u_6	a_2	b_2	1
u_7	a_2	b_1	1
u_8	a_4	b_2	1
u_9	a_3	b_4	1
u_{10}	a_2	b_5	1

Table 14.10. The reduced matrix corresponding to the decision table provided in Table 14.9.

$\mathcal{M}'$	u_1	u_2	u_3	u_4
u_5	$b_{b_4}^{b_1}$	$b_{b_4}^{b_2}$	$a_{a_2}^{a_1}, b_{b_4}^{b_3}$	$a_{a_3}^{a_1}, b_{b_4}^{b_1}$
u_6	$a_{a_2}^{a_1}, b_{b_2}^{b_1}$	$a_{a_2}^{a_1}$	$b_{b_3}^{b_2}$	$a_{a_3}^{a_2}, b_{b_2}^{b_1}$
u_7	$a_{a_2}^{a_1}$	$a_{a_2}^{a_1}, b_{b_2}^{b_1}$	$b_{b_3}^{b_1}$	$a_{a_3}^{a_2}$
u_8	$a_{a_4}^{a_1}, b_{b_2}^{b_1}$	$a_{a_4}^{a_1}$	$a_{a_4}^{a_2}, b_{b_3}^{b_2}$	$a_{a_4}^{a_3}, b_{b_2}^{b_1}$
u_9	$a_{a_3}^{a_1}, b_{b_4}^{b_1}$	$a_{a_3}^{a_1}, b_{b_4}^{b_2}$	$a_{a_3}^{a_2}, b_{b_4}^{b_3}$	$b_{b_4}^{b_1}$
u_{10}	$a_{a_2}^{a_1}, b_{b_5}^{b_1}$	$a_{a_2}^{a_1}, b_{b_5}^{b_2}$	$b_{b_5}^{b_3}$	$a_{a_3}^{a_2}, b_{b_5}^{b_1}$

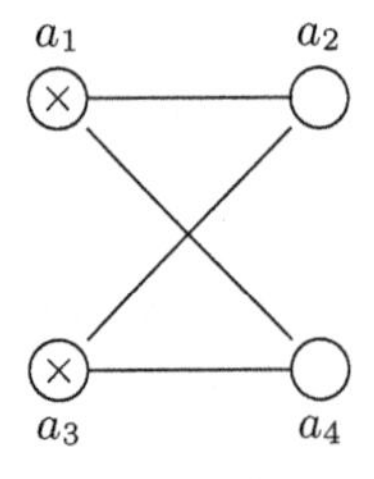

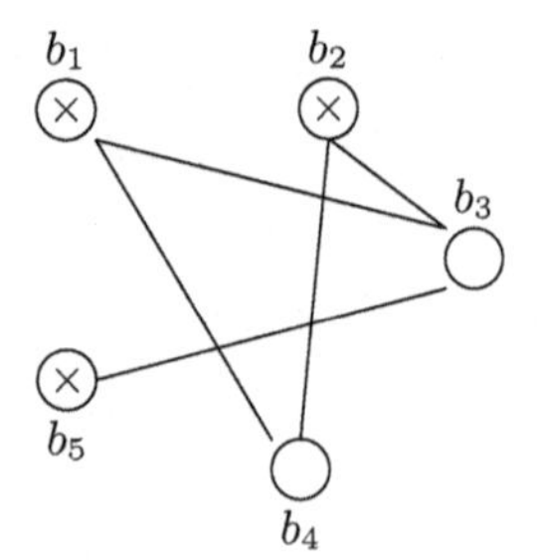

Fig. 14.1. Coloring of attribute value graphs constructed in Example 14.4.14.

$$c_b(b_1) = c_b(b_2) = c_b(b_5) = 1$$
$$c_b(b_3) = c_b(b_4) = 2.$$

Given these clustering functions, one can construct a new decision table using Definition 14.4.13. The result is provided in Table 14.11. □

Table 14.11. The reduced table corresponding to graphs shown in Figure 14.1.

a^c	b^c	d
1	1	0
2	2	0
1	2	1
2	1	1

Observe that discretization and symbolic attribute value grouping can be simultaneously used in decision systems including both real-value and symbolic attributes.

14.5 Minimal Decision Rules

In this section, techniques for constructing minimal rules for decision systems will be considered.

Definition 14.5.1. *Given a decision table $\mathcal{A}$, a* minimal decision rule *(w.r.t. $\mathcal{A}$) is a rule which is* TRUE *in $\mathcal{A}$ and which becomes* FALSE *in $\mathcal{A}$ if any elementary descriptor from the left-hand side of the rule is removed.*[8] □

The minimal number of elementary descriptors in the left-hand side of a minimal decision rule defines the largest subset of a decision class. Accordingly, information included in the conditional part of any minimal decision rule is sufficient for predicting the decision value of all objects satisfying this part of the rule. The conditional parts of minimal decision rules define the largest object sets relevant for approximating decision classes. The conditional parts of minimal decision rules can be computed using prime implicants.

To compute the set of all minimal rules w.r.t. to a decision system $\mathcal{A} = \langle U, A, d\rangle$, we proceed as follows.

For any object $x \in U$:

1. Construct a decision-relative discernibility function f_x^r by considering the row corresponding to object x in the decision-relative discernibility matrix for $\mathcal{A}$
2. Compute all prime implicants of f_x^r
3. On the basis of the prime implicants, create minimal rules corresponding to x. To do this, consider the set $A(I)$ of attributes corresponding to

[8] A decision rule $\varphi \Rightarrow \psi$ is TRUE in $\mathcal{A}$ if and only if $\|\varphi\|_{\mathcal{A}} \subseteq \|\psi\|_{\mathcal{A}}$ (see Chapter 3, Section 3.6).

Table 14.12. Decision table considered in Example 14.5.2

Object	L	W	C	S
1	7.0	large	green	no
2	7.0	large	blue	no
3	4.0	medium	green	yes
4	4.0	medium	red	yes
5	5.0	medium	blue	no
6	4.5	medium	green	no
7	4.0	large	red	no

Table 14.13. $\{L, W\}$-reduction considered in Example 14.5.2

Objects	L	W	S
1, 2	7.0	large	no
3,4	4.0	medium	yes
5	5.0	medium	no
6	4.5	medium	no
7	4.0	large	no

propositional variables in I, for each prime implicant I, and construct a rule:

$$\left(\bigwedge_{a \in A(I)} (a = a(x)) \right) \Rightarrow d(x).$$

The following example illustrates the idea.

Example 14.5.2. Consider the decision system $\mathcal{A}$ whose decision table is provided in Table 14.12. Table 14.12 contains the values of conditional attributes of vehicles (L, W, C, standing for *Length, Width*, and *Color*, respectively), and a decision attribute S standing for *Small* which allows one to decide whether a given vehicle is small.

This system has exactly one decision-relative reduct consisting of attributes L and W. The $\{L, W\}$-reduction of $\mathcal{A}$ is shown as Table 14.13.

Table 14.13 results in the following set of non-minimal decision rules:

$$(L = 7.0) \wedge (W = \mathsf{large}) \Rightarrow (S = \mathsf{no})$$
$$(L = 4.0) \wedge (W = \mathsf{medium}) \Rightarrow (S = \mathsf{yes})$$
$$(L = 5.0) \wedge (W = \mathsf{medium}) \Rightarrow (S = \mathsf{no})$$
$$(L = 4.5) \wedge (W = \mathsf{medium}) \Rightarrow (S = \mathsf{no})$$
$$(L = 4.0) \wedge (W = \mathsf{large}) \Rightarrow (S = \mathsf{no}).$$

Table 14.14. Reduced decision-relative discernibility matrix from Example 14.5.2

	3	4
1	L, W	L, W, C
2	L, W, C	L, W, C
5	L, C	L, C
6	L	L, C
7	W, C	W

To obtain the minimal decision rules, we apply the construction provided above, for $x \in \{1, \ldots, 7\}$.

1. The decision-relative discernibility functions $f_1^r, \ldots, f_7^r$ are constructed on the basis of the reduced discernibility matrix shown in Table 14.14:

$$
\begin{aligned}
f_1^r &\equiv (L \vee W) \wedge (L \vee W \vee C) \equiv (L \vee W) \\
f_2^r &\equiv (L \vee W \vee C) \wedge (L \vee W \vee C) \equiv (L \vee W \vee C) \\
f_3^r &\equiv (L \vee W) \wedge (L \vee W \vee C) \wedge (L \vee C) \wedge L \wedge (W \vee C) \\
&\equiv (L \wedge W) \vee (L \wedge C) \\
f_4^r &\equiv (L \vee W \vee C) \wedge (L \vee W \vee C) \wedge (L \vee C) \wedge (L \vee C) \wedge W \\
&\equiv (L \wedge W) \vee (C \wedge W) \\
f_5^r &\equiv (L \vee C) \wedge (L \vee C) \equiv (L \vee C) \\
f_6^r &\equiv L \wedge (L \vee C) \equiv L \\
f_7^r &\equiv (W \vee C) \wedge W \equiv W.
\end{aligned}
$$

2. The following prime implicants are obtained from formulas $f_1^r, \ldots, f_7^r$:

   f_1^r: L, W
   f_2^r: L, W, C
   f_3^r: $L \wedge W$, $L \wedge C$
   f_4^r: $L \wedge W$, $C \wedge W$
   f_5^r: L, C
   f_6^r: L
   f_7^r: W.

3. Based on the prime implicants, minimal decision rules are created for objects $1, \ldots, 7$. For instance, from prime implicants L and W corresponding to f_1^r, the following minimal decision rules are generated based on object 1:

$$
\begin{aligned}
&(L = 7.0) \Rightarrow (S = \mathsf{no}) \\
&(W = \mathsf{large}) \Rightarrow (S = \mathsf{no}).
\end{aligned}
$$

On the basis of object 3 and prime implicants $L \wedge W$ and $L \wedge C$ for f_3^r we obtain the rules:

$$(L = 4.0) \wedge (W = \mathsf{medium}) \Rightarrow (S = \mathsf{yes})$$
$$(L = 4.0) \wedge (C = \mathsf{green}) \Rightarrow (S = \mathsf{yes}).$$

Similarly, minimal decision rules can easily be obtained for all other formulas. □

In practice, the number of minimal decision rules can be large. One then tries to consider only subsets of these rules or to drop some conditions from minimal rules.

Remark 14.5.3. The main challenge in inducing rules from decision systems lies in determining which attributes should be included in the conditional parts of the rules. Using the strategy outlined above, the minimal rules are computed first. Their conditional parts describe the largest object sets with the same generalized decision value in a given decision system. Although such minimal decision rules can be computed, this approach can result in a set of rules of unsatisfactory classification quality. Such rules might appear too general or too specific for classifying new objects. This depends on the data analyzed. Techniques have been developed for the further tuning of minimal rules. □

14.6 Rough Classifiers

In Section 12.4 a general scheme for constructing rule-based classifiers has been presented using the idea of information granules. Let us now discuss more precisely how classifiers can be used to specify rough concepts, where one deals with positive, negative and unknown information.

In general, a classifier can be understood as an entity[9] providing one with a relation $C(\bar{x}, y)$, where $\bar{x}$ is a vector of values of attributes of an object to be classified and y is a decision.

From the rough set perspective we are interested in interpreting the results of the classifier by means of $C^+(\bar{x}, y)$ and $C^-(\bar{x}, y)$ with $C^\pm(\bar{x}, y)$ containing, as usual, objects classified neither to $C^+(\bar{x}, y)$ nor to $C^-(\bar{x}, y)$.

Consider an object characterized by vector $\bar{v}$ of values of attributes. There are the following cases:

1. $\neg\exists y. C(\bar{v}, y)$, e.g., no rule of the classifier is applicable to input arguments $\bar{v}$. In this case it is natural to assume that the classifier cannot classify

[9] For example, an information granule or a software component.

the input object, thus for any y, it is unknown whether $C(\bar{v}, y)$ holds or not, i.e., for any y we have that neither $C^+(\bar{v}, y)$ nor $C^-(\bar{v}, y)$ holds (thus for any decision y we have $C^{\pm}(\bar{v}, y)$)

2. $\exists y.C(\bar{v}, y)$, e.g., there is a rule of the classifier, which is applicable to $\bar{v}$. In this case:
   a) if the decision is unique, i.e., we have that
   $$\exists d. \{C(\bar{v}, d) \wedge \forall y.[C(\bar{v}, y) \rightarrow y = d]\},$$
   then one can assume that $C^+(\bar{v}, d)$ holds, and we also have that for all $d' \neq d$, $C^-(\bar{v}, d')$.
   b) if all decisions are from the set $\{d_1, \ldots, d_k\}$, where $k > 1$, i.e.,
   $$\exists d_1, \ldots, d_k. \{C(\bar{v}, d_1) \wedge \ldots \wedge C(\bar{v}, d_k) \wedge \\ \forall y.[C(\bar{v}, y) \rightarrow (y = d_1 \vee \ldots \vee y = d_k)]\},$$
   then one can assume that $C^{\oplus}(\bar{v}, d_1), \ldots, C^{\oplus}(\bar{v}, d_k)$ hold and for all $d \notin \{d_1, \ldots, d_k\}$ we have that $C^-(\bar{v}, d)$.

Consider two CAKE granules, Cl denoting the original classifier and RCl denoting its interpretation as a rough relation. Observe that Cl and RCl deliver relation $C(\bar{x}, z)$, where z denotes the result of the classifier. The following formulas are used in constructing CAKE diagrams for RCl

$$\exists d. \{Cl.C^+(\bar{x}, d) \wedge \forall y.[Cl.C^+(\bar{x}, y) \rightarrow y = d] \wedge z = d\} \tag{14.6}$$
$$\exists d. \{Cl.C^+(\bar{x}, d) \wedge \forall y.[Cl.C^+(\bar{x}, y) \rightarrow y = d] \wedge z \neq d\}. \tag{14.7}$$

The following implications reflect case 2(a):

$$\forall \bar{x}, z.[(14.6) \rightarrow RCl.C^+(\bar{x}, z)]$$
$$\forall \bar{x}, z.[(14.7) \rightarrow RCl.C^-(\bar{x}, z)].$$

Let for some $k > 1$, $D = \{d_1, \ldots, d_k\}$ and let $x \in D$ stands for $x = d_1 \vee \ldots \vee x = d_k$. In the case 2(b) we need the following formulas:

$$\exists d_1 \ldots d_k. \left[\bigwedge_{d \in D} Cl.C^+(\bar{x}, d) \wedge \forall y.[Cl.C^+(\bar{x}, y) \rightarrow y \in D] \wedge z \in D \right] \tag{14.8}$$

$$\exists d_1 \ldots d_k. \left[\bigwedge_{d \in D} Cl.C^+(\bar{x}, d) \wedge \forall y.[Cl.C^+(\bar{x}, y) \rightarrow y \in D] \wedge z \notin D \right]. \tag{14.9}$$

The following implications reflect then the case 2(b):

$$\forall \bar{x}, z.[(14.8) \rightarrow RCl.C^{\oplus}(\bar{x}, z)]$$
$$\forall \bar{x}, z.[(14.9) \rightarrow RCl.C^-(\bar{x}, z)].$$

The first implication is irrelevant, since in CAKE we only represent positive and negative facts. Moreover, $RCl.C^{\oplus}(\bar{x}, z)$ can be computed on the basis of the second implication and the fact that $RCl.C^{\oplus}(\bar{x}, z) \equiv \neg RCl.C^{-}(\bar{x}, z)$.

The corresponding CAKE diagram is shown in Figure 14.2.

It us worth emphasizing that in the above schema we assume that information supplied by the classifier is available only via the relation $C(\bar{x}, y)$. In particular, prioritization of the rules and voting is already done by the classifier.

Note also that one could develop other strategies for interpreting classifiers' results as rough sets. One possibility is to consider the majority voting decision as the proper one. Another possibility is to prioritize the rules or use more sophisticated solutions.

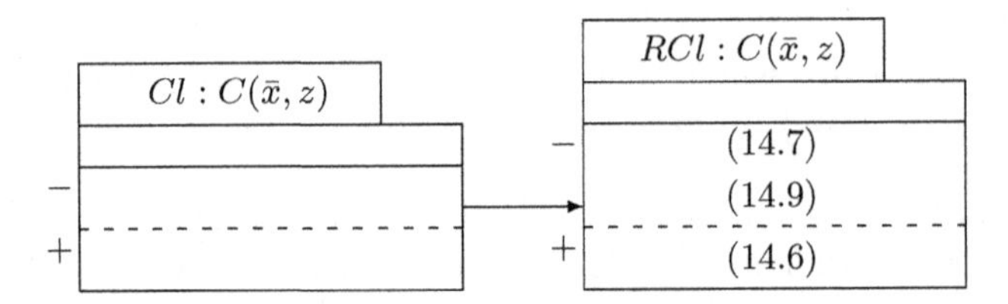

Fig. 14.2. CAKE interpretation of a classifier.

14.7 Example: Learning of Concepts

Given that one has all the techniques described in the previous sections at one's disposal, an important task is to induce definitions of concepts from training data where the representation of the definition is as efficient and high quality as possible. These definitions may then be used as classifiers for the induced concepts. In this section, we will show how a concept can be induced and then integrate it with the CAKE representation techniques described in an earlier chapter.

Consider the example described in Section 7.6, where a concept of traffic congestion was approximated by a combination of rough and crisp knowledge. Approximations of some of the concepts used, such as *Speed* or *Distance*, were assumed to be given. Let us now show how classifiers of these features can automatically be generated on the basis of training data and machine learning techniques.

Let us concentrate on the concept of *Distance*. Recall that the rough relation $Distance(x, y, z)$ denotes the approximate distance between vehicles x and

Table 14.15. Training data considered in Example 14.7.

Object	*SL*	*VS*	*W*	*AD*	*Distance*
1	70	60	rain	3.0	small
2	70	70	sun	5.0	medium
3	50	60	rain	5.0	small
4	50	60	sun	9.0	medium
5	30	15	rain	9.0	large
6	30	10	sun	5.0	large
7	70	60	rain	15.0	large
8	50	40	rain	15.0	large

y, where $z \in \{\text{small}, \text{medium}, \text{large}, \text{unknown}\}$. Below we simplify the definition somewhat, and consider $Distance(x, z)$ which denotes that the distance between x and the vehicle directly preceding x is z.[10] Assume that sample training data has been gathered in a decision table which is provided in Table 14.15, where:[11]

- SL stands for the "speed limit" on a considered road segment
- VS stands for the "vehicle speed"
- W stands for "weather conditions"
- AD stands for "actual distance" between a given vehicle and its predecessor on the road.

For the sake of simplicity we concentrate on generating rules to determine whether the distance between two objects is small.

On the basis of the training data one can compute a discernibility matrix. Since we are interested in rules for the decision small only, it suffices to consider a simplified discernibility matrix with columns labelled by objects 1 and 3, as these are the only two objects where the corresponding decision is small. The resulting discernibility matrix is shown in Table 14.16.

The discernibility matrix gives rise to the following discernibility functions:

$$\begin{aligned}
f_1 \equiv\ & (VS \lor W \lor AD) \land (SL \lor W \lor AD) \land (SL \lor VS \lor AD) \\
& \land (SL \lor VS \lor W \lor AD) \land AD \land (SL \lor VS \lor AD) \\
\equiv\ & AD \\
f_3 \equiv\ & (SL \lor VS \lor W) \land (W \lor AD) \land (SL \lor VS \lor AD) \\
& \land (SL \lor VS \lor W) \land (SL \lor AD) \land (VS \lor AD)
\end{aligned}$$

[10] In fact, here we consider a distance to be small if it causes a dangerous situation, and to be large if the situation is safe.

[11] Of course, real-life sample data would consist of hundreds or thousands of examples.

Table 14.16. Discernibility matrix of Table 14.15 for decision small.

Object	1	3
2	VS, W, AD	SL, VS, W
4	SL, W, AD	W, AD
5	SL, VS, AD	SL, VS, AD
6	SL, VS, W, AD	SL, VS, W
7	AD	SL, AD
8	SL, VS, AD	VS, AD

$$\equiv (W \wedge AD) \vee (SL \wedge AD) \vee (VS \wedge AD) \vee (SL \wedge VS \wedge W).$$

Based on the discernibility functions, one can easily find prime implicants and obtain the following rules for the decision small:[12]

$$\begin{aligned}
&(AD = 3.0) \Rightarrow (Distance = \mathsf{small}) \\
&(W = \mathsf{rain}) \wedge (AD = 5.0) \Rightarrow (Distance = \mathsf{small}) \\
&(SL = 50) \wedge (AD = 5.0) \Rightarrow (Distance = \mathsf{small}) \\
&(VS = 60) \wedge (AD = 5.0) \Rightarrow (Distance = \mathsf{small}) \\
&(SL = 50) \wedge (VS = 60) \wedge (W = \mathsf{rain}) \Rightarrow (Distance = \mathsf{small}).
\end{aligned} \tag{14.10}$$

Based on the rules generated above, one can construct a knowledge structure and its corresponding representation as a CAKE diagram which corresponds to the situation described. Assume that incoming information about objects (vehicles) is stored in the extensional database as the relation $Obj(ob, sl, vs, w, ad)$ associated with the structure similar to that of Table 14.15, except that no decision column is given.[13] Such data may come directly from sensors such as cameras or be fused from a number of sensors via a merging and pre-processing phase.

The resulting CAKE diagram is provided in Figure 14.3. Of course, the diagram should also contain rules for the decisions medium and large which can also be obtained through the machine learning process. Also, as shown in Section 14.6, one should add rules and voting policies allowing one to compute the boundary and negative regions of considered concepts.

The resulting representational structure allows the user to ask questions about newly observed vehicles and their status regarding distance to the vehicle in front of them. The qualitative relation *Distance* is tightly grounded with the environment through the use of training data collected via sensors. The relation can in turn be used in the definition of more abstract relations in additional knowledge structures.

[12] In practical applications one would have to discretize AD before creating rules.
[13] This is decided on the basis of the rules obtained.

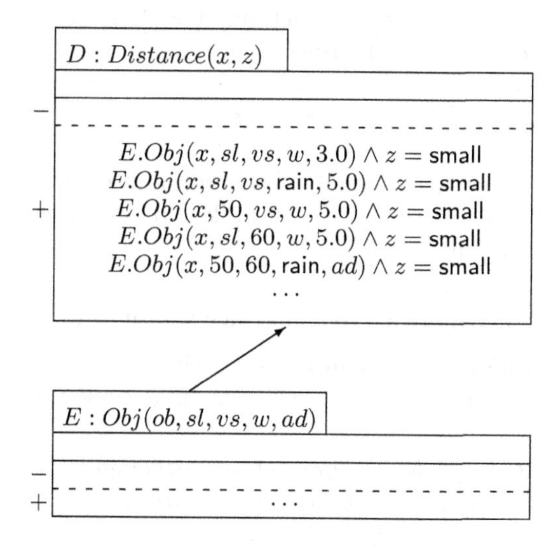

Fig. 14.3. CAKE diagram for *Distance*.

14.8 Association Rules

In this section, we show how rough set techniques can be used to extract association rules from information systems. Association rules, which play an important role in the field of data mining, provide associations among attributes. A real number from the interval [0,1] is assigned to each rule and provides a measure of the confidence of the rule. The following example will help to illustrate this.

Example 14.8.1. Consider the information table provided in Table 14.17.

Each row in the table represents items bought by a customer. For instance, customer 1 bought bread and milk, whereas customer 4 bought milk and jam. An association rule that can be extracted from the above table is: *a customer who bought bread also bought milk.* This is represented by

$(Bread = \text{yes}) \Rightarrow (Milk = \text{yes})$.

Table 14.17. Information table considered in Example 14.8.1.

Customer	*Bread*	*Milk*	*Jam*	*Beer*
1	yes	yes	no	no
2	yes	yes	yes	yes
3	yes	yes	yes	no
4	no	yes	yes	no

Since all customers who bought bread actually bought milk too, the confidence of this rule is 1. Now consider the rule

$$(Bread = \mathsf{yes}) \wedge (Milk = \mathsf{yes}) \Rightarrow (Jam = \mathsf{yes})$$

stating that a customer who bought bread and milk bought jam also. Since three customers bought both bread and milk and two of them bought jam, the confidence of this rule is 2/3. □

We now formalize this approach to confidence measures for association rules.

Recall that by a template we mean a conjunction of elementary descriptors, i.e., expressions of the form $a = v$, where a is an attribute and $v \in V_a$.

Definition 14.8.2. *Let $\mathcal{A}$ be an information system and T be a template. By $support_{\mathcal{A}}(T)$ we denote the number of objects satisfying T.* □

Definition 14.8.3. *Let $\mathcal{A}$ be an information system and $T = D_1 \wedge \ldots \wedge D_m$ be a template. By an* association rule generated from T, *we mean any expression of the form*

$$\bigwedge_{D_i \in P} D_i \Rightarrow \bigwedge_{D_j \in Q} D_j,$$

where $\{P, Q\}$ is a partition of $\{D_1, \ldots, D_m\}$.

By a confidence of an association rule $\phi \equiv \bigwedge_{D_i \in P} D_i \Rightarrow \bigwedge_{D_j \in Q} D_j$ *we mean the coefficient*

$$confidence_{\mathcal{A}}(\phi) = \frac{support_{\mathcal{A}}(D_1 \wedge \ldots \wedge D_m)}{support_{\mathcal{A}}(\bigwedge_{D_i \in P} D_i)}.$$

□

There are two basic steps used in methods aimed at generating association rules. (Below s and c stand for support and confidence thresholds w.r.t. a given information system $\mathcal{A}$).

1. Generate as many templates $T = D_1 \wedge \ldots \wedge D_k$ as possible, such that $support_{\mathcal{A}}(T) \geq s$ and $support_{\mathcal{A}}(T \wedge D_i) < s$, for any descriptor D_i different from all descriptors $D_1, \ldots, D_k$.
2. Search for a partition $\{P, Q\}$ of T, for each T generated in the previous step, satisfying

   a) $support_{\mathcal{A}}(P) < \dfrac{support_{\mathcal{A}}(T)}{c}$

b) P has the shortest length among templates satisfying (a).

Each such partition leads to an association rule of the form $P \Rightarrow Q$ whose confidence is greater than c.

The second step, which is crucial to the process of extracting association rules, can be solved using rough set methods.

Definition 14.8.4. *Let $T = D_1 \wedge D_2 \wedge \ldots \wedge D_m$ be a template such that $support_{\mathcal{A}}(T) \geq s$. For a given confidence threshold $c \in [0,1]$, the association rule $\phi \equiv P \Rightarrow Q$ is called* c-irreducible *if $confidence_{\mathcal{A}}(P \Rightarrow Q) \geq c$ and for any association rule $\phi' \equiv P' \Rightarrow Q'$ such that P' is a sub-formula of P, we have $confidence_{\mathcal{A}}(P' \Rightarrow Q') < c$.* □

The problem of searching for c-irreducible association rules from a given template is equivalent to the problem of searching for α-reducts (see Definition 14.2.13), for some α, in a decision table.

Definition 14.8.5. *Let $\mathcal{A}$ be an information system and $T = D_1 \wedge D_2 \wedge \ldots \wedge D_m$ be a template. By a* characteristic table *for T w.r.t. $\mathcal{A}$, we understand a decision system $\mathcal{A}|_T = \langle U, A|_T, d\rangle$, where*

1. *$A|_T = \{a_{D_1}, a_{D_2}, ..., a_{D_m}\}$ is a set of attributes corresponding to the descriptors of T such that*

$$a_{D_i}(u) = \begin{cases} 1 & \text{if the object } u \text{ satisfies } D_i, \\ 0 & \text{otherwise.} \end{cases}$$

2. *the decision attribute d determines if the object satisfies a template T, i.e.,*

$$d(u) = \begin{cases} 1 & \text{if the object } u \text{ satisfies } T, \\ 0 & \text{otherwise.} \end{cases}$$

□

The following theorem provides the relationship between association rules and approximations of reducts.

Theorem 14.8.6. *For a given information system $\mathcal{A} = \langle U, A\rangle$, a template $T = D_1 \wedge D_2 \wedge \ldots \wedge D_m$ and a set of descriptors $P \subseteq \{D_1, \ldots, D_m\}$, the association rule*

$$\bigwedge_{D_i \in P} D_i \quad \Rightarrow \quad \bigwedge_{D_j \in \{D_1,\ldots,D_m\}-P} D_j$$

is

1. *a* 1*-irreducible association rule from* T *if and only if* $\bigcup_{D_i \in P} \{a_{D_i}\}$ *is a decision-relative reduct of* $\mathcal{A}|_T$.
2. *a* c*-irreducible association rule from* T *if and only if* $\bigcup_{D_i \in P} \{a_{D_i}\}$ *is an* α*-reduct of* $\mathcal{A}|_T$*, where*

$$\alpha = 1 - \left[\left(\frac{1}{c} - 1 \right) \Big/ \left(\frac{|U|}{support_{\mathcal{A}}(T)} - 1 \right) \right].$$

□

The problem of searching for the shortest association rules is NPTIME-hard.

The following example illustrates the main ideas used in the method for searching for association rules.

Table 14.18. Information table $\mathcal{A}$ considered in Example 14.8.7.

$\mathcal{A}$	a_1	a_2	a_3	a_4	a_5	a_6	a_7	a_8	a_9
u_1	0	1	1	1	80	2	2	2	3
u_2	0	1	2	1	81	0	aa	1	aa
u_3	0	2	2	1	82	0	aa	1	aa
u_4	0	1	2	1	80	0	aa	1	aa
u_5	1	1	2	2	81	1	aa	1	aa
u_6	0	2	1	2	81	1	aa	1	aa
u_7	1	2	1	2	83	1	aa	1	aa
u_8	0	2	2	1	81	0	aa	1	aa
u_9	0	1	2	1	82	0	aa	1	aa
u_{10}	0	3	2	1	84	0	aa	1	aa
u_{11}	0	1	3	1	80	0	aa	2	aa
u_{12}	0	2	2	2	82	0	aa	2	aa
u_{13}	0	2	2	1	81	0	aa	1	aa
u_{14}	0	3	2	2	81	2	aa	2	aa
u_{15}	0	4	2	1	82	0	aa	1	aa
u_{16}	0	3	2	1	83	0	aa	1	aa
u_{17}	0	1	2	1	84	0	aa	1	aa
u_{18}	1	2	2	1	82	0	aa	2	aa

Example 14.8.7. Consider the information table $\mathcal{A}$ with 18 objects and 9 attributes presented in Table 14.18.

Consider the template

$$T = (a_1 = 0) \wedge (a_3 = 2) \wedge (a_4 = 1) \wedge (a_6 = 0) \wedge (a_8 = 1). \tag{14.11}$$

It is easily seen that $support_{\mathcal{A}}(T) = 10$. The new constructed decision table $\mathcal{A}|_T$ is presented in Table 14.19.

Table 14.19. Decision table $\mathcal{A}|_T$ considered in Example 14.8.7.

$\mathcal{A}\|_T$	a_{D_1} $(a_1 = 0)$	a_{D_2} $(a_3 = 2)$	a_{D_3} $(a_4 = 1)$	a_{D_4} $(a_6 = 0)$	a_{D_5} $(a_8 = 1)$	d
u_1	1	0	1	0	0	0
u_2	1	1	1	1	1	1
u_3	1	1	1	1	1	1
u_4	1	1	1	1	1	1
u_5	0	1	0	0	1	0
u_6	1	0	0	0	1	0
u_7	0	0	0	0	1	0
u_8	1	1	1	1	1	1
u_9	1	1	1	1	1	1
u_{10}	1	1	1	1	1	1
u_{11}	1	0	1	1	0	0
u_{12}	1	0	0	1	0	0
u_{13}	1	1	1	1	1	1
u_{14}	1	1	0	0	0	0
u_{15}	1	1	1	1	1	1
u_{16}	1	1	1	1	1	1
u_{17}	1	1	1	1	1	1
u_{18}	0	1	1	1	0	0

Table 14.20. Reduced discernibility matrix for $\mathcal{A}|_T$ form Example 14.8.7.

$\mathcal{M}(\mathcal{A}\|_T)$	u_2, u_3, u_4, u_8, u_9 $u_{10}, u_{13}, u_{15}, u_{16}, u_{17}$
u_1	$a_{D_2}, a_{D_4}, a_{D_5}$
u_5	$a_{D_1}, a_{D_3}, a_{D_4}$
u_6	$a_{D_2}, a_{D_3}, a_{D_4}$
u_7	$a_{D_1}, a_{D_2}, a_{D_3}, a_{D_4}$
u_{11}	$a_{D_1}, a_{D_3}, a_{D_5}$
u_{12}	$a_{D_2}, a_{D_3}, a_{D_5}$
u_{14}	$a_{D_3}, a_{D_4}, a_{D_5}$
u_{18}	a_{D_1}, a_{D_5}

The reduced discernibility matrix $\mathcal{A}|_T$ is provided as Table 14.20, where for simplicity, the second column represents, in fact, ten columns with identical contents, labeled by $u_2, u_3, u_4, u_8, u_9, u_{10}, u_{13}, u_{15}, u_{16}, u_{17}$, respectively.

Given the discernibility matrix, one can easily compute the discernibility function for $\mathcal{A}|_T$:

$$\begin{aligned} f_{\mathcal{A}_T}(a_{D_1}, a_{D_2}, a_{D_3}, a_{D_4}, a_{D_5}) \equiv\ & (a_{D_2} \vee a_{D_4} \vee a_{D_5}) \\ & \wedge(a_{D_1} \vee a_{D_3} \vee a_{D_4}) \\ & \wedge(a_{D_2} \vee a_{D_3} \vee a_{D_4}) \\ & \wedge(a_{D_1} \vee a_{D_2} \vee a_{D_3} \vee a_{D_4}) \end{aligned}$$

$$\begin{aligned}
&\wedge(a_{D_1} \vee a_{D_3} \vee a_{D_5}) \\
&\wedge(a_{D_2} \vee a_{D_3} \vee a_{D_5}) \\
&\wedge(a_{D_3} \vee a_{D_4} \vee a_{D_5}) \\
&\wedge(a_{D_1} \vee a_{D_5}),
\end{aligned}$$

where D_i denotes the i-th conjunct of (14.11).

The discernibility function has the following prime implicants: $a_{D_3} \wedge a_{D_5}$, $a_{D_4} \wedge a_{D_5}$, $a_{D_1} \wedge a_{D_2} \wedge a_{D_3}$, $a_{D_1} \wedge a_{D_2} \wedge a_{D_4}$, $a_{D_1} \wedge a_{D_2} \wedge a_{D_5}$, $a_{D_1} \wedge a_{D_3} \wedge a_{D_4}$. This gives rise to the reducts: $\{a_{D_3}, a_{D_5}\}$, $\{a_{D_4}, a_{D_5}\}$, $\{a_{D_1}, a_{D_2}, a_{D_3}\}$, $\{a_{D_1}, a_{D_2}, a_{D_4}\}$, $\{a_{D_1}, a_{D_2}, a_{D_5}\}$, $\{a_{D_1}, a_{D_3}, a_{D_4}\}$. Thus, there are 6 association rules with confidence 1, i.e., 1-irreducible:

$$\begin{aligned}
D_3 \wedge D_5 &\Rightarrow D_1 \wedge D_2 \wedge D_4 \\
D_4 \wedge D_5 &\Rightarrow D_1 \wedge D_2 \wedge D_3 \\
D_1 \wedge D_2 \wedge D_3 &\Rightarrow D_4 \wedge D_5 \\
D_1 \wedge D_2 \wedge D_4 &\Rightarrow D_3 \wedge D_5 \\
D_1 \wedge D_2 \wedge D_5 &\Rightarrow D_3 \wedge D_4 \\
D_1 \wedge D_3 \wedge D_4 &\Rightarrow D_2 \wedge D_5.
\end{aligned}$$

For confidence 0.9, we look for α-reducts for the decision table $\mathcal{A}|_T$, where

$$\alpha = 1 - \left(\frac{1}{0.9} - 1\right) / \left(\frac{18}{10} - 1\right) \approx 0.86.$$

Hence, we look for a set of descriptors that covers at least $\lceil (18 - 10) * (\alpha) \rceil = \lceil 8 * 0.86 \rceil = 7$ elements of the discernibility matrix $\mathcal{M}(\mathcal{A}|_{\mathbf{T}})$. One can see that the following sets of descriptors: $\{D_1, D_2\}$, $\{D_1, D_3\}$, $\{D_1, D_4\}$, $\{D_1, D_5\}$, $\{D_2, D_3\}$, $\{D_2, D_5\}$, $\{D_3, D_4\}$ have nonempty intersections with exactly 7 members of the discernibility matrix $\mathcal{M}(\mathcal{A}|_T)$. Consequently, the 0.9-irreducible association rules obtained from those sets are the following:

$$\begin{aligned}
D_1 \wedge D_2 &\Rightarrow D_3 \wedge D_4 \wedge D_5 \\
D_1 \wedge D_3 &\Rightarrow D_2 \wedge D_4 \wedge D_5 \\
D_1 \wedge D_4 &\Rightarrow D_2 \wedge D_3 \wedge D_5 \\
D_1 \wedge D_5 &\Rightarrow D_2 \wedge D_3 \wedge D_4 \\
D_2 \wedge D_3 &\Rightarrow D_1 \wedge D_4 \wedge D_5 \\
D_2 \wedge D_5 &\Rightarrow D_1 \wedge D_3 \wedge D_4 \\
D_3 \wedge D_4 &\Rightarrow D_1 \wedge D_2 \wedge D_5.
\end{aligned}$$

The technique illustrated by this example can be applied in finding useful dependencies between attributes in complex application domains. In particular, one could use such dependencies in constructing robust classifiers conforming to the laws of the underlying reality. □

14.9 Bibliographic Notes

Machine learning has a long tradition in artificial intelligence. The books [124, 128] provide a good introduction to the subject.

A variety of methods for computing reducts and their applications can be found in [109, 146, 158, 162, 163, 190, 202, 234]. The fact that the problem of finding a minimal reduct of a given information system is NPTIME-hard has been proved in [190].

As we mentioned, there exists a number of good heuristics that compute sufficiently many reducts in an acceptable time. Moreover, a successful methodology, based on different reducts, has been developed for solution of many problems like attribute selection, decision rule generation, association rule generation, discretization of real-valued attributes and symbolic value grouping. For further readings the reader is referred to [9, 181, 210] (attribute selection); [131, 134, 136, 187] (discretization); [132, 133] (discretization of data stored in relational databases); [135] (reduct approximation and association rules). In particular, parts of Section 14.4.1, including the MD-heuristics are based on [131].

There have been also developed methods for approximation of compound concepts based on rough sets, hierarchical learning, and ontology approximation (see, e.g., [14, 11, 12, 13, 138, 140, 139, 185, 188, 193, 194]).

Many of the presented results have been implemented in the RSES and ROSETTA software systems (see `http://logic.mimuw.edu.pl/~rses/` for RSES and `http://rosetta.lcb.uu.se/general/` for ROSETTA), see also [10, 15, 16, 99, 204, 219].

15

UAV Learning Process: A Case Study

15.1 Introduction

In his chapter, the process of inducing classifiers from actual data is discussed. This is done using a case study roughly related to the example considered in Section 14.7. A classifier will be constructed for the concept *Dangerous*. Recall that this concept is intended to represent potentially dangerous traffic situations.

It is assumed that the associated decision attribute *Danger* used in the decision table representing raw data for the *Dangerous* concept can be assigned the values low, moderate, and high. We consider the danger to be high if the vehicular and environmental attributes contribute together to potentially dangerous traffic situations. Similarly, traffic situations are very safe if the danger is considered to be low and relatively safe (or dangerous) if the danger is considered to be moderate. Vehicular attributes could be the speed of a vehicle and its distance to the vehicle in front of it. Environmental attributes could be the condition of roads, and maximum speed limits for particular roads.

15.2 Acquisition of Data

The first step in constructing suitable classifiers depends on the acquisition of raw and relevant sample data from diverse sources. In the initial stage of acquisition, such data could be acquired from a number of different sources. Sensors on a UAV such as the camera sensor could be used to collect video streams of traffic situations over diverse areas. Image processing techniques would have to be used to extract relevant feature data to be stored as attributes in a decision table. Files about road structure, traffic signals, and maximum and minimum speed limits could be used along with data gathered

P. Doherty et al.: *Knowledge Representation Techniques*, Studfuzz **202**, 311–321 (2006)
www.springerlink.com

about previous accidents in specific areas of interest. Once all the data is collected, an initial set of attributes must be agreed upon and specific vehicle and traffic situations should be stored as *cases* in a *case file*. Each case would be represented as a row in a very large decision table. Each case would also have to be annotated with a proper value for the decision attribute *Danger*, to specify which of the cases are judged to be dangerous or not. The annotations would most probably be supplied by human experts. The following aspects make the data collection and annotation process relatively complex:

- the number of cases in a case file are usually large (several thousands of cases are a rule rather than exception)
- the set of cases in the case file should be representative for the concept under consideration
- raw data acquired in the acquisition process is usually noisy and incomplete.

Experts, when annotating, would normally mark all cases where there is danger and all unmarked cases would be treated as safe. This creates a potential problem since the ratio of unsafe to safe cases is usually quite small. Not only are the number of cases quite large, but the complexity of the data for each case is quite high since it is often gathered from raw signal data, images, video, etc. Raw data is generally noisy, incomplete and inconsistent. Noise in a case file can be caused by the use of inaccurate sensors or an expert making a mistake in annotation or storing of data in the table. Incompleteness results from the impossible task of acquiring all possible scenarios and situations that might contribute to dangerous traffic situations. Inconsistency might arise if two information sources (e.g., sensors, experts) provide contradictory information about the same case.

15.3 Preprocessing

The data tables consisting of case files generated in the data acquisition phase are generally quite large and contain a great amount of redundant and unstructured data. The purpose of the preprocessing phase is to take raw data tables, transform them into representations as decision tables and then try to reduce their size without compromising the high quality data implicitly represented in the tables. The potential quality of data has to be preserved in order to successfully generate classifiers of high quality.

The processes involved in taking a raw data table and transforming it into a high quality compact decision table are considered in the following sections.

15.3.1 From a Case File to a Local Case File

The main goal in the preprocessing stage is to transfer raw case files into a decision table where each row contains information about a case and a corresponding decision. Attempts are then made to reduce the table size by eliminating irrelevant information from various cases. The resulting reduced case descriptions are called *local cases*. Local cases are extracted from the original cases stored in the case file. In the example considered here, cases may contain images. After image processing one can reduce the initial images describing traffic situations to local neighborhoods of pairs of vehicles in images and the distance between each pair. The extraction processes used in transforming initial cases to local cases is generally a complex process but is required in order to reduce the search space for relevant features. Unfortunately, there is no uniform approach assuring the proper extraction of data.

15.3.2 From a Local Case File to a Reduced Local Case File

Assuming a decision table consisting of local cases is obtained, the next preprocessing steps depend on reducing the number of local cases to a table of *reduced local cases*. One might view a reduced local case as a form of a prototype which permits the grouping of similar local cases into one or more reduced local cases. There are various techniques that may be used to do this. One possibility is to introduce a tolerance relation on local cases and identify all similar local cases. The main goal would be to identify a *dominant set* of representative local cases, i.e., the minimal set of objects such that all other objects in the table are similar to at least one object in the dominant set.

15.4 Attribute Selection and Value Set Reduction

Given a decision table consisting of reduced local cases, the techniques of value set reduction and attribute selection can be used to obtain more compact representations of the original decision table without compromising the quality of the data. These techniques not only remove certain attributes from cases, but may even introduce new attributes summarizing other attributes that are removed. Rows of similar cases may also be grouped into a smaller number of rows.

In the example under consideration, the local cases in our tables contain both real-valued and symbolic attributes. Usually, real valued attributes are discretized and symbolic attributes are grouped. This leads to the reduction of attribute values as was shown in Chapter 14. The result of this combination of techniques is that the original decision table is reduced which often leads to

Table 15.1. Decision table containing training data.

Object	*SL*	*VS*	*RC*	*AD*	*Danger*
1	70.0	60.0	dry	small	moderate
2	76.0	60.0	wet	small	high
3	74.0	60.0	snow	small	high
4	72.0	60.0	ice	small	high
5	70.0	80.0	dry	small	high
6	50.0	60.0	dry	medium	moderate
7	50.0	60.0	wet	medium	moderate
8	50.0	60.0	snow	medium	high
9	30.0	10.0	snow	large	moderate
10	30.0	10.0	dry	large	low
11	50.0	50.0	dry	medium	low
12	50.0	40.0	wet	medium	low
13	50.0	60.0	ice	medium	high

a higher quality of classifiers. New attributes can be obtained by combination of existing attributes in the original table. For example, the difference between the speed limit and the actual vehicle speed limit might be a relevant new feature that can be generated and used in classification while removing the two original attributes. Rough set techniques for attribute selection, value set reduction and construction of minimal rules are considered in Chapter 14. In the following sections, we will apply these techniques to the example.

Let us assume that both the data acquisition and preprocessing phases have been completed and a decision table for the concept *Dangerous* traffic situation has been generated. The decision table that provides the initial training data is shown in Table 15.1, where[1]

- SL stands for "speed limit"
- VS stands for "vehicle speed"
- RC stands for "road condition"
- AD stands for "actual distance"
- *Danger* indicates how dangerous the distance is, when particular values of attributes SL, VS, RC and AD are given.

Consider the attributes SL and VS. Their meaning suggests that their difference can be used as a new attribute. Thus we add a new attribute while removing two existing ones with a reduction of one column and hopefully the same or even increased quality in the data. The new attribute DIF is defined as

$$DIF(x) = SL(x) - VS(x)$$

[1] Observe that this decision table is consistent.

Table 15.2. Decision system obtained from Table 15.1 after introducing a new attribute DIF.

Object	*DIF*	*RC*	*AD*	*Danger*
1	10.0	dry	small	moderate
2	16.0	wet	small	high
3	14.0	snow	small	high
4	12.0	ice	small	high
5	-10.0	dry	small	high
6	-10.0	dry	medium	moderate
7	-10.0	wet	medium	moderate
8	-10.0	snow	medium	high
9	20.0	snow	large	moderate
10	20.0	dry	large	low
11	0.0	dry	medium	low
12	10.0	wet	medium	low
13	-10.0	ice	medium	high

for any object x. The attributes SL, VS are then replaced by DIF. The newly modified data table is presented in Table 15.2.

Value set reduction will now be used to group values in the value sets V_{RC}, V_{AD} associated with the attributes RC, AD. Here we apply the techniques shown in Section 14.4. For attribute RC we introduce Boolean variables R^d_w, R^d_s, R^d_i, R^i_w, R^i_s, R^s_w, where d, w, s, i stand for dry, wet, snow and ice, respectively. For attribute AD we introduce Boolean variables A^m_s, A^l_m, A^l_s, where l, m, s stand for large, medium and small, respectively.

The required reduced matrix is shown in Table 15.3.

After computing the corresponding Boolean function from entries of Table 15.3, the following single prime implicant can be constructed:

$$R^d_w \wedge R^d_s \wedge R^d_i \wedge R^s_w \wedge R^i_w \wedge A^m_s \wedge A^l_s \wedge A^l_m. \quad (15.1)$$

The corresponding attribute value graph for the prime implicant (see Section 14.4) used in the value set reduction process is shown in Figure 15.1.

Hence, we can group values of RC as follows:

{dry}, {wet}, {snow, ice}

and values of AD remain ungrouped.

The modified decision table which results from the above value set reduction is shown in Table 15.4. The new decision table is consistent. Note that there is a reduction in table size from 13 to 11 rows.

Table 15.3. The relevant fragment of the matrix constructed grouping values of attributes RC and AD.

	1	6	7	9	10	11	12
2	R_w^d	R_w^d, A_s^m	A_s^m	R_w^s, A_s^l	R_w^d, A_s^l	R_w^d, A_s^m	A_s^m
3	R_s^d	R_s^d, A_s^m	R_w^s, A_s^m	A_s^l	R_s^d, A_s^l	R_s^d, A_s^m	R_w^s, A_s^m
4	R_i^d	R_i^d, A_s^m	R_w^i, A_s^m	R_s^i, A_s^l	R_i^d, A_s^l	R_i^d, A_s^m	R_w^i, A_s^m
5	$\emptyset$	A_s^m	R_w^d, A_s^m	R_s^d, A_s^l	A_s^l	A_s^m	R_w^d, A_s^m
8	R_s^d, A_s^m	R_s^d	R_w^s	A_m^l	R_s^d, A_m^l	R_s^d	R_w^s
10	A_s^l	A_m^l	R_w^d, A_m^l	R_s^d	$\emptyset$	$\emptyset$	$\emptyset$
11	A_s^m	$\emptyset$	R_w^d	R_s^d, A_m^l	$\emptyset$	$\emptyset$	$\emptyset$
12	A_s^m	R_w^d	$\emptyset$	R_w^s, A_m^l	$\emptyset$	$\emptyset$	$\emptyset$
13	R_i^d, A_s^m	R_i^d	R_w^i	R_s^i, A_m^l	R_i^d, A_m^l	R_i^d	R_w^i

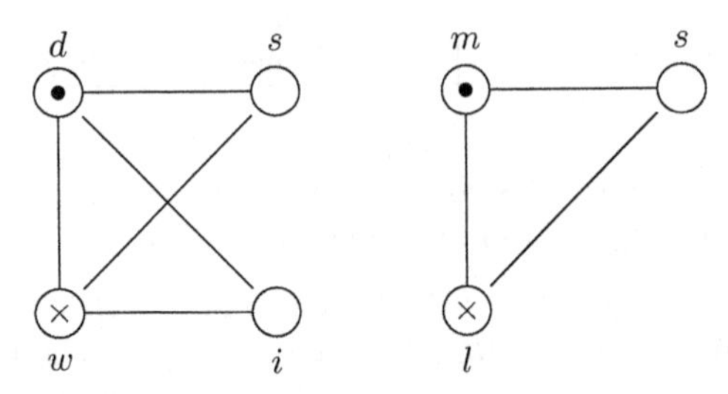

Fig. 15.1. Coloring of the attribute value graph corresponding to prime implicant (15.1).

To further reduce the number of attribute values, we can apply another value set reduction technique, the discretization process, to the new real-valued attribute DIF. The attribute DIF defines six cuts (see Section 14.4):

$$\langle DIF, -5.0\rangle, \langle DIF, 5.0\rangle, \langle DIF, 11.0\rangle, \langle DIF, 13.0\rangle, \langle DIF, 15.0\rangle, \langle DIF, 18.0\rangle.$$

To extract the optimal set of cuts we apply the method based on prime implicants, described in Section 14.4.1. First we introduce Boolean variables $p_1^{DIF}, \ldots, p_6^{DIF}$ corresponding to intervals $[-10.0, 0.0)$, $[0.0, 10.0)$, $[10.0, 12.0)$, $[12.0, 14.0)$, $[14.0, 16.0)$, $[16.0, 20.0)$, respectively. After performing steps 2-4 of the method, we obtain a discernibility formula whose single prime implicant is

$$p_1^{DIF} \wedge p_2^{DIF} \wedge p_3^{DIF} \wedge p_6^{DIF}.$$

Hence we obtain the following minimal set of cuts:

$$\langle DIF, -5.0\rangle, \langle DIF, 5.0\rangle, \langle DIF, 11.0\rangle, \langle DIF, 18.0\rangle.$$

Table 15.4. Decision system obtained from Table 15.1 after grouping values of *RC* and *AD*.

Object	*DIF*	*RC*	*AD*	*Danger*
1	10.0	dry	small	moderate
2	16.0	wet	small	high
3, 4	14.0, 12.0	snow, ice	small	high
5	-10.0	dry	small	high
6	-10.0	dry	medium	moderate
7	-10.0	wet	medium	moderate
8, 13	-10.0	snow, ice	medium	high
9	20.0	snow, ice	large	moderate
10	20.0	dry	large	low
11	0.0	dry	medium	low
12	10.0	wet	medium	low

The corresponding discretization of Table 15.4 is shown in Table 15.5, where i1, i2, i3, i4, i5 denote intervals $(-\infty, -5)$, $[-5, 5)$, $[5, 11)$, $[11, 18)$, $[18, +\infty)$, respectively.

Table 15.5. Decision system obtained from Table 15.4 after discretization of *DIF*.

Object	*DIF'*	*RC*	*AD*	*Danger*
1	i3	dry	small	moderate
2	i4	wet	small	high
3, 4	i4	snow, ice	small	high
5	i1	dry	small	high
6	i1	dry	medium	moderate
7	i1	wet	medium	moderate
8, 13	i1	snow, ice	medium	high
9	i5	snow, ice	large	moderate
10	i5	dry	large	low
11	i2	dry	medium	low
12	i3	wet	medium	low

The original decision table has now been reduced. In the next section, we will show how to generate minimal decision rules which are then used to construct a classifier.

15.5 Decision Rule Generation

Section 14.5 of Chapter 14 provides a method for generating minimal decision rules from decision tables. This method will now be used to generate minimal decision rules for the newly constructed decision table given in Table 15.5.

The decision attribute *Danger* has three possible values: low, moderate, and high. In order to generate a complete classifier, three sets of minimal decision rules, one set for each decision value, would have to be generated.

In order to obtain minimal decision rules, we first construct the decision-relative discernibility functions for all objects with the decision high. These discernibility functions and their reductions to Blake canonical form[2] (see Section 2.5.3) are shown below, where Boolean variables D, R, W, A correspond to attributes DIF', W, RC, AD, respectively:

$$f_2^r \equiv (D \vee R) \wedge (D \vee R \vee A) \wedge (D \vee A) \equiv (D \vee R) \wedge (D \vee A)$$
$$f_{3,4}^r \equiv (D \vee R) \wedge (D \vee R \vee A) \wedge (D \vee A) \equiv (D \vee R) \wedge (D \vee A)$$
$$f_5^r \equiv D \wedge A \wedge (R \vee A) \wedge (D \vee R \vee A) \wedge (D \vee A) \equiv D \wedge A$$
$$f_{8,13}^r \equiv (D \vee R \vee A) \wedge R \wedge (D \vee A) \wedge (D \vee R) \equiv R \wedge (D \vee A).$$

The following prime implicants are obtained from formulas $f_2^r, f_{3,4}^r, f_5^r, f_{8,13}^r$:

f_2^r : $D \wedge R, D \wedge A$
$f_{3,4}^r$: $D \wedge R, R \wedge A$
f_5^r : $D \wedge A$
$f_{8,13}^r$: $R \wedge D, R \wedge A$.

On the basis of the prime implicants generated above, the following set of minimal decision rules for the decision high can be generated. For clarity, the rules are labeled by the sets of objects used in their construction:

$$
\begin{aligned}
2: &\quad (DIF' = \mathsf{i4}) \wedge (RC = \mathsf{wet}) \Rightarrow (Danger = \mathsf{high}) \\
2: &\quad (DIF' = \mathsf{i4}) \wedge (AD = \mathsf{small}) \Rightarrow (Danger = \mathsf{high}) \\
3,4: &\quad (DIF' = \mathsf{i4}) \wedge (RC \in \{\mathsf{snow}, \mathsf{ice}\}) \Rightarrow (Danger = \mathsf{high}) \\
3,4: &\quad (DIF' = \mathsf{i4}) \wedge (AD = \mathsf{small}) \Rightarrow (Danger = \mathsf{high}) \\
5: &\quad (DIF' = \mathsf{i1}) \wedge (AD = \mathsf{small}) \Rightarrow (Danger = \mathsf{high}) \\
8,13: &\quad (DIF' = \mathsf{i1}) \wedge (RC \in \{\mathsf{snow}, \mathsf{ice}\}) \Rightarrow (Danger = \mathsf{high}) \\
8,13: &\quad (RC \in \{\mathsf{snow}, \mathsf{ice}\}) \wedge (AD = \mathsf{medium}) \Rightarrow (Danger = \mathsf{high}).
\end{aligned}
$$

In the rule generation phase, it is often the case that duplicates of rules are generated, as is the case above. In the considered case the fourth rule is redundant, since it is the same as the second one. Thus we obtain the following set of rules for the decision high:

$$(DIF' = \mathsf{i4}) \wedge (RC = \mathsf{wet}) \Rightarrow (Danger = \mathsf{high})$$
$$(DIF' = \mathsf{i4}) \wedge (AD = \mathsf{small}) \Rightarrow (Danger = \mathsf{high})$$
$$(DIF' = \mathsf{i4}) \wedge (RC \in \{\mathsf{snow}, \mathsf{ice}\}) \Rightarrow (Danger = \mathsf{high})$$

[2] Recall that sentences in Blake canonical form are disjunctions of prime implicates.

$(DIF' = \mathsf{i1}) \wedge (AD = \mathsf{small}) \Rightarrow (Danger = \mathsf{high})$
$(DIF' = \mathsf{i1}) \wedge (RC \in \{\mathsf{snow}, \mathsf{ice}\}) \Rightarrow (Danger = \mathsf{high})$
$(RC \in \{\mathsf{snow}, \mathsf{ice}\}) \wedge (AD = \mathsf{medium}) \Rightarrow (Danger = \mathsf{high}).$

Further reduction of the rule set can be made by applying a number of other techniques. For example, one can drop some conditions from minimal decision rules. The classification quality of known objects (training data) decreases, but one often improves the classification quality of new objects, especially in the case of noisy training data.

Another technique could depend on generating tolerance spaces for the value sets of attributes in the decision system and then using them in the comparison of minimal decision rules. For example, assume that a tolerance space on the RC's value set is provided and that values wet and {snow, ice} are identified as being in the same neighborhood of the provided tolerance relation. Then one can conclude that the first rule is identified with the third one. Hence, one of these rules can be removed from the minimal decision rule set for the decision high. Using this technique we obtain the following rules for decisions moderate and low:

$(DIF' = \mathsf{i3}) \wedge (RC = \mathsf{dry}) \Rightarrow (Danger = \mathsf{moderate})$
$(DIF' = \mathsf{i3}) \wedge (AD = \mathsf{small}) \Rightarrow (Danger = \mathsf{moderate})$
$(DIF' = \mathsf{i1}) \wedge (RC = \mathsf{dry}) \wedge (AD = \mathsf{medium}) \Rightarrow (Danger = \mathsf{moderate})$
$(DIF' = \mathsf{i1}) \wedge (RC = \mathsf{wet}) \Rightarrow (Danger = \mathsf{moderate})$
$(DIF' = \mathsf{i5}) \wedge (RC \in \{\mathsf{snow}, \mathsf{ice}\}) \Rightarrow (Danger = \mathsf{moderate})$
$(RC \in \{\mathsf{snow}, \mathsf{ice}\}) \wedge (AD = \mathsf{large}) \Rightarrow (Danger = \mathsf{moderate})$
$(DIF' = \mathsf{i5}) \wedge (RC = \mathsf{dry}) \Rightarrow (Danger = \mathsf{low})$
$(RC = \mathsf{dry}) \wedge (AD = \mathsf{large}) \Rightarrow (Danger = \mathsf{low})$
$(DIF' = \mathsf{i2}) \Rightarrow (Danger = \mathsf{low}).$

15.6 From Decision Rules to Classifiers

Techniques for constructing classifiers in terms of decision rules are provided in Chapters 12 and 14. The rough set interpretation of classifiers is described in Section 14.6.

In order to create a reasonable classifier from a set of decision rules one should provide a matching mechanism, i.e., a method which is to be used to decide if a given object matches (exactly or at least to some acceptable degree) the lefthand side of any rule in the rule set. This can be done by supplying tolerance spaces for the individual value sets associated with the conditional attributes in the rules. Since a classifier may be viewed as an information granule, (see Section 12.4) such tolerances can be lifted to the classifier level, as is shown in Chapter 13.

Consider testing data contained in Table 15.6.

Table 15.6. Test data for the classifier considered in Section 15.6

Object	*DIF(DIF′)*	*RC*	*AD*	*Danger*
o1	11 (i4)	snow	medium	high
o2	-7 (i1)	dry	UNKNOWN	UNKNOWN
o3	UNKNOWN	ice	medium	high
o4	20 (i5)	snow	large	moderate
o5	0 (i2)	dry	large	low

For objects stored in Table 15.6 we have the following facts, where $C(\mathsf{o}, \mathsf{d})$ states that object o, which is to be classified relatively to *Danger*, has the value d computed by using the decision rules for *Danger*:

$$C^{+}(\mathsf{o1}, \mathsf{high}), C^{-}(\mathsf{o1}, \mathsf{moderate}), C^{-}(\mathsf{o1}, \mathsf{low})$$
$$C^{\pm}(\mathsf{o2}, \mathsf{high}), C^{\pm}(\mathsf{o2}, \mathsf{moderate}), C^{\pm}(\mathsf{o2}, \mathsf{low})$$
$$C^{+}(\mathsf{o3}, \mathsf{high}), C^{-}(\mathsf{o3}, \mathsf{moderate}), C^{-}(\mathsf{o3}, \mathsf{low})$$
$$C^{-}(\mathsf{o4}, \mathsf{high}), C^{+}(\mathsf{o4}, \mathsf{moderate}), C^{-}(\mathsf{o4}, \mathsf{low})$$
$$C^{-}(\mathsf{o5}, \mathsf{high}), C^{-}(\mathsf{o5}, \mathsf{moderate}), C^{+}(\mathsf{o5}, \mathsf{low}).$$

We conclude this section by observing that any classifier is an information granule. Thus it presents its results in the form of relations that can be used in the intensional layer of databases for higher level reasoning.

It is worthwhile mentioning here that we have also developed hierarchical learning methods for compound concepts approximation using sensory data and domain knowledge represented by ontology of concepts. One can consider as examples of such concepts *dangerous situation on the road* or concepts related to identification of behavioral patterns on the road (e.g., *overtaking manouver*). The interested reader is referred to the literature on ontology approximation cited in Section 3.9.

This short case study shows how, beginning with sensor data, one can use the techniques described in this book to generate decision rules, classifiers and then rough relations which may be stored and used for grounded inference at higher layers in a robotics architecture.

References

1. S. Abiteboul, R. Hull, and V. Vianu. *Foundations of Databases.* Addison-Wesley Pub. Co., 1996.
2. W. Ackermann. Untersuchungen über das Eliminationsproblem der mathematischen Logik. *Mathematische Annalen*, 110:390–413, 1935.
3. R. Agrawal, H. Mannila, R. Srikant, H. Toivonen, and A. Verkano. Fast discovery of association rules. In Fayyad et al. [72], pages 307–328.
4. K.R. Apt, H. Blair, and A. Walker. Towards a theory of declarative knowledge. In Minker [125], pages 89–148.
5. K.R. Apt and R. Bol. Logic programming and negation: A survey. *Journal of Logic Programming*, 19-20:9–71, 1994.
6. F. Baader and B. Hollunder. How to prefer more specific defaults in terminological default logic. In *Proc. 13th IJCAI*, pages 669–674, San Mateo, CA, 1993. Morgan Kaufmann.
7. F. Bancilhon and R. Ramakrishnan. An amateur's introduction to recursive query processing strategies. In M. Stonebraker, editor, *Readings in Database Systems*, pages 507–555, Los Altos, CA, 1988. Morgan Kaufmann, Inc.
8. D. Batens, C. Mortensen, G. Priest, and Jean-Paul van Bendegem, editors. *Frontiers of Paraconsistent Logic.* Taylor and Francis, 2000.
9. J. Bazan. A comparison of dynamic and non-dynamic rough set methods for extracting laws from decision system. In Polkowski and Skowron [162], pages 321–365.
10. J. Bazan, R. Latkowski, and M. Szczuka. DIXER - Distributed executor for rough set exploration system. In Ślęzak et al. [204], pages 362–371.
11. J. Bazan, H. S. Nguyen, J. F. Peters, A. Skowron, and M. Szczuka. Rough set approach to pattern extraction from classifiers. In A. Skowron and M. Szczuka, editors, *Proceedings of the Workshop on Rough Sets in Knowledge Discovery and Soft Computing at ETAPS 2003, April 12-13, 2003*, volume 82(4) of *Electronic Notes in Computer Science*, pages 20–29. Elsevier, Amsterdam, Netherlands, 2003.
12. J. Bazan, J. F. Peters, and A. Skowron. Behavioral pattern identification through rough set modelling. In Ślęzak et al. [204], pages 688–697.
13. J. Bazan and A. Skowron. Classifiers based on approximate reasoning schemes. In Dunin-Kęplicz et al. [67], pages 191–202.

14. J. Bazan and A. Skowron. On-line elimination of non-relevant parts of complex objects in behavioral pattern identification. In Pal et al. [144], pages 720–725.
15. J. Bazan, M. Szczuka, Wojna. A, and M. Wojnarski. On the evolution of rough set exploration system. In Tsumoto et al. [219], pages 592–601.
16. J.G. Bazan, M.S. Szczuka, and J. Wróblewski. A new version of rough set exploration system. In J.J. Alpigini, J.F. Peters, A. Skowron, and N. Zhong, editors, *Proceedings of the 3rd International Conference on Rough Sets and Current Trends in Computing RSCTC*, volume 2475 of *LNAI*, pages 397–404. Springer-Verlag, 2002.
17. N.D. Belnap. A useful four-valued logic. In G. Eptein and J.M. Dunn, editors, *Modern Uses of Many Valued Logic*, pages 8–37. Reidel, 1977.
18. M. Ben-Jacob and M. Fitting. Stratified, weak stratified, and three-valued semantics. *Fundamenta Informaticae*, 13:19–33, 1990.
19. P. Besnard. *An Introduction to Default Logic.* Springer-Verlag, 1989.
20. P. Besnard and T. Schaub. Possible world semantics for default logic. *Fundamenta Informaticae*, 21:39–66, 1994.
21. J. Bicarregui, T. Dimitrakos, D. Gabbay, and T. Maibaum. Interpolation in practical formal development. *Journal of the IGPL*, 9(2):247–259, 2001.
22. L. Bolc and P. Borowik. *Many-Valued Logics, 1. Theoretical Foundations.* Springer, Berlin, 1992.
23. G. Booch, I. Jacobson, and J. Rumbaugh. *The Unified Modeling Language Reference Manual.* Addison-Wesley Pub. Co., 1998.
24. G. Booch, I. Jacobson, and J. Rumbaugh. *The Unified Modeling Language User Guide.* Addison-Wesley Pub. Co., 1998.
25. M. Brandon and N. Rescher. *The Logic of Inconsistency.* Basil Blackwell, Oxford, 1978.
26. G. Brewka. Comulative default logic: In defense of non-monotonic inference rules. *Artificial Intelligence J.*, 50:183–205, 1991.
27. G. Brewka. *Non-Monotonic Reasoning: Logical Foundations of Commonsense.* Cambridge University Press, 1991.
28. G. Brewka. Adding priorities and specificity to default logic. In *European Workshop on Logics in AI (JELIA-94)*, volume 1409 of *Lecture Notes in Artificial Intelligence*, pages 247–260. Springer, 1994.
29. R.R. Brooks and S.S. Iyengar. *Multi-Sensor Fusion.* Prentice-Hall PTR, Upper Saddle River, NJ, 1998.
30. F.M. Brown. *Boolean Reasoning.* Kluwer Academic Publishers, Dordrecht, 1990.
31. D. Busch. Sequent formalizations of three-valued logic. In P. Doherty, editor, *Partiality, Modality and Nonmonotonicity*, pages 45–75, Stanford, California, 1996. CSLI Publications.
32. M. Cadoli. *Tractable Reasoning in Artificial Intelligence*, volume 941 of *LNAI.* Springer-Verlag, Berlin Heidelberg, 1995.
33. W. Carnielli, M. Coniglio, and I. D'Ottaviano, editors. *Paraconsistency: the Logical Way to the Inconsistent.* Marcel Dekker, New York, 2002.
34. G. Cattaneo. Abstract approximation spaces for rough theories. In Polkowski and Skowron [162], pages 59–98.
35. A.K. Chandra and D. Harel. Structure and complexity of relational queries. In *Proc. IEEE Conf. on Foundations of Computer Science*, pages 333–347, 1980.
36. A.K. Chandra and D. Harel. Horn clause queries and generalizations. *Journal of Logic Programming*, 2(1):1–15, 1985.

37. E.F Codd. A relational model of data for large shared databanks. *Communications of the ACM*, 13(6):377–387, 1970.
38. R.M. Colomb. *Deductive Databases and their Applications*. Taylor & Francis, 1998.
39. G. Conte, S. Duranti, and T. Merz. Dynamic 3D path following for an autonomous helicopter. In *Proc. of the IFAC Symp. on Intelligent Autonomous Vehicles*, 2004.
40. C.J. Date. *The database Relational Model: A Retrospective review and Analysis*. Addison-Wesley pub. Co., 2000.
41. C.J. Date and H. Darwen. *The Guide to the SQL Standard: A User's Guide to the Standard SQL*. Addison-Wesley Pub. Co., 1997.
42. P. Degrace and L.H. Stahl. *The Olduvai Imperative: Case and the State of Software Engineering Practice*. Yourdon, 1993.
43. J. P. Delgrande, T. Schaub, and W. K. Jackson. Alternative approaches to default logic. *Artificial Intelligence J.*, 70:167–237, 1994.
44. E. W Dijkstra. *A Discipline of Programming*. Prentice-Hall, 1976.
45. P. Doherty. Advanced Research with Autonomous Unmanned Aerial Vehicles. In *Proc. of the Int. Conf. on the Principles of Knowledge Representation and Reasoning*, pages 731–732, 2004.
46. P. Doherty, M. Grabowski, W. Łukaszewicz, and A. Szałas. Towards a framework for approximate ontologies. *Fundamenta Informaticae*, 57(2-4):147–165, 2003.
47. P. Doherty, G. Granlund, K. Kuchcinski, E. Sandewall, K. Nordberg, E. Skarman, and J. Wiklund. The WITAS unmanned aerial vehicle project. In *Proc. of the European Conf. on Artificial Intelligence*, pages 747–755, 2000.
48. P. Doherty, P. Haslum, F. Heintz, T. Merz, T. Persson, and B. Wingman. A Distributed Architecture for Autonomous Unmanned Aerial Vehicle Experimentation. In *Proc. of the Int. Symp. on Distributed Autonomous Robotic Systems*, pages 221–230, 2004.
49. P. Doherty, J. Kachniarz, and A. Szałas. Meta-queries on deductive databases. *Fundamenta Informaticae*, 40(1):17–30, 1999.
50. P. Doherty, J. Kachniarz, and A. Szałas. Using contextually closed queries for local closed-world reasoning in rough knowledge databases. In Pal et al. [145], pages 219–250.
51. P. Doherty and W. Łukaszewicz. A non-monotonic logic with explicit defaults. *Journal of Applied Non-Classical Logics*, 2:9–48, 1992.
52. P. Doherty, W. Łukaszewicz, A. Skowron, and A. Szałas. Approximation transducers and trees: A technique for combining rough and crisp knowledge. In Pal et al. [145], pages 189–218.
53. P. Doherty, W. Łukaszewicz, and A. Szałas. Computing circumscription revisited. *Journal of Automated Reasoning*, 18(3):297–336, 1997. See also 14th International Joint Conference on AI (IJCAI'95), Morgan Kaufmann Pub. Inc.
54. P. Doherty, W. Łukaszewicz, and A. Szałas. General domain circumscription and its effective reductions. *Fundamenta Informaticae*, 36(1):23–55, 1998.
55. P. Doherty, W. Łukaszewicz, and A. Szałas. Declarative PTIME queries for relational databases using quantifier elimination. *Journal of Logic and Computation*, 9(5):739–761, 1999.
56. P. Doherty, W. Łukaszewicz, and A. Szałas. Computing strongest necessary and weakest sufficient conditions of first-order formulas. *International Joint Conference on AI (IJCAI'2001)*, pages 145 - 151, 2000.

57. P. Doherty, W. Łukaszewicz, and A. Szałas. Efficient reasoning using the local closed-world assumption. In A. Cerri and D. Dochev, editors, *Proc. 9th Int. Conference AIMSA 2000*, volume 1904 of *LNAI*, pages 49–58. Springer-Verlag, 2000.
58. P. Doherty, W. Łukaszewicz, and A Szałas. CAKE: A computer-aided knowledge engineering technique. In F. van Harmelen, editor, *Proc. 15th European Conference on Artificial Intelligence, ECAI'2002*, pages 220–224, Amsterdam, 2002. IOS Press.
59. P. Doherty, W. Łukaszewicz, and A Szałas. Information granules for intelligent knowledge structures. In G. Wang, Q. Liu, Y. Yao, and A. Skowron, editors, *Proceedings of 9th Internatinal Conference on Rough Sets, Fuzzy Sets, Data Mining and Granular Computing*, volume 2639 of *LNCS*, pages 405–412. Springer-Verlag, 2003.
60. P. Doherty, W. Łukaszewicz, and A Szałas. Tolerance spaces and approximative representational structures. In *Proceedings 26th German Conference on Artificial Intelligence*, LNCS, pages 475–489. Springer-Verlag, 2003.
61. P. Doherty, W. Łukaszewicz, and A. Szałas. Approximate databases and query techniques for agents with heterogenous perceptual capabilities. In *Proceedings of the 7th International Conference on Information Fusion, FUSION'2004*, pages 175–182, 2004.
62. P. Doherty, W. Łukaszewicz, and A. Szałas. Approximative query techniques for agents with heterogeneous ontologies and perceptive capabilities. In D. Dubois, Ch. Welty, and M-A. Williams, editors, *Proceedings of 9th International Conference on the Principles of Knowledge Representation and Reasoning, KR'2004*, pages 459–468. AAAI Press, 2004.
63. P. Doherty, M. Magnusson, and A. Szałas. Approximate databases: A support tool for approximate reasoning. *To appear*, 2006.
64. P. Doherty and A. Szałas. On the correspondence between approximations and similarity. In S. Tsumoto, R. Slowinski, J. Komorowski, and J.W. Grzymala-Busse, editors, *Proceedings of 4th International Conference on Rough Sets and Current Trends in Computing, RSCTC'2004*, volume 3066 of *LNAI*, pages 143–152. Springer-Verlag, 2004.
65. G. Dong and J. Li. Interestingness of discovered association rules in terms of neighborhood-based unexpectedness. In X. Wu, Ramamohanarao K., and K.B. Korb, editors, *Proceedings of Research and Development in Knowledge Discovery and Data Mining, PAKDD-98*, volume 1384 of *LNAI*, pages 72–86, Heidelberg, 1998. Springer-Verlag.
66. W.P. Dowling and J.H. Galier. Linear-time algorithms for testing the satisfiability of propositional Horn formulae. *Journal Logic Programming*, 1:267–284, 1984.
67. B. Dunin-Kęplicz, A. Jankowski, A. Skowron, and M. Szczuka, editors. *Monitoring, Security, and Rescue Tasks in Multiagent Systems (MSRAS'2004)*. Advances in Soft Computing. Springer, Heidelberg, Germany, 2005.
68. H-D. Ebbinghaus and J. Flum. *Finite Model Theory*. Springer-Verlag, Heidelberg, 1995.
69. H-D. Ebbinghaus, J. Flum, and W. Thomas. *Mathematical Logic*. Springer-Verlag, Heidelberg, 1994.
70. D. Etherington. *Reasoning with Incomplete Information*. Research Notes in Artificial Intelligence. Pitman, London, 1988.

71. O. Etzioni, K. Golden, and D.S. Weld. Sound and efficient closed-world reasoning for planning. *Artificial Intelligence*, 89:113–148, 1997.
72. U.M. Fayyad, G. Piatetsky-Shapiro, P. Smyth, and R. Uthurusamy, editors. *Advances in Knowledge Discovery and Data Mining.* The AAAI Press/The MIT Press, Cambridge, MA., 1996.
73. M. Fitting. A Kripke/Kleene semantics for logic programs. *Journal of Logic Programming*, 2:295–312, 1985.
74. Peter A. Flach and Antonis Kakas. Abduction and induction in AI. *Logic Journal of the IGPL*, 6(4):651–656, 1998.
75. D. M. Gabbay and H. J. Ohlbach. Quantifier elimination in second-order predicate logic. In B. Nebel, C. Rich, and W. Swartout, editors, *Principles of Knowledge representation and reasoning, KR 92*, pages 425–435. Morgan Kaufmann, 1992.
76. H. Garcia-Molina, J. Ullnman, and J. Widom. *Database Systems Implementation.* Prentice Hall, 1999.
77. P. Gärdenfors. *Conceptual spaces: The geometry of thought.* MIT Press, Cambridge, Mass., 2000.
78. M.R Garey and D.S. Johnson. *Computers and Intractability: A Guide to the Theory of NP-completeness.* W.H. Freeman & Co., 1979.
79. G. Gottlob. Complexity results for nonmonotonic logics. *J. Logic Computat.*, 2:397–425, 1992.
80. G. Graefe. Query evaluation techniques for large databases. *ACM Computeing Surveys*, 25(2):73–170, 1993.
81. S. Greco, B. Matarazzo, and R. Słowiński. Fuzzy similarity relation as a basis for rough approximations. In L. Polkowski and A. Skowron, editors, *Rough Sets and Current Trends in Computing*, volume 1424 of *LNAI*, pages 283–289, Berlin, 1998. Springer Verlag.
82. S. Greco, B. Matarazzo, and R. Słowiński. Rough set processing of vague information using fuzzy similarity relations. In C. S. Calude and G. Paun, editors, *Finite Versus Infinite - Contributions to an Eternal Dilemma*, volume 2, pages 149–173, London, 2000. Springer Verlag.
83. Y. Gurevich. Toward a logic tailored for computational complexity. In Richter M.M. et al., editor, *Computation and Proof Theory*, volume 1104 of *LNM*, pages 175–216. Springer Verlag, 1984.
84. Y. Gurevich and S. Shelah. Fixed-point extensions of first-order logic. *Annals of Pure and Applied Logic*, 32:265–280, 1986.
85. J.Y. Halpern. *Reasoning about uncertainty.* MIT Press, Cambridge, Mass., 2003.
86. C. Hartshorne, P. Weiss, and A.W. Burks, editors. *The Collected Papers of Charles Sanders Peirce.* Harvard University Press, 1958.
87. T. Hastie, R. Tibshirani, and J. Friedman. *The Elements of Statistical Learning: Data Mining, Inference, and Prediction.* Springer–Verlag, Heidelberg, 2001.
88. S. Hirano, M. Inuiguchi, and S. Tsumoto, editors. *Proceedings of International Workshop on Rough Set Theory and Granular Computing, RSTGC'01.* Bulletin of the International Rough Set Society 5(1-2). IRSS, Shimane, 2001.
89. J.E. Hopcroft and J.D. Ullman. *Introduction to Automata Theory, Languages and Computation.* Addison-Wesley Pub., Co., 1979.
90. M.N. Huhns and M.P. Singh, editors. *Readings in Agents.* Morgan Kaufmann, San Mateo, 1998.

91. N. Immerman. Relational queries computable in polynomial time. *Information and Control*, 68(1-3):86–104, 1986.
92. N. Immerman. *Descriptive Complexity.* Springer-Verlag, New York, Berlin, 1998.
93. J. Kachniarz and A. Szałas. On a static approach to verification of integrity constraints in relational databases. In E. Orłowska and A. Szałas, editors, *Relational Methods for Computer Science Applications*, pages 97–109. Springer Physica-Verlag, 2001.
94. A. Kakas, R.A. Kowalski, and F. Toni. The role of logic programming in abduction. In *Handbook of Logic programming.* Oxford University Press, 1994.
95. A. Kakas, R.A. Kowalski, and F. Toni. The role of abduction in logic programming. In D.M. Gabbay, C.J. Hogger, and J.A. Robinson, editors, *Handbook of Logic in Artificial Intelligence and Logic Programming.* Oxford University Press, 1995.
96. P.C. Kanellakis. Elements of relational database theory. In J. Van Leeuwen, editor, *Handbook of Theoretical Computer Science*, pages 1074–1156, Amsterdam, 1991. Elsevier.
97. H. Kautz and B. Selman. Knowledge compilation and theory approximation. *Journal of the ACM*, 43(2):193–224, 1996.
98. H. A. Kautz and B Selman. Hard problems for simple default logic. *Artificial Intelligence J.*, 49:243–279, 1991.
99. J. Komorowski, A. Øhrn, and A. Skowron. Rosetta and other software systems for rough sets. In W. Klösgen and J. Żytkow, editors, *Handbook of Data Mining and Knowledge Discovery*, pages 554–559. Oxford University Press, 2000.
100. J. Komorowski, Z. Pawlak, L. Polkowski, and A. Skowron. Rough sets: A tutorial. In Pal and Skowron [146], pages 3–98.
101. R. Kowalski. A proof procedure using connection graphs. *Journal of the ACM*, 22:572–595, 1975.
102. K. Krawiec, R. Słowiński, and D. Vanderpooten. Learning decision rules from similarity based rough approximations. In Polkowski and Skowron [163], pages 37–54.
103. V. Lifschitz. Computing circumscription. In *Proc. 9th IJCAI*, pages 229–235, Palo Alto, CA, 1985. Morgan Kaufmann.
104. V. Lifschitz. On the satisfiability of circumscription. *Artificial Intelligence J.*, 28:17–27, 1986.
105. V. Lifschitz. On the declarative semantics of logic programs with negation. In Minker [125], pages 177–192.
106. V. Lifschitz. Pointwise circumscription. In M. Ginsberg, editor, *Readings in Nonmonotonic Reasoning*, pages 179–193. Morgan Kaufmann, Palo Alto, CA, 1988.
107. V. Lifschitz. Circumscription. In D. M. Gabbay, C. J. Hogger, and J. A. Robinson, editors, *Handbook of Artificial Intelligence and Logic Programming*, volume 3, pages 297–352. Oxford University Press, 1991.
108. F. Lin. On strongest necessary and weakest sufficient conditions. In A.G. Cohn, F. Giunchiglia, and B. Selman, editors, *Proc. 7th International Conf. on Principles of Knowledge Representation and Reasoning, KR2000*, pages 167–175. Morgan Kaufmann Pub., Inc., 2000.
109. T. Y. Lin and N. Cercone. *Rough sets and data mining. Analysis of imprecise data.* Kluwer Academic Publishers, Boston, 1997.

110. T.Y. Lin. Granular computing on binary relations I, II. In Polkowski and Skowron [162], pages 107–140.
111. M. Luck, P. McBurney, O. Shehory, and S. Willmott. *Agent Technology: Computing as Interaction. A Roadmap for Agent Based Computing.* University of Southampton on behalf of AgentLink III, Southampton, UK, 2005.
112. J. Łukasiewicz. *Die logischen Grundlagen der Wahrscheinlichkeitsrechnung.* Polska Akademia Umiejetnosci, Krakow, 1913.
113. J. Łukasiewicz. Philosophische bemerkungen zu mehrwertigen systemen des aussagenkalküls. *Société des Sciences et des Lettres de Varsovie*, 23:55–77, 1930.
114. W. Łukaszewicz. Two results on default logic. In *Proc. 9th International Joint Conf. on Artificial Intelligence, IJCAI-85*, pages 459–461, San Francisco, Ca., 1985. Morgan Kaufmann Pub., Inc.
115. W. Łukaszewicz. Considerations on default logic: An alternative approach. *Computational Intelligence*, 4:1–16, 1988.
116. W. Łukaszewicz. *Non-Monotonic Reasoning - Formalization of Commonsense Reasoning.* Ellis Horwood Series in Artificial Intelligence. Ellis Horwood, 1990.
117. M. Marcotty and H. Ledgard. *The World of Programming Languages.* Springer-Verlag, Berlin, 1986.
118. V. W. Marek and M. Truszczyński. *Nonmonotonic Logic.* Springer-Verlag, 1993.
119. J. McCarthy. Circumscription: A form of non-monotonic reasoning. *Artificial Intelligence J.*, 13:27–39, 1980.
120. J. McCarthy. Applications of circumscription to formalizing commonsense knowledge. *Artificial Intelligence J.*, 28:89–116, 1986.
121. J. McCarthy. Approximate objects and approximate theories. In A.G. Cohn, F. Giunchiglia, and B. Selman, editors, *Proc. 7th International Conf. on Principles of Knowledge Representation and Reasoning, KR2000*, pages 519–526. Morgan Kaufmann Pub., Inc., 2000.
122. T. Merz. Building a System for Autonomous Aerial Robotics Research. In *Proc. of the IFAC Symp. on Intelligent Autonomous Vehicles*, 2004.
123. T. Merz, S. Duranti, and G. Conte. Autonomous landing of an unmanned aerial helicopter based on vision and inertial sensing. In *Proc. of the 9th International Symposium on Experimental Robotics*, 2004.
124. D. Michie, D.J. Spiegelhalter, and C.C. Taylor, editors. *Machine learning, Neural and Statistical Classification.* Ellis Horwood, New York, 1994.
125. J. Minker, editor. *Foundations of Deductive Databases and Logic Programming.* Morgan Kaufmann Publishers, Los Altos, CA., 1988.
126. J. Minker. Logic and databases: Past, present, and future. *AI Magazine*, 18(3), 1997.
127. M. Minsky. *The Society of Mind.* Simon & Schuster, Inc., 1986.
128. T.M. Mitchell. *Machine Learning.* Mc Graw-Hill, Portland, 1997.
129. R. C. Moore. Possible-world semantics for autoepistemic logic. In *Proc. 1st Nonmonotonic Reasoning Workshop*, pages 344–354, New Paltz, NY, 1984.
130. R. C. Moore. Semantical considerations on nonmonotonic logic. *Artificial Intelligence J.*, 25:75–94, 1985.
131. H.S. Nguyen. *Discretization of Real Value Attributes, Boolean Reasoning Approach, Ph.D. Thesis.* Warsaw University, Warsaw, 1997.

132. H.S. Nguyen. Efficient SQL-learning method for data mining in large data bases. In *Proceedings of the 16th International Joint Conference on Artificial Intelligence (IJCAI'99)*, pages 806–811, 1999.
133. H.S. Nguyen. On efficient handling of continuous attributes in large data bases. *Fundamenta Informaticae*, 48(1):61–81, 2001.
134. H.S. Nguyen and S.H. Nguyen. Pattern extraction from data. *Fundamenta Informaticae*, 34:129–144, 1998.
135. H.S. Nguyen and S.H. Nguyen. Rough sets and association rule generation. *Fundamenta Informaticae*, 40(4), 1999.
136. H.S. Nguyen and A. Skowron. Quantization of real value attributes. In *Proceedings of the Second Joint Annual Conference on Information Sciences, Wrightsville Beach, North Carolina, 1995, USA*, pages 34–37, 1995.
137. H.S. Nguyen, A. Skowron, and J. Stepaniuk. Granular computing: A rough set approach. *Computational Intelligence*, 17:514–544, 2001.
138. S. H. Nguyen, J. Bazan, Skowron. A., and H. S. Nguyen. Layered learning for concept synthesis. In J.F. Peters and A. Skowron, editors, *Transactions on Rough Sets I: Journal Subline*, volume 3100 of *Lecture Notes in Computer Science*, pages 187–208. Springer, Heidelberg, Germany, 2004.
139. T. T. Nguyen. Eliciting domain knowledge in handwritten digit recognition. In Pal et al. [144], pages 762–767.
140. T. T. Nguyen and A. Skowron. Rough set approach to domain knowledge approximation. In Wang et al. [232], pages 221–228.
141. A. Nonnengart, H.J. Ohlbach, and A. Szałas. Elimination of predicate quantifiers. In H.J. Ohlbach and U. Reyle, editors, *Logic, Language and Reasoning. Essays in Honor of Dov Gabbay, Part I*, pages 159–181. Kluwer, 1999.
142. A. Nonnengart and A. Szałas. A fixpoint approach to second-order quantifier elimination with applications to correspondence theory. In E. Orłowska, editor, *Logic at Work: Essays Dedicated to the Memory of Helena Rasiowa*, volume 24 of *Studies in Fuzziness and Soft Computing*, pages 307–328. Springer Physica-Verlag, 1998.
143. E. Orłowska, editor. *Incomplete Information, Rough Set Analysis*, volume 13 of *Advances in Fuzziness and Soft Computing*. Physica-Verlag, Heidelberg, 1998.
144. S.K. Pal, S. Bandoyopadhay, and S. Biswas, editors. *First International Conference on Pattern Recognition and Machine Intelligence (PReMI'05) December 18-22, 2005, Indian Statistical Institute, Kolkata*, volume 3776 of *Lecture Notes in Computer Science*. Springer-Verlag, Heidelberg, Germany, 2005.
145. S.K. Pal, L. Polkowski, and A. Skowron, editors. *Rough-Neural Computing: Techniques for Computing with Words*. Cognitive Technologies. Springer–Verlag, Heidelberg, 2004.
146. S.K. Pal and A. Skowron, editors. *Rough Fuzzy Hybridization: A New Trend in Decision–Making*. Springer-Verlag, Singapore, 1999.
147. C.H. Papadimitriou. *Computational Complexity*. Addison-Wesley Pub., Co., 1994.
148. Z. Pawlak. Rough sets. *International Journal of Computer and Information Sciences*, 11:341–356, 1982.
149. Z. Pawlak. Rough logic. *Bull. Polish Acad. Sci. Tech*, 35(5-6):253–258, 1987.
150. Z. Pawlak. *Rough Sets. Theoretical Aspects of Reasoning about Data*. Kluwer Academic Publishers, Dordrecht, 1991.
151. Z. Pawlak, J.F. Peters, A. Skowron, Z. Suraj, S. Ramanna, and M. Borkowski. Rough measures: Theory and applications. In Hirano et al. [88], pages 177–183.

152. J. F. Peters, A. Skowron, P. Synak, and S. Ramanna. Rough sets and information granulation. In O. Kaynak T. Bilgic, D. Baets, editor, *Tenth International Fuzzy Systems Association World Congress (IFSA 2003), Istanbul, Turkey, June 30 - July 2, 2003*, volume 2715 of *Lecture Notes in Artificial Intelligence*, pages 370–377. Springer-Verlag, Heidelberg, Germany, 2003.
153. J.F. Peters, S. Ramanna, A. Skowron, J. Stepaniuk, Z. Suraj, and M. Borkowski. Sensor fusion: A rough granular approach. In *Proc. of the Joint 9th International Fuzzy Systems Association World Congress and 20th NAFIPS International Conference, Vancouver, Canada*, pages 1367–1371, 2001.
154. J.F. Peters, A. Skowron, and J. Stepaniuk. Rough granules in spatial reasoning. In *Proc. of the Joint 9th International Fuzzy Systems Association World Congress and 20th NAFIPS International Conference, Vancouver, Canada*, pages 1355–1360, 2001.
155. P-O. Pettersson. Using Randomized Algorithms for Helicopter Path Planning. *Lic. Thesis Linköping University.* To appear, 2006.
156. P-O. Pettersson and P. Doherty. Probabilistic Roadmap Based Path Planning for an Autonomous Unmanned Aerial Vehicle. In *Proc. of the ICAPS-04 Workshop on Connecting Planning Theory with Practice*, 2004.
157. L. Polkowski. *Rough Sets: Mathematical Foundations.* Advances in Fuzziness and Soft Computing. Physica-Verlag, Heidelberg, 2002.
158. L. Polkowski, Y.Y. Lin, and S. Tsumoto, editors. *Rough Set Methods and Applications: New Developments in Knowledge Discovery in Information Systems*, volume 56 of *Studies in Fuzziness and Soft Computing.* Springer Physica-Verlag, Heidelberg, 2000.
159. L. Polkowski and A. Skowron. Rough mereological approach to knowledge–based distributed AI. In J.K. Lee, J. Liebowitz, and J.M. Chae, editors, *Proceedings of the 3rd World Congress on Expert Systems*, pages 774–781. Cognizant Communication Corporation, 1996.
160. L. Polkowski and A. Skowron. Rough mereology: A new paradigm for approximate reasoning. *International Journal of Approximate Reasoning*, 15(4):333–365, 1996.
161. L. Polkowski and A. Skowron, editors. *First International Conference on Rough Sets and Soft Computing, RSCTC'98.* LNAI 1424. Springer-Verlag, Heidelberg, 1998.
162. L. Polkowski and A. Skowron, editors. *Rough Sets in Knowledge Discovery 1: Methodology and Applications*, volume 17 of *Studies in Fuzziness and Soft Computing.* Physica-Verlag, Heidelberg, 1998.
163. L. Polkowski and A. Skowron, editors. *Rough Sets in Knowledge Discovery 2: Applications, Case Studies and Software Systems*, volume 18 of *Studies in Fuzziness and Soft Computing.* Physica-Verlag, Heidelberg, 1998.
164. L. Polkowski and A. Skowron. Towards adaptive calculus of granules. In Zadeh and Kacprzyk [238], pages 201–227.
165. L. Polkowski and A. Skowron. Rough mereology in information systems. a case study: Qualitative spatial reasoning. In Polkowski et al. [158], pages 89–135.
166. L. Polkowski and A. Skowron. Rough mereological calculi of granules: A rough set approach to computation. *Computational Intelligence*, 17:472–492, 2001.
167. L. Polkowski and A. Skowron. Rough-neuro computing. In W. Ziarko and Y.Y. Yao, editors, *Third International Conference on Rough Sets and Soft*

Computing, RSCTC'00, Banff, Canada, October, LNAI 2005, pages 57–64, Heidelberg, 2001. Springer-Verlag.

168. L. Polkowski, A. Skowron, and J. Żytkow. Rough foundations for rough sets. In T. Y. Lin and A. M. Wildberger, editors, *Soft Computing: Rough Sets, Fuzzy Logic, Neural Networks, Uncertainty Management, Knowledge Discovery*, pages 55–58. Simulation Councils, Inc., San Diego, CA, USA, 1995.
169. G. Priest, R. Routley, and J. Norman, editors. *Paraconsistent Logic. Essays on the Inconsistent*. Philosophia Verlag, München, 1989.
170. T. Przymusiński. Well-founded semantics coincides with three-valued stable semantics. *Fundamenta Informaticae*, XIII:445–463, 1990.
171. W. V. O. Quine. *The Web of Belief*. McGraw-Hill, 2nd edition edition, 1978.
172. A. Radzikowska. A three-valued approach to default logic. *Journal of Applied Non-Classical Logics*, 6:149–190, 1996.
173. R. Ramakrishnan, editor. *Applications of Logic Databases*. Kluwer, 1995.
174. M Reingruber and W.W. Gregory. *The Data Modeling Handbook: A Best-Practice Approach to Building Quality Data Models*. John Wiley & Sons, 1994.
175. R. Reiter. On closed world data bases. In H. Gallaire and J. Minker, editors, *Logic and Data Bases*, pages 55–76. Plenum Press, 1978.
176. R. Reiter. A logic for default reasoning. *Artificial Intelligence J.*, 13:81–132, 1980.
177. R. Reiter and G. Criscuolo. Some representational issues in default reasoning. *Int. J. Computers and Mathematics*, 9:1–13, 1983.
178. N. Rescher. *Many-Valued Logic*. McGraw Hill, New York, 1969.
179. T. Schaub. *The Automation of Reasoning with Incomplete Information*. Springer-Verlag, 1997.
180. H. Simmons. The monotonous elimination of predicate variables. *Journal of Logic and Computation*, 4:23–68, 1994.
181. A. Skowron. Synthesis of adaptive decision systems from experimental data. In A. Aamodt and J. Komorowski, editors, *Proc. of the 5th Scandinavian Conference on Artificial Intelligence (SCAI'95)*, pages 220–238, Amsterdam, 1995. IOS Press.
182. A. Skowron. Approximate reasoning by agents. In B. Dunin-Keplicz and E. Nawarecki, editors, *2nd International Workshop of Central and Eastern Europe on Multi-Agent Systems*, volume 2296 of *LNAI*, pages 3–14. Springer-Verlag, 2001.
183. A. Skowron. Approximate reasoning by agents in distributed environments. In N. Zhong, J. Liu, S. Ohsuga, and J. Bradshaw, editors, *Proceedings of the 2nd Pacific-Asia Conference on Intelligent Agent Technology*, pages 28–39. World Scientific, 2001.
184. A. Skowron. Toward intelligent systems: Calculi of information granules. In Hirano et al. [88], pages 9–30.
185. A. Skowron. Approximate reasoning in distributed environments. In N. Zhong and J. Liu, editors, *Intelligent Technologies for Information Analysis*, pages 433–474. Springer, Heidelberg, Germany, 2004.
186. A. Skowron. Rough sets in perception-based computing (keynote talk). In *First International Conference on Pattern Recognition and Machine Intelligence (PReMI'05) December 18-22, 2005, Indian Statistical Institute, Kolkata*, pages 21–29, 2005.

187. A. Skowron and H. S. Nguyen. Boolean reasoning scheme with some applications in data mining. In *Proceedings of the 3-rd European Conference on Principles and Practice of Knowledge Discovery in Databases*, volume 1704 of *Lecture Notes in Computer Science*, pages 107–115, Berlin, 1999. Springer Verlag.
188. A. Skowron and J. Peters. Rough sets: Trends and challenges. In Wang et al. [232], pages 25–34. (plenary talk).
189. A. Skowron and L. Polkowski. Rough mereological foundations for design, analysis, synthesis, and control in distributive systems. *Information Sciences International Journal*, 104(1-2):129–156, 1998.
190. A. Skowron and C. Rauszer. The discernibility matrices and functions in information systems. In Słowiński [205], pages 331–362.
191. A. Skowron and J. Stepaniuk. Tolerance approximation spaces. *Fundamenta Informaticae*, 27:245–253, 1996.
192. A. Skowron and J. Stepaniuk. Information granules: Towards foundations of granular computing. *International Journal of Intelligent Systems*, 16/1:57–86, 2001.
193. A. Skowron and J. Stepaniuk. Information granules and rough-neurocomputing. In Pal et al. [145], pages 43–84.
194. A. Skowron and J. Stepaniuk. Ontological framework for approximation. In Ślęzak et al. [203], pages 718–727.
195. A. Skowron, J. Stepaniuk, J. F. Peters, and R. Swiniarski. Calculi of approximation spaces in distributed environments. In D. Ślęzak, E. Menasalvas-Ruiz, Ch.-J. Liau, and M. Szczuka, editors, *Proceedings of the International workshop on Rough Sets and Soft Computing in Intelligent Agent and Web Technologies in Conjunction with the 2005 IEEE/WIC/ACM International Conference on Web Intelligence and Intelligent Agent Technology, Compiègne, France, September 19, 2005, 17-24*, pages 17–24. Compiègne University, Compiègne, France, 2005.
196. A. Skowron, J. Stepaniuk, J. F. Peters, and R. Swiniarski. Calculi of approximation spaces. *Fundamenta Informaticae*, 2006. (to appear).
197. A. Skowron, J. Stepaniuk, and J.F. Peters. Extracting patterns using information granules. In T. Terano, T. Nishida, A. Namatame, S. Tsumoto, Y. Ohsawa, and T. Washio, editors, *New Frontiers in Artificial Intelligence, Joint JSAI Workshop Post Proceedings*, LNAI 2253, pages 359–363. Springer-Verlag, Heidelberg, 2001.
198. A. Skowron, J. Stepaniuk, and S. Tsumoto. Information granules for spatial reasoning. *Bulletin of the International Rough Set Society*, 3/4:147–154, 1999.
199. A. Skowron, R. Swiniarski, and P. Synak. Approximation spaces and information granulation. In J. F. Peters and A. Skowron, editors, *Transactions on Rough Sets III: Journal Subline*, volume 3400 of *Lecture Notes in Computer Science*, pages 175–189. Springer, Heidelberg, Germany, 2005.
200. A. Skowron and P. Synak. Complex patterns. *Fundamenta Informaticae*, 60(1-4):351–366, 2004.
201. A. Skowron and N. Zhong, editors. *Proceedings of the 7-th International Workshop on Rough Sets, Fuzzy Sets, Data Mining, and Granular-Soft Computing (RSFDGrC'99)*, volume 1711 of *Lecture Notes in Artificial Intelligence*. Springer-Verlag, Heidelberg, 1999.
202. D. Ślęzak. *Approximate Decision Reducts, Ph. D. Thesis.* Warsaw University, Warsaw, 2002.

203. D. Ślęzak, G. Wang, M. Szczuka, I. Duentsch, and Y. Y. Yao, editors. *Proceedings of the 10th International Conference on Rough Sets, Fuzzy Sets, Data Mining, and Granular Computing (RSFDGrC'2005), Regina, Canada, 2005, Part I*, volume 3641 of *Lecture Notes in Artificial Intelligence*. Springer-Verlag, Heidelberg, Germany, 2005.
204. D. Ślęzak, J. T. Yao, J. F. Peters, Ziarko W., and X. Hu, editors. *Proceedings of the 10th International Conference on Rough Sets, Fuzzy Sets, Data Mining, and Granular Computing (RSFDGrC'2005), Regina, Canada, August 31-September 3, 2005, Part II*, volume 3642 of *Lecture Notes in Artificial Intelligence*. Springer-Verlag, Heidelberg, Germany, 2005.
205. R. Słowiński, editor. *Intelligent Decision Support - Handbook of Applications and Advances of the Rough Sets Theory*. Kluwer Academic Publishers, Dordrecht, 1992.
206. R. Słowiński and D. Vanderpooten. Similarity relation as a basis for rough approximations. In P. Wang, editor, *Advances in Machine Intelligence & Soft Computing*, pages 17–33, Raleigh NC, 1997. Bookwrights.
207. J. Stepaniuk. Knowledge discovery by application of rough set models. In Polkowski et al. [158], pages 137–233.
208. J. Stillman. It's not my default: The complexity of membership problems in restricted default logic. In *Proc. 8th National Conference on AI (AAAI-90)*, pages 571–578, Menlo Park, CA, 1990. AAAI Press.
209. P. Stone. *Leyered Learning in Multiagent Systems: A Winning Approach to Robotic Soccer*. MIT Press, 2000.
210. R. Swiniarski and A. Skowron. Rough set methods in feature selection and extraction. *Pattern Recognition Letters*, 24(6), 2003. 833–849.
211. A. Szałas. On the correspondence between modal and classical logic: An automated approach. *Journal of Logic and Computation*, 3:605–620, 1993.
212. A. Szałas. On an automated translation of modal proof rules into formulas of the classical logic. *Journal of Applied Non-Classical Logics*, 4:119–127, 1994.
213. A. Tarski. A lattice-theoretical theorem and its applications. *Pacific Journal of Mathematics*, 5:285–309, 1955.
214. B. Thalheim. *Entity-Relationship Modeling: Foundations of Database Technology*. Springer Verlag, 2000.
215. The Autonomous UAV Technologies Lab. Linkoping University, http://www.ida.liu.se/~patdo/auttek/.
216. The WITAS UAV Project. http://www.ida.liu.se/ext/witas.
217. H. Toivonen, M. Klemettinen, P. Ronkainen, K. Hätönen, and H. Mannila. Pruning and grouping discovered association rules. In *Proceedings of the MLnet Familiarization Workshop on Statistics, Machine Learning and Knowledge Discovery in Databases*, pages 47–52, Heraklion, Crete, Greece, 1995.
218. S. Tsumoto. Modelling diagnostic rules based on rough sets. In Polkowski and Skowron [161], pages 475–482.
219. S. Tsumoto, R. Słowiński, J. Komorowski, and J. Grzymała-Busse, editors. *Proceedings of the 4th International Conference on Rough Sets and Current Trends in Computing (RSCTC'2004), Uppsala, Sweden, June 1-5, 2004*, volume 3066 of *Lecture Notes in Artificial Intelligence*. Springer-Verlag, Heidelberg, Germany, 2004.
220. J. D. Ullman and J. Widom. *Database Systems*. Prentice Hall, 1997.
221. J.D. Ullman. Implementation of logical query languages for databases. *ACM Trans. on database Systems*, 10(3):289–321, 1985.

222. J.D Ullman. *Principles of Database and Knowledge Base Systems*, volume I. Computer Science Press, Rockville, 1988.
223. J.D Ullman. *Principles of Database and Knowledge Base Systems. The New Technologies*, volume II. Computer science Press, Rockville, 1989.
224. A. Urquhart. Many-valued logic. In D.M. Gabbay and F. Guenthner, editors, *Handbook of Philosophical Logic*, volume 3, pages 71–116, Dordrecht, 1986. Reidel.
225. M.H. Van Emden and R.A. Kowalski. The semantics of predicate logic as a programming language. *Journal of the ACM*, 23(4):733–742, 1976.
226. A. Van Gelder. Negation as failure using tight derivations for general logic programs. In *IEEE Symp. on Logic Programming*, pages 127–139, 1988.
227. A. Van Gelder, K.A. Ross, and J.S. Schlipf. The well-founded semantics for general logic programs. In *ACM Symp. on Principles of Database Systems*, pages 221–230, 1988.
228. M.Y. Vardi. The complexity of relational query languages. In *Proc. ACM SIGACT Symp. on the Theory of Computing*, pages 137–146, 1982.
229. A. Vitória, C.V. Damásio, and J. Małuszyński. From rough sets to rough knowledge bases. *Fundamenta Informaticae*, 57(2-4):215–246, 2003.
230. A. Vitória, C.V. Damásio, and J. Małuszyński. Query answering for rough knowledge bases. In G. Wang, Q. Liu, Y. Yao, and A. Skowron, editors, *Proceedings of 9th Internatinal Conference on Rough Sets, Fuzzy Sets, Data Mining and Granular Computing*, volume 2639 of *LNCS*, pages 197–204. Springer-Verlag, 2003.
231. A. Vitória, C.V. Damásio, and J. Małuszyński. Toward rough knowledge bases with quantitative measures. In S. Tsumoto, R. Slowinski, J. Komorowski, and J.W. Grzymala-Busse, editors, *Proceedings of 4th International Conference on Rough Sets and Current Trends in Computing, RSCTC'2004*, volume 3066 of *LNAI*, pages 153–158. Springer-Verlag, 2004.
232. G. Wang, Q. Liu, Y. Y. Yao, and A. Skowron, editors. *Proceedings of the 9-th International Conference on Rough Sets, Fuzzy Sets, Data Mining, and Granular Computing (RSFDGrC'2003), Chongqing, China, May 26-29, 2003*, volume 2639 of *Lecture Notes in Artificial Intelligence*. Springer-Verlag, Heidelberg, Germany, 2003.
233. X. Wang, B. De Baets, and E. Kerre. A comparative study of similarity measures. *Fuzzy Sets and Systems*, 73(2):259–268, 1995.
234. J. Wróblewski. *Adaptive Methods of Object Classification, Ph. D. Thesis*. Warsaw University, Warsaw, 2002.
235. M. Wzorek and P. Doherty. Preliminary Report: Reconfigurable Path Planning for an Autonomous Unmanned Aerial Vehicle. In *Proceedings of the 24th Annual Workshop of the UK Planning and Scheduling Special Interest Group (PlanSIG-05)*, 2005.
236. L.A. Zadeh. Fuzzy logic = computing with words. *IEEE Trans. on Fuzzy Systems*, 4:103–111, 1996.
237. L.A. Zadeh. A new direction in AI: Toward a computational theory of perceptions. *AI Magazine*, 22(1):73–84, 2001.
238. L.A. Zadeh and J. Kacprzyk, editors. *Computing with Words in Information/Intelligent Systems*, volume 1-2. Physica-Verlag, Heidelberg, 1999.

Index